THE ROLE OF ORGANIC MATTER IN MODERN AGRICULTURE

Developments in Plant and Soil Sciences

1. J. Monteith and C. Webb, eds.,
    Soil Water and Nitrogen in Mediterranean-type Environments. 1981. ISBN 90-247-2406-6
2. J.C. Brogan, ed.,
    Nitrogen Losses and Surface Run-off from Landspreading of Manures. 1981. ISBN 90-247-2471-6
3. J.D. Bewley, ed.,
    Nitrogen and Carbon Metabolism. 1981. ISBN 90-247-2472-4
4. R. Brouwer, I. Gašparíková, J. Kolek and B:C. Loughman, eds.,
    Structure and Function of Plant Roots. 1981. ISBN 90-247-2510-0
5. Y.R. Dommergues and H.G. Diem, eds.,
    Microbiology of Tropical Soils and Plant Productivity. 1982. ISBN 90-247-2624-7
6. G.P. Robertson, R. Herrera and T. Rosswall, eds.,
    Nitrogen Cycling in Ecosystems of Latin America and the Caribbean. 1982. ISBN 90-247-2719-7
7. D. Atkinson et al., eds.,
    Tree Root Systems and their Mycorrhizas. 1983. ISBN 90-247-2821-5
8. M.R. Sarić and B.C. Loughman, eds.,
    Genetic Aspects of Plant Nutrition. 1983. ISBN 90-247-2822-3
9. J.R. Freney and J.R. Simpson, eds.,
    Gaseous Loss of Nitrogen from Plant-Soil Systems. 1983. ISBN 90-247-2820-7
10. United Nations Economic Commission for Europe.
    Efficient Use of Fertilizers in Agriculture. 1983. ISBN 90-247-2866-5
11. J. Tinsley and J.F. Darbyshire, eds.,
    Biological Processes and Soil Fertility. 1984. ISBN 90-247-2902-5
12. A.D.L. Akkermans, D. Baker, K. Huss-Danell and J.D. Tjepkema, eds.,
    Frankia Symbioses. 1984. ISBN 90-247-2967-X
13. W.S. Silver and E.C. Schröder, eds.,
    Practical Application of Azolla for Rice Production. 1984. ISBN 90-247-3068-6
14. P.G.L. Vlek, ed.,
    Micronutrients in Tropical Food Crop Production. 1985. ISBN 90-247-3085-6
15. T.P. Hignett, ed.,
    Fertilizer Manual. 1985. ISBN 90-247-3122-4
16. D. Vaughan and R.E. Malcolm, eds.,
    Soil Organic Matter and Biological Activity. 1985. ISBN 90-247-3154-2
17. D. Pasternak and A. San Pietro, eds.,
    Biosalinity in Action: Bioproduction with Saline Water. 1985. ISBN 90-247-3159-3.
18. M. Lalonde, C. Camiré and J.O. Dawson, eds.,
    Frankia and Actinorhizal Plants. 1985. ISBN 90-247-3214-X
19. H. Lambers, J.J. Neeteson and I. Stulen, eds.,
    Fundamental, Ecological and Agricultural Aspects of Nitrogen Metabolism in Higher Plants. 1986.
    ISBN 90-247-3258-1
20. M.B. Jackson, ed.
    New Root Formation in Plants and Cuttings. 1986. ISBN 90-247-3260-3
21. F.A. Skinner and P. Uomala, eds.,
    Nitrogen Fixation with Non-Legumes. 1986. ISBN 90-247-3283-2
22. A. Alexander, ed.
    Foliar Fertilization. 1986. ISBN 90-247-3288-3.
23. H.G. v.d. Meer, J.C. Ryden and G.C. Ennik, eds.,
    Nitrogen Fluxes in Intensive Grassland Systems. 1986. ISBN 90-247-3309-x.
24. A.U. Mokwunye and P.L.G. Vlek, eds.,
    Management of Nitrogen and Phosporus Fertilizers in Sub-Saharan Africa. 1986.
    ISBN 90-247-3312-x
25. Y. Chen and Y. Avnimelech, eds.,
    The Role of Organic Matter in Modern Agriculture. 1986. ISBN 90-247-3360-x
26. S.K. De Datta and W.H. Patrick Jr., eds.,
    Nitrogen Economy of Flooded Rice Soils. 1986. ISBN 90-247-3361-8

The Role of Organic Matter in Modern Agriculture

Edited by

Y. CHEN
The Hebrew University of Jerusalem
Rehovot, Israel

and

Y. AVNIMELECH
The Technion
Haifa, Israel

1986 **MARTINUS NIJHOFF PUBLISHERS**
a member of the KLUWER ACADEMIC PUBLISHERS GROUP
DORDRECHT / BOSTON / LANCASTER

Distributors

for the United States and Canada: Kluwer Academic Publishers, 190 Old Derby Street, Hingham, MA 02043, USA
for the UK and Ireland: Kluwer Academic Publishers, MTP Press Limited, Falcon House, Queen Square, Lancaster LA1 1RN, UK
for all other countries: Kluwer Academic Publishers Group, Distribution Center, P.O. Box 322, 3300 AH Dordrecht, The Netherlands

Library of Congress Cataloging in Publication Data

The Role of organic matter in modern agriculture.

    (Developments in plant and soil sciences ; v.   )
    Includes bibliographies.
    1. Organic wastes as fertilizer. 2. Crops and soils. 3. Humus. 4. Soil microbiology. 5. Plant growing media. I. Chen, Y. II. Avnimelech, Y. III. Series.
S654.R65 1986   631.8'6   86-9754
ISBN-13:978-94-010-8470-3

ISBN-13:978-94-010-8470-3   e-ISBN-13: 978-94-009-4426-8
DOI: 10.1007/978-94-009-4426-8

Preface

The use of organic residues as a means of maintaining and increasing soil fertility is of long-standing. This tradition has been somewhat neglected since the introduction of mineral fertilizers at low cost. More and more farmers and scientists are now showing renewed interest in the proper and effective use of organic residues, composts and other recycled organic additives. The role and function of organic amendments in modern agricultural systems have become topics of major interest in the scientific and agricultural communities. Research work on residue disposal has provided new concepts on the interaction between organic components and soils as well as new handling technologies (e.g. pelletizing of organic residues). The trend to conserve energy has led scientists to study the minimal tillage system, to find ways of replacing conventional inorganic fertilizers with natural organic products or microbial preparations, and to develop new composting methods. The drive to achieve higher yields in commercial greenhouse farming has led to a search for optimum substrates as growth media and for improved management techniques. This has led to the introduction of organic substitutes for peat, notably those originating from agricultural wastes.

Another important aspect is the current interest in organic farming, where use of synthetic chemicals is avoided or prohibited. An increasing percentage of the population in highly developed countries is willing to pay premium prices for food produced on soils where inorganic fertilizers and other agricultural chemicals have not been used.

These considerations led us to the conclusion that this book is timely, and that it fills a void on a subject that lacks integrated scientific information. The volume has been designed to serve as a reference text on select topics from the broad field of soil organic matter. Undoubtedly, there are relevant subjects that are not adequately covered. Notwithstanding, we feel that the book will be of great interest to crop and soil scientists, to the general agronomists, and to environmentalists throughout the world.

Following a brief introduction (Chapter 1), the subject matter is conveniently divided into five sections. The first section (Chapters 2, 3, and 4) covers nitrogen and phosphorus-supply to plants as influenced by soil organic matter and biological transformations. The second section (Chapters 5 and 6) deals primarily with the effects of soil organic matter and redox potentials on micronutrient transport and availability to plants. Chapters 7 and 8 comprise the third section and focus on interactions between microorganisms and soil organic matter, with emphasis on soil fertility. Biofertilizers and biocontrol agents are also thoroughly discussed. The fourth section (Chapters 9 and 10) deals with the effects of soil organic matter and applied sewage sludge on soil structure and fertility, with emphasis on the effects of digested sewage sludge on chemical and biological soil properties. The fifth and last section (Chapters 11 and 12) consists of two comprehensive reviews on the use of peat and compost as container substrates in horticulture, including discussions on the suppression of soil-borne pathogens by composts and on various agricultural wastes that have been recently used as peat substitutes in container growth media.

Appreciation and gratitude is expressed to all chapter contributers who have made the publication of this book possible.

The Hebrew University of Jerusalem, 1985   Y. CHEN
and
The Technion, Haifa, 1985   Y. AVNIMELECH

Contributors

Avnimelech, Y., Department of Soils and Fertilizers, Faculty of Agricultural Engineering, The Technion, Haifa, Israël.
Broadbent, F.E., Department of Land, Air and Water Resources, University of Cal California, Davis, California, USA.
Chen, Y., Department of Soil and Water Sciences, The Hebrew University of Jerusalem, Rehovot, 76100, Israël.
Clay, D.E., Soil and Water Management Research Unit, University of Minnesota, St. Paul, Minnesota 55108, USA.
Clapp, C.E., Soil and Water Management Research Unit, University of Minnesota, St. Paul, Minnesota 55108, USA.
Feijtel, T.C., Louisiana State University, Baton Rouge, Louisiana 70803, USA.
Hadar, Y., Department of Plant Pathology and Microbiology, The Hebrew University of Jerusalem, Rehovot, 76100, Israël.
Hayes, M.H.B., Department of Chemistry, University of Birmingham, Edgbastan, Birmingham, United Kingdom.
Henis, Y., Department of Plant Pathology and Microbiology, The Hebrew University of Jerusalem, Rehovot, 76100, Israël.
Hoitink, H.A.J., Department of Plant Pathology, Ohio State University, Ohio Agricultural Research and Development Centre, Wooster, Ohio 44691, USA.
Inbar, Y., Department of Soil and Water Sciences, The Hebrew University of Jerusalem, Rehovot, 76100, Israël.
Kuter, G.A., Department of Plant Pathology, Ohio State University, Ohio Agricultural Research and Development Centre, Wooster, Ohio 44691, USA.
Larson, W.E., Soil and Water Management Research Unit, University of Minnesota St. Paul, Minnesota 55108, USA.
Olson, S.R., USDA-ARS, Agronomy, Colorado State University, Fort Collins, Colorado, USA.
Patrick Jr, W.H., Louisiana State University, Baton Rouge, Louisiana 70803, USA.
Raviv, M., Mewe-Yaar Experimental Station, ARO, Doar Haifa, Israël.
Reddy, K.R., Institute of Food and Agricultural Sciences, University of Florida, Sanford, Florida 32771, USA.
Stark, S.A., Soil and Water Management Research Unit, University of Minnesota, St. Paul, Minnesota 55108, USA.

Stevenson, F.J., Department of Agronomy, University of Illinois, Urbana, Illinois 61801, USA.

Terry, R.E., Department of Agronomy and Horticulture, Brigham Young University, Provo, Utah 84602, USA.

Contents

Preface   V

List of contributors   VII

Introduction
  1. Y. Avnimelech, **Organic residues in modern agriculture**   1

      1.1   Introduction   1
      1.2   Supply of nutrients by organic additives   2
      1.3   Effects of organic additives on the physical properties of the
              – soil   4
      1.4   Effects of organic amendments on yield   5
      1.5   Conclusions   8
      1.6   References   9

**Section 1: Nitrogen and phosphorus supply to plants by organic matter and
their transformations**   11

  2. F.E. Broadbent, **Effects of organic matter on nitrogen and phosphorus
     supply to plants**   13

      2.1   Introduction   13
      2.2   Nitrogen   13
      2.2.1   Forms of organic N in soil   13
      2.2.2   N mineralization   14
      2.2.2.1 Organic soils   14
      2.2.2.2 Mineral soils   15
      2.2.3   Available nitrogen   15
      2.2.3.1 Chemical extractants   15
      2.2.3.2 Incubation procedures   16
      2.2.3.3 Field estimates of N availability   17
      2.2.3.4 Crop uptake of soil N   18
      2.2.4   Environmental influences   18
      2.2.5   Forest soils   18
      2.3   Phosphorus   19
      2.3.1   Nature of soil organic P   19

X

Contents

2.3.2	P mineralization	19
2.3.2.1	Phosphatases	20
2.3.2.2	Effect of liming	20
2.3.3	Soil organisms in relation to P availability	20
2.4	Organic amendments	21
2.4.1	Animal manures	21
2.4.2	Other organic amendments	22
2.5	Conclusions	23
2.6	References	23

3. S.R. Olsen, The role of organic matter and ammonium in producing high corn yields — **29**

3.1	Introduction	29
3.2	Review of literature	31
3.2.1	Beneficial effect on yields of organic manures	31
3.2.2	Ammonium and nitrate as a nitrogen source	32
3.2.3	Energy requirements for NH_4^+ -grown and NO_3^- -grown — plants and for the combined sources	35
3.3	Results	40
3.4	Discussion	49
3.5	References	51

4. R.E. Terry, Nitrogen transformations in Histosols — **55**

4.1	Introduction	55
4.2	Histosol subsidence	55
4.3	Nitrogen mineralization	56
4.4	Nitrification	60
4.5	Nitrogen in drainage water of organic soils	61
4.6	Denitrification	64
4.7	Conclusions	68
4.8	References	69

Section II: Effects of soil organic matter and redox on micronutrients availability to plants — **71**

5. Y. Chen and F.J. Stevenson, Soil organic matter interactions with trace elements — **73**

5.1	Introduction	73
5.2	Importance of complexes of Fe, Mn, Zn and Cu with humic — substances to agriculture	73
5.2.1	Physiological and biochemical functions	73
5.2.2	Transport of micronutrients to plant roots	76
5.2.3	Supply of micronutrients to higher plants	79
5.3	Nature of organic complexing agents in soil	81
5.3.1	Defined biochemical compounds	81

5.3.2	Humic substances	83
5.3.2.1	Extraction and fractionation	83
5.3.2.2	Mechanisms of metal ion binding by humic and fulvic acids	86
5.3.2.3	Solubility characteristics	88
5.3.2.4	Metal ion binding capacity	89
5.3.2.5	Reduction properties	89
5.4	Use of micronutrient-enriched organic wastes and naturally — occuring metal organic complexes as soil amendments	89
5.4.1	Iron-organo complexes	90
5.4.1.1	Polyflavenoids and lignosulfonates	90
5.4.1.2	Manure and composts	91
5.4.1.3	Sewage sludge	92
5.4.1.4	Coal, lignite and peat	92
5.4.2	Zinc-, copper- and manganese- organo complexes	94
5.5	Stability constants of metal complexes with humic and fulvic — acids	95
5.5.1	General considerations	96
5.5.2	Modeling approaches	98
5.5.2.1	Macromolecule as the central group	99
5.5.2.2	Metal ion as the central group	104
5.5.2.3	Polynuclear complexes	108
5.6	Summary and conclusions	108
5.7	References	109

6. **K.R. Reddy, T.C. Feijtel and W.H. Patrick Jr, Effect of soil redox conditions on microbial oxidation of organic matter** — **117**

6.1	Introduction	117
6.2	Sources and types of organic matter	118
6.2.1	Soil organic matter	119
6.2.2	Root exudates	119
6.2.3	Added substrates	120
6.3	Role of inorganic redox couples on microbial respiration	120
6.3.1	Aerobic respiration	122
6.3.2	Facultative anaerobic respiration	125
6.3.3	Nitrate respiration	125
6.3.4	Manganese respiration	127
6.3.5	Iron respiration	128
6.3.6	Anaerobic respiration	130
6.3.7	Sulfate respiration	130
6.3.7.1	Fermentation	132
6.4	Kinetics of microbial organic matter oxidation	133
6.4.1	Rate of reaction	133

Contents

6.4.2	Soil and environmental factors	136
6.4.2.1	Soil moisture	136
6.4.2.2	Oxidant supply	137
6.4.2.3	Temperature	138
6.4.3	Substrate factors	139
6.5	Influence of aerobic/anaerobic respiration of organic matter — on soil biochemical processes	141
6.5.1	Nitrogen	142
6.5.2	Phosphorus	143
6.5.3	Potassium, calcium and magnesium	144
6.5.4	Sulfur	144
6.5.5	Micronutrients and heavy metals	145
6.6	Agronomic and environmental significance	147
6.7	References	148

Section III: Soil microorganisms, biofertilizers and biocontrol agents: their interactions with soil organic matter and effects on soil fertility   **157**

7. Y. Henis, Soil microorganisms, soil organic matter and soil fertility   **159**

7.1	Introduction	159
7.2	The living fraction (plants, animals, microorganisms)	160
7.3	The dead fraction: Fresh organic matter	161
7.4	The natural soil organic matter: Humus	161
7.5	The role of soil microorganisms in phosphorus availability — to higher plants	163
7.6	Subsoil, humus and fertility	164
7.7	Biological transformation of microbial residues in soil	164
7.8	Nitrogen fixation and soil organic matter	165
7.9	Rhizosphere microflora, organic matter and soil fertility	165
7.10	Soil organic matter and plant diseases	165
7.11	Conclusions	166
7.12	References	166

8. Y. Hadar, The role of organic matter in the introduction of biofertilizers and biocontrol agents to soils   **169**

8.1	Introduction	169
8.2	*Rhizobium*	169
8.3	*Azotobacter* and other free living bacteria	171
8.4	Vesicular-arbuscular mycorrhizae	172
8.5	Ectomycorrhiza	173
8.6	Systems for biological control of soilborne plant pathogens	174
8.7	Conclusions	175
8.8	References	176

Section IV: Effects of soil organic matter and applied sewage sludge on soil structure and fertility   **181**

Contents

9. M.H.B. Hayes, Soil organic matter extraction, fractionation, structure and effects on soil structure   **183**

9.1	Introduction	183
9.2	Structure of humus materials	185
9.3	Extraction of humic substances	185
9.4	Extraction of soil polysaccharides	186
9.5	Fractionation of humic substances	187
9.6	Fractionation of soil polysaccharides	188
9.7	Primary structures of humic substances	188
9.8	Secondary and terrtiary structures of humic substances	191
9.9	General conclusions from studies of humic structures	192
9.10	Structures of soil polysaccarides	194
9.11	Interactions of humus materials with soil inorganic — components	194
9.12	Humic substances - clay interactions	195
9.13	Soil polysaccharide - clay interactions	197
9.14	Humus - oxyhydroxide interactions	199
9.15	Humus and soil aggregates	199
9.16	References	204

10. C.E. Clapp, S.A. Stark, D.E. Clay and W.E. Larson, Sewage sludge organic matter and soil properties   **209**

10.1	Introduction	209
10.2	Effect of sewage sludge organic matter on soil physical — properties	210
10.2.1	Bulk density	210
10.2.2	Aggregation and aggregate stability	215
10.2.3	Porosity and pore size distribution	220
10.3	Hydraulic conductivity	224
10.3.1	Moisture retention	226
10.3.2	Effect of sewage sludge organic matter on soil chemical — properties	226
10.3.3	Carbon	232
10.3.4	Nitrogen	232
10.3.5	pH	237
10.3.6	Cation exchange capacity	237
10.3.7	Electrical conductivity	238
10.3.8	Phosphorus	238
10.3.9	Metals	241
10.3.10	Redox potential	242
10.4	Effect of sewage sludge organic matter on soil biological — properties	242
10.4.1	Microorganisms	243

Contents

10.4.2	Macrofauna	245
10.4.3	Plants	246
10.5	References	248

Section V: The use of peat and composts as container media   **255**

11. M.Raviv, Y. Chen and Y. Inbar, Peat and peat subsitutes as growth media for container-grown plants   **257**

11.1	Historical review	257
11.2	General introduction	257
11.2.1	Physical properties	258
11.2.2	Chemical properties	259
11.2.3	Other properties	259
11.3	Physical characteristics	259
11.3.1	Bulk and particle density	260
11.3.1.1	Particle and pore size distribution	261
11.3.2	Porosity and aeration	263
11.3.3	Water holding capacity (water retention curve)	263
11.4	Chemical characteristics	266
11.4.1	Carbon/Nitrogen (C/N) ratio	266
11.4.2	Cation exchange capacity	267
11.4.3	Substrate pH	269
11.4.4	Nutrients availability in organic substrates	269
11.5	Biological characteristics	270
11.5.1	Decomposition rate	270
11.5.2	Effects of decomposition products	271
11.5.2.1	Enzymatic activity	271
11.5.2.2	Growth regulating activity	272
11.6	Organic materials used as growth media	273
11.6.1	Peat	273
11.6.2	Peat substitutes	275
11.6.2.1	Bark	275
11.6.2.2	Sawdust and woodchips	277
11.6.2.3	Sewage sludge and municipal composts	279
11.6.2.4	Treated animal excreta	279
11.6.2.5	Other organic materials	280
11.7	References	280

12. H.A.J. Hoitink and G.A. Kuter, Effects of composts in growth media on soilborne pathogens   **289**

12.1	Introduction	289
12.2	The composting process	290
12.3	Maturity of composts	292
12.4	Chemical properties	293
12.5	Physical properties	295
12.6	Biological properties	297
12.7	References	302

1. Organic residues in modern agriculture

Y. AVNIMELECH

1.1. Introduction

Manures and composts have been used as a means to increase soil fertility and crop production all along the history of farming. Organic residues were the only means of adding nitrogen and the most important means to add other nutrients until development of chemical fertilizers production and distribution systems. Presently, the chemical industry provides concentrated inorganic fertilizers that are easily handled and distributed, and that can supply the need for any nutrient element. This development offsets the use of organic residues as a sole source for nutrients and in some cases eliminates the use of manures and compost to a point where these materials are accumulating and not being used. In many places organic residues are becoming more of a problem rather than an asset. Whereas the old Jewish laws dealt with the ownership of manure found on public properties, modern laws deal with the responsibility for the disposal of manure.

These tendencies are now being questioned by a growing number of scientists and farmers. Evidence and considerations for and against the economical usage of organic residues are the topic of this symposium and volume.

One way to evaluate the economical value of inorganic fertilizer elements in organic residues is to consider the quantities of these elements as equivalent to those supplied by inorganic fertilizers. The assumptions made here are that all the nutrients supplied with organic residue will be released to the soil solution within a short time and that there are no interactions between the organic matter and the nutrient chemistry or availability. These assumptions are probably wrong, yet they will be used here as a first approximation.

Allison (1) considered the fertilizer value of barnyard and poultry manure as $ 2.0 and $ 5.0 per ton, respectively. With these values it was considered that most farm manures (at least cow and pig manures) are too low in nutrients to justify hauling and spreading. The fertilizer prices have changed since, yet hauling and spreading are more expensive. The fertilizer value of barnyard manure is now about 4 times higher than its value in 1962 ($7.6/ton), yet hauling (assuming a 30

km distance) and spreading will cost about 10-12 $ per ton. The value of the nu-
trients as such cannot justify the use of manures. The use of organic residues can
only be justified under two sets of conditions. The first is the case when the residue has a significant negative price, i.e., it has to be disposed of and a price is
given for its hauling. The other set of conditions holds if the organic residue applied has properties and functions above those of an equivalent amount of fertilizer.

1.2. Supply of nutrients by organic additives

The analysis made above was based on the assumption that nutrient elements
in organic additives and those found in fertilizer salts are equivalent. This assumption may be wrong due to a number of mechanisms and circumstances. A major
difference between organic and inorganic sources of nutrients is the rate of nutrient release. Most fertilizers are soluble and the nutrient elements are released
upon application to the soil. This instantaneous release is often a disadvantage.
Ammonia is volatilized from surface-applied fertilizers (10), excessive nitrates are
leached below the root zone and a large portion of the applied phosphorus is fixed.

Slow release fertilizers are being developed now in order to overcome the
above-mentioned problems (14). Manures and composts are, in essence, slow release fertilizers. Nutrients are slowly released from the added organic materials
through the microbially induced mineralization process. The rates of the release
process are controlled through the properties of added organic material and
those of the soil. A labile organic material will decompose faster than a stabilized
material, yet, this does not necessarily mean that nutrient release will be faster.
The decomposition process often leads to an effective binding of nutrients by the
developing and growing biomass. A fast decomposition leads to a high rate of nutrient release only when the organic substrate is rich in nutrients (low C:N and
C:P ratios). Different formulations of manures and composts, different composting techniques and durations, as well as different application levels, technique
and timing, would have an effect on the rate of nutrient release. A stabilised
residue will support a slow rate of nutrient supply while the addition of fresh
residue, rich in nutrients, will lead to a fast supply of nutrients. In similar material
having high C:N and C:P ratios, nutrients will be first immobilized by the growing
biomass and only later released from the manure to the soil. The rates of organic
matter decomposition and nutrient release are unfortunately not measured in
many research works dealing with the nutritive value of manures and composts.
Without this information it is impossible to compare results and draw any general
conclusions.

The net release of nutrients from organic matter is a function of decomposition rates of the different organic matter fractions and of the uptake of nutrients

by the growing biomass. Theory, experimental techniques and simulation models to describe this process are available (*e.g.*, Gasser, (11) Paul and Juma (20), Van Veen *et al.* (27)). It is possible to develop a predictive model for the net release of nutrients (N, P, *etc.*) from a given manure or a compost as a function of time. Based on such models, there is a possibility to design and produce organic additives made to release nutrients at a predetermined rate, as needed by the crop, a characteristic that is yet the unachieved goal of the controlled release fertilizer industry.

Inorganic ammonium applied with fertilizers is nitrified in the soil within a few weeks (4) and thus, the major inorganic component of nitrogen in the soil is usually NO_3. This leads to the rapid leaching of nitrate and often to losses through denitrification. The lack of a stable ammonium supply has an effect on the growing plant. Olsen (22) has reviewed research works indicating that a stable supply of ammonium is essential for achievement of high yields. Slow nutrient release from manure and composts provides a stable supply of ammonium. Manures are suggested, in that work, to be a stable source of ammonium and thus to support maximal yields.

Phosphorus availability can also be increased through addition of organic residues *via* several mechanisms. First, the slow release of inorganic phosphorus during the decomposition of the organic matter provides a continuous supply of phosphorus with a minimal exposure to the different fixation mechanisms. Moreover, the presence of organic matter in the soil effectively decreases the phosphorus fixation by the soil through the acidifying and chelation mechanisms. The decomposing organic matter releases organic acids and CO_2, both tending to lower the soil pH in neutral and basic soils and thus to raise phosphorus solubility (4). The added organic matter and its decomposition products have significant chelation capacity, lowering the activity of the polyvalent cations (Ca, Fe, Al) that form insoluble salts with phosphorus (4).

Another possibly very significant effect is the presence of labile soluble organic phosphorus in the soil solution. In studies related to aquatic systems, it was found that a large fraction (90-99%) of the phosphorus determined with the molybdenum blue method is actually not soluble orthophosphate (18, 24, 25) but polyphosphates and easily hydrolisable organic phosphates. These labile phosphates are hydrolised in the acidic molybdate reagent and are thus included in the conventional orthophosphate determination. An increase in apparent phosphorus solubility was found in soils amended with manure. This increase was due to the fact that a large fraction of the soluble phosphorus was actually soluble organic phosphorus (Avnimelech, unpublished data). The availability to plants and mobility in soils of these phosphorus compounds needs further investigation. It seems that this fraction of the soluble phosphorus may be very significant as to the management of phosphorus supply to plants.

Trace metals availability and solubility are affected by organic matter mainly through chelation (6). The solubility of trace metals is often too low to support optimal plant growth ([e.g., iron solubility in a neutral soil is about 10^{-18} mole/l (19)]. The chelating agents supplied with the organic additives are able to raise the levels of trace metals in the soil significantly [e.g., chelated iron concentration in sanitary landfill leachates is in the order of 10^{-2} mole/l (23)].

1.3. Effects of organic additives on the physical properties of the soil

Organic amendments are known to have favourable effects on soil physical properties. One of the effects most commonly cited is the improvement of the soil structure. Soil structure is a complex function involving cohesion and adhesion between individual particles and among clusters, geometric arrangement and orientation of particles and clusters as well as the stability of any given arrangement under varying conditions of moisture and compaction. The different soil parameters combined under the term structure affect processes such as infiltration, wind and water erosion, root growth, distribution of air and water filled pores, energy consumption for soil cultivation, seed germination, etc. As such, it is very difficult to obtain a simple answer as to the value of the organic amendment effect on soil structure.

One of the measurable functions of soil structure is the bulk density of the soil. Continuous cultivation tends to raise the bulk density, i.e., to compact the soil and thus reduce infiltration, aeration or root growth and to raise the energy needed for cultivation (13).

Khaleel et al. (17) surveyed results of 42 field experiments dealing with the effects of manures and composts on soil properties. A highly significant correlation was found between the increase in soil organic carbon induced by manure application (ΔC) and the lowering in percents of bulk density (ΔBD) of the soil:

$$\Delta BD = 3.99 + 6.62 \; \Delta C \quad R^2 = 0.69**$$

** 0.01 confidence level

The data on the beneficial effect of organic matter on the soil bulk density include a report from the Rothamsted plots where the bulk density of the soil in plots receiving only fertilizers since 1852 was 1.52 compared to a density of 1.29 in plots amended with manure (16). An interesting finding is given by Petterson and Von Wistinghausen (21) reporting that the subsoil was compacted in plots receiving only inorganic fertilizers for a period of 20 years. The subsoil in the manured plots had a better structure and a lower bulk density. Such an effect on the deep subsoil layers would indicate that organic fractions migrate downward and are active below the plowed layer of the soil. Such migration could be due to leaching of soluble organic matter or via bioturbation mechanism such as the movement of earth worms.

Effects of organic amendments on other functions of the soil structure, such as infiltration rate and aggregate stability, have also been reported.

Water holding capacity of soils especially that of sandy soils, is raised by the addition of organic matter. Khaleel *et al.* (17) found a correlation, based on numerous published works, between the increase in field capacity (ΔFC) and the increase in organic carbon (ΔC) in plots treated with organic residues:

$$\Delta FC = Exp[1.09 + 2.141\Delta C - 0.409(\Delta C)^2 - 0.0167(\% \text{ sand}) + \\ + 0.0004(\% \text{ sand})^2] \quad R^2 = 0.81$$

1.4. Effects of organic amendments on yield

The application of organic materials stimulate the growth and activity of heterotrophic microbial population. This in turn may affect plant growth through either the supply of biochemically important substances or through the effects of the saprophytic population on soil microorganisms causing plant diseases. The two mechanisms mentioned above are discussed in this volume (15).

It can be concluded that the application of organic materials has a potential to increase plant yields to an extent above that based on the application of fertilizer equivalent nutrients. One manifestation of such a potential is the effect of organic amendments on the crop yield *vs.* fertilizer application curve, brought schematically in Figure 1.1. The crop yield increases with the increase in fertilizer application until a saturation value above which the yield does not increase, but often decreases (Curve a). If the application of organic amendments contributes to plant growth through mechanisms other than nutrient application then a different response curve (Curve b) and a higher saturation value are obtained.

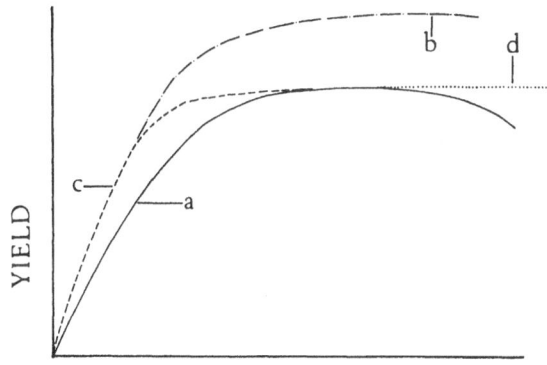

FERTILIZER RATE

Fig. 1.1. Potential effects of manure on nutrient efficiency (schematic).
Curve a: Yield response to inorganic fertilizer; Curve b: Increased efficiency of nutrient uptake and higher maximal yield; Curve c: Increased efficiency of nutrients. Curve d: Prevention of yield decline at high nutrient concentration.

In cases where nutrient elements supplied with the organic matter are of a higher availability, the response curve is steeper yet the saturation value is the same as that obtained for the inorganic fertilizer application (Curve c). An additional potential effect on the yield curve is the prevention of the yield reduction found with excessive amounts of fertilizers (Curve d).

Evidence for the existence of each of these curves can be found in the literature. Cooke (7), reviewing the experience accumulated in England, states that until recently most experiments yielded Type a curves, *i.e.*, the manures had no effect on yields above that of the nutrients included in the organic matter. Since 1970 data has accumulated that fit both Curve c -- indicating a more efficient nutrient application -- and Curve b -- indicating that the manures had effects above those of nutrient application. Similar results are brought by De Haan (8). Avnimelech *et al.* (2) conducted a pot experiment with 4 compacted clay-loam subsoils sampled from 80-120 cm deep layer. A set of sequential treatments have been tested (Table 1.1). The first treatment, nitrogen (130 ppm NH_4-N), had a significant positive effect in 3 of the soils in the first crop grown (oats). The other fertilizer treatments given on top of the added nitrogen did not have any significant

Table 1.1. Effects of fertilizers and manure on yield (g dry matter/3 kg soil; averages of 5 replicates) in compacted clay loam subsoils.

Treatment	Soil location			
	Yesodot	Kochav	Bet Nir	Saad
	First crop: Oats			
Control	0.67	0.38	0.59	0.42
N	0.70	0.74	0.77	0.60
N P K	0.78	0.76	0.78	0.62
N P K + microelements	0.65	0.73	0.71	0.72
N P K + microelements + manure	1.24	1.29	1.32	1.12
	Second crop:Clover			
Control	0.23	0.21	0.22	0.27
N	0.27	0.24	0.23	0.33
N P K	0.25	0.24	0.25	0.31
N P K + microelements	0.20	0.25	0.23	0.30
N P K + microelements + manure	1.19	0.89	0.84	0.66

effect. Application of composted farmyard manure at a rate of 4% (wet manure: dry soil) had a highly significant effect, doubling the first crop and tripling the second crop yield. The effect of the manure does not seem to be due to the addition of nutrients. The manure was applied on top of a rather high level of nutrients, to which the response was very mild. It seems that the soils used, all very low in organic matter content and biological activity, responded to other effects of the manure. At least one of those effects was a very clear change of the soil structure. The soils tested had all very compacted and dense aggregates. These were changed only by manure addition, making the aggregates less dense and friable (Moduli of rapture were reduced to about 50% of the original values). It seems that the change of the soil aggregates from the compacted dense ones to friable porous aggregates allowed root penetration, water and solute diffusion into and out of the aggregates and had thus improved the properties of the soil as a growing medium.

An important feature of the experiment described here is its irreproducibility. Additional experiments with the same soils, yet with different manures and composts, and apparently different incubation conditions of the organic materials with the same soil, failed to yield similar results. This irreproducibility is a typical feature as to the state of the art of the use of organic amendments. In some cases one finds a clear and at times a very dramatic increase in yields and soil fertility following the application of manure or composts. Yet, on the other hand, quite often no response is found, especially in relation to plant yields. Several explanations can be proposed to explain this irreproducibility. First, the application of organic materials is a name attributed to a complex sequence that may have different starting points (composition, physical properties of manure, temperature, moisture of soil manure complex, *etc.*) and many reactions. Raveh and Avnimelech (23) found that the decomposition of municipal refuse can follow two different routes, depending on the temperature and aeration conditions. One route leads to the formation of organic acids as the main product and the other to the formation of humic compounds. The decomposition sequence seemed to follow the route dictated during the initial stages of the process. It is possible that even with the application of the same organic product, different decomposition sequences in the soil will lead to completly different results. It may be that different products are needed for different soils, crops or sets of conditions. One example is the probable contradiction between the efficiency of the added organic material of improved soil structure through the production of polysaccharides and its role as a source of nitrogen to the crop. The first role is better achieved by an organic material having a wide C:N ratio (Somani and Saxena (26)) while the second needs a material with a narrow C:N ratio.

A different explanation as to the variable effects of manures and composts on soil fertility is given by Olsen (22). According to this explanation, an important role of the organic matter is to supply ammonium to high yielding crops. Medium or low yielding crops will not respond to the ammonium supply and thus, the re-

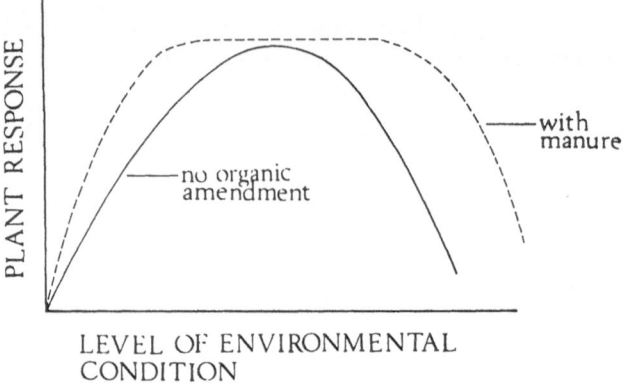

LEVEL OF ENVIRONMENTAL
CONDITION

Fig. 1.2. Schematic representation of the effects of manure on yield of sub or over optimal
levels of environmental conditions.

sponse to the organic amendment will be significant only for the high yielding
crop. Another very general explanation is provided by Flaig *et al.* (9). The role of
the organic matter according to the suggested concept is to "buffer" the system,
i.e., to improve the growing conditions of the plant when a given parameter is
below or above the optimal range. The effect of the organic matter can be represented schematically as in Figure 1.2. As demonstrated, the organic amendment is
not effective when growing conditions are optimal. Yet, it helps to maintain the
maximal yield even when conditions are at either side of the optimal range.

1.5. Conclusions

The major role of organic wastes application in primitive agriculture was to
restore the soil nutrient balance. This role is marginal in modern agriculture. The
addition of nutrients *per se* with the organic wastes will hardly cover the expenses
of the process. Yet, the addition of organic matter to soils has potentially other
functions. The nutrients added with the organic residue are released gradually and
are thus less sensitive to leaching, volatilization or fixation. The gradual release of
ammonium seems to be an important physiological function. In addition, nutrients are often released in a more available form such as chelated metals or organic phosphorus.

The added organic matter induced changes in the physical properties of soils.
A marked improvement in the structure of soils is often detected. This effect

seems to be of prime importance in intensive agriculture. Soil aeration becomes a limiting factor under the conditions of a generous nutrient and water supply and can only be ensured by proper soil structure. In addition, modern agriculture is almost equivalent to the use of heavy machinery, leading to the compaction of the soil. Maintaining a proper soil structure seems to be of major importance under such conditions.

An additional wide spectrum of functions is revealed when biological and bio-chemical effects of the organic matter are reviewed. The biological activity induced by the organic matter has the potential to antagonize root diseases and supply different growth factors.

With all the potential functions, one is faced with a lot of inconsistent information concerning the performance of organic residues in the field. Moreover, we do not have answers to many questions that are essential to the design of composting systems, to the use of differing raw materials or as to the application of organic residues to different soils and crops. The important mechanisms of manure activity are not known to a point where they can be quantified. Even the characterization of manures and composts is quite vague and there is no clear idea as to what properties should be tested, evaluated or graded.

1.6. References

1 Allison F.E. 1973 Soil organic matter and its role in crop production. Elsevier Sci. Pub. Co. (Amsterdam) 637 p.
2 Avimelech Y., Polachek Y., Moses A. and Cohen E. 1971 The effects of manures on the fertility of compacted soils. Hasadeh 51, 316-318 (Hebrew).
3 Baver L.D. 1963 The effect of organic matter on soil structure. *In* Organic Matter and Soil Fertility Pontif. Acad. Sci. Scripta Varia 32, 383-413. North Holland Publ. Co Amsterdam.
4 Black C.A. 1968 Soil-plant Relationship, John Wiley and Sons Inc. N.Y. Second Edition, 792 p.
5 Broadbent F.E. 1968 Turnover of nitrogen in soil organic matter. *In* Organic Matter and Soil Fertility Pontif. Acad. Sci. Scripta Variam 32, 61-82. North Holland Publ. Co. Amsterdam.
6 Chen Y. and Stevenson F.J. 1986 Soil organic matter interactions with trace elements. *In* The Role of Organic Matter in Modern Agriculture. Eds. Y. Chen and Y. Avnimelech. Martinus Nijhoff, Dordrecht, pp 73-115.
7 Cooke G.W. 1979 Some priorities for Britisch soil science. Soil Sci. 30, 187-213.
8 De Haan S. 1977 Humus, its formation, its relation with the mineral part of the soil and its significance for soil productivity. *In* Soil Organic Matter Studies, Interna. Atomic Energy Commission, Vienna, STI/PUB/438. Vol 1, 21-30.
9 Flaig W., Nagar B., Sochtig H. and Tietjen C. 1977 Organic materials and soil productivity. FAO Soil Bull. 35, Rome 1977. 119 p.
10 Freney J.R. and Simpson J.R. (Eds), 1983 Gaseous Loss of Nitrogen from Plant-Soil Systems. Martinus Nijhoff, The Hague. 317 p.
11 Gasser J.K.R. (Ed). 1979 Modelling nitrogen from farm waste, Applied Sci. Publ. Ltd. London 195 p.

10

12 Gattani P.D., Jain S.V. and Seth S.P. 1976 Effects of continuous use of chemical fertilizers and manures on soil physical and chemical properties. J. Indian Soc. Soil Sci. 24, 284-289.

13 Hafez A.A.R. 1974 Comparative changes in soil physical properties induced by admixtures of manures from various domestic animals. Soil Sci. 118, 53-59.

14 Hauke R.D. 1972 Synthetic slow release fertilizers and fertilizer amendments. In C.A.I. Goring and Hamaker J.W. (Eds.) Organic chemicals in the soil environment. Vol 2, pp. 633-690. Marcel Dekker, N.Y.

15 Hoitink H.A.J. and Kuter G.A. 1986 Effects of composts in growth media on soilborne plant pathogens. In The Role of Organic Matter in Modern Agriculture. Eds. Y. Chen and Y. Avnimelech. Martinus Nijhoff, Dordrecht. pp. 289-306.

16 Jenkinson D.S. and Johnston A.E. 1977 Soil organic matter in the Hoosfield continuous barley experiment. Report, Rothamsted Exp. Sta. 1976, Part 2, 87-101.

17 Khaleel R., Reddy K.R. and Overcash M.R. 1981 Changes in soil physical properties due to organic waste application : A review. J. Environm. Quality 110, 133-141.

18 Kuenzler E.F. and Ketchum B.H. 1962 Rate of phosphorus uptake by *Phaeodactylum tricornutum*. Biol. Bull. 123, 134-145.

19 Lindsay W.L. 1979 Chemical Equilibria in Soils. John Wiley and sons, N.Y. 449 p.

20 Paul E.A. and Juma N.G. 1981 Mineralization and immobilization of soil nitrogen by microorganisms. In Terrestrial Nitrogen Cycles. Processes, Ecosystem Strategies and Management Impacts. Eds. F.E. Clark and T. Rosswall. Ecological Bull. (Stockholm) 33, 179-195.

21 Pettersson B.D. and von Wistinghausen E. 1979 Effects of organic and inorganic fertilizers on soils and crops. Results of long term experiment in Sweden. Misc. Publications. Woods End Agric. Inst. No. 1, 44 p.

22 Olsen S.R. 1986 The role of organic matter and ammonium in producing high corn yields. In The Role of Organic Matter in Modern Agriculture. Eds. Y. Chen and Y. Avnimelech. Martinus Nijhoff, Dordrecht. pp. 29-54.

23 Raveh A. and Avnimelech Y. 1979 Leaching of pollutants from sanitary landfills models. J. Water Pollution Control Fed. 51, 2705-2716.

24 Rigler F. 1966 Radiobiological analysis of inorganic phosphorus in lakewater. Int. Ver. Theor. Angew. Limnol. Verh. 16, 465-470.

25 Rigler F. 1968 Further observations inconsistent with the hypothesis that the molybdenum blue method measures orthophosphate in lake water. Limnol. Oceanogr. 13, 7-13.

26 Somani L.L. and Saxena S.N. 1975 Effects of some organic matter sources on nutrient availability, humus build-up, soil physical properties and wheat yield under field conditions. Annals of Arid Zone 14, 149-158.

27 Van Veen J.A., McGill W., Hunt H.W., Frissel M.J. and Cole C.V. 1981 Simulation models of the terrestrial nitrogen cycle. In Terrestrial Nitrogen Cycles. Processes, Ecosystem Strategies and Management Impacts. Eds. F.E. Clark and T. Rosswall. Ecological Bull. (Stockholm) 33, 25-48.

28 Weil R.R. and Kroontje Y. 1979 Physical conditions of a Davidson clay loam after five years heavy poultry manure application J. Environ. Qual. 8, 387-392.

Section 1

Nitrogen and phosphorus supply to plants by organic matter and their transformations

2. Effects of organic matter on nitrogen and phosphorus supply to plants

F.E. BROADBENT

2.1. Introduction

Nitrogen is unique among major nutrient elements in that soil reserves are almost entirely in the organic form. The quantity and nature of soil organic matter accordingly, have a very significant influence on the availability of this element to growing plants. While many soils contain important amounts of organic phosphorus, the influence of organic matter on phosphorus availability is not as great as in the case of nitrogen. While carbon and nitrogen are usually mineralized in fairly constant proportions, phosphorus release from the organic form is not nearly as well correlated with that of carbon (85).

2.2. Nitrogen

2.2.1. Forms of organic N in soil

Characterization of forms of N in soil hydrolyzates shows the presence of a considerable amount of ammonium-ion derived primarily from amides, amino acids, amino sugars, and small amounts of nucleic acids (9, 24). Typically, a substantial fraction, often as much as half the total, cannot be identified in any of these forms. On the basis of analyses of $6N$ HCl hydrolyzates of 14 proteins, Greenfield (32) concluded that the unaccounted-for N originates mainly from non-amino N in arginine, tryptophan, lysine and proline. He believes most soil N is in amino acids, some of which form complexes with other organic materials such as tannins and polysaccharides. Several investigators (42, 80) have discussed the concept of an active or labile fraction of soil organic matter from which most mineralizable N is derived, and numerous attempts have been made to characterize such an active fraction and to estimate its size. The organic matter of mineral soils may differ from that of organic soils, as suggested by a study of organic N distribution in selected peats and peat fractions (79), where it was found that amino acid and amino sugar contents of peat soils appears to be characteristic of individual bogs, in contrast to the situation with mineral soils.

Y. Chen and Y. Avnimelech (eds.), The Role of Organic Matter in Modern Agriculture.
ISBN 90-247-3360-X.
© 1986, Martinus Nijhoff Publishers, Dordrecht.

Swift and Posner (82) found N and P contents of high molecular weight fractions of soil humic acids to be higher than in lower molecular weight fractions, and suggested that the degradation process leads to preferential loss of N and P from the low molecular weight fractions, which presumably would comprise a part of the active or labile pool. Freney and Simpson (27) studied mineralization of labeled and indigenous N in soils of different N contents after long periods of equilibration with added $^{15}NH_4^+$. They found the non-distillable acid-soluble fraction of soil N was decomposed to greater extent than other labeled N fractions, and that labeled N was more available than indigenous N. They concluded that chemical fractionation may not be an especially useful technique for estimating mineralizable N unless it can be combined with a technique for separating recently formed organic N from older forms. Several investigators have observed progressive stabilization of N recently incorporated into the organic fraction (13, 45, 66). Cropping of 12 prairie surface soils in Saskatchewan was found by Campbell and Souster (14) to result in the loss of 31-56% of the potentially available N.

Michrina et al. (56) determined molecular weight distribution in two extracts used to sample the pool of available soil N, 0.01 M CaCl$_2$ and 0.01 M NaHCO$_3$, and found that 78-87% of the humic substances in the CaCl$_2$ extract had molecular weights less than 1000 daltons, whereas in the NaHCO$_3$ extract 42-83% of the humic substances were in the 1000 to 21,000 range. Of the latter, 43-92% was protein and 8-30% was ninhydrin-positive. In another study particle size fractionation of soils in relation to N availability showed that an increasingly greater proportion of N within size fractions was mineralized as particle size decreased (15). However, Robinson (70) concluded that use of soil fractions smaller than 2 mm would not be advantageous for use in incubation procedures to estimate mineralizable N.

2.2.2. N mineralization

2.2.2.1. Organic soils

Conversion of organic N to the mineral form is usually rapid in peat and muck soils, often in excess of N requirements of crop plants. On this account there has been little work done with organic soils to evaluate their N mineralization potential. Studies of N release rates have more often been based on concern for the possibility of water pollution in aquifers into which drainage from organic soils is discharged (84). Net release of 500-600 kg N ha^{-1}yr^{-1} or more from organic soils has been reported (33, 84), and even these values probably underestimate total N mineralization by a considerable margin because of the conditions in peat soils which are very favorable to denitrification (33). Rate of N mineralization, like subsidence, is related to depth of water table (87). Avnimelech (5) observed a linear rate of nitrate production in peat samples incubated aerobically. Guthrie and Duxbury (33) reported that rate of N mineralized in drained columns of peat soils was essentially independent of incubation time. The data of Terry (84) for six Histosols in Florida likewise show rates of N release which are nearly linear. He estimated that for a Pahokee muck soil 686 kg N ha^{-1} is mineralized for each cm of soil lost to microbial oxidation. Reddy (67) estimated that 1.3 to 4.2% of the total soil organic N was mineralized per year in Florida Histosols. In his experiments amounts of N mineralized ranged as high as 1230 kg N ha^{-1}yr^{-1} for virgin soils.

2.2.2.2. Mineral soils

In soils with low or moderate concentrations of organic matter the rate of N release to plants is thought to be controlled by the size of an active organic pool. Jansson (42) postulated this pool to be in equilibrium with a larger, more inert source of organic N. In reality, it seems probable that organic N in soils represents a continuum in terms of availability, rather than two discrete pools of definable size, but for purposes of estimating N supplying capacity of soils the concept is useful.

2.2.3. Available nitrogen

Three general approaches to estimation of that portion of soil organic N which is readily subject to mineralization are the use of chemical extractants, incubation procedures and field measurement of N uptake by crops.

2.2.3.1. Chemical extractants

Many attempts have been made to devise suitable extractants for soil N which can be used for predicting N-supplying capacity (22, 47, 77). The extensive literature on this subject has been reviewed by Dahnke and Vasey (22), but a few more recent developments will be mentioned here. Fox and Piekielek (26) compared eight methods for obtaining an index of N availability with field uptake of N by corn at eight locations in two seasons. Boiling 0.01 M CaCl$_2$ and 0.01 M NaHCO$_3$-extractable N were both significantly correlated with field response at the 1% level, and total soil N and autoclave-extractable ammonium-N were correlated at the 5% level. Sahrawat (71) found that organic C and total N were as well correlated with N-supplying capacity of 39 wetland rice soils as were any other of six chemical methods or two incubation methods. He proposed that mineralizable N and organic C could be determined simultaneously by distillation and measurement of the ammonium-N released by the oxidative action of chromic acid in the Walkley-Black procedure (72). Kawaguchi and Kyuma (44) surveyed a total of 505 soils from six southern Asian countries and found a highly significant correlation between acid-hydrolyzable N and the quantity of ammonium-N produced by anaerobic incubation for two weeks.

2.2.3.2. Incubation procedures

In recent years considerable attention has been focused on incubation procedures and on attempts to determine biomass N. These are based on the premise that a labile fraction of soil organic matter exists which can be readily converted to mineral N. Stanford and Smith (80) designated the size of this labile pool by the symbol N_o and assumed that decomposition of this organic N proceeds by first order kinetics according to the equation $N_o - N_m = N_o \exp(-kt)$, where N_m is quantity of N mineralized during time t. In practice several values of N_m are measured at various times and a curve fitted to the points, from which N_o and k are estimated by an iterative process after selection of a reasonable value of N_o as a first estimate. In making their first estimate Stanford and Smith (80) implicitly assumed a hyperbolic model of the form $N_m = t/(a + N_o t)$, where N_o is the y-asymptote of the hyperbola. Other investigators (50, 76, 83) have suggested modifications in the procedure or of the equation to permit more accurate estimates of N_o and of the rate constant k. For example, in a study of chaparral soils in Southern California (50) it was found that a modification of the transformed equation of the $\log(N_o - N_m) = \log N_o - kt^b$ predicted N mineralization values which agreed within $\pm 10\%$ with independent estimates of N mineralization based on N-balance equations. Talpaz *et al.* (83) analyzed the data of Stanford and Smith (80) by a nonlinear regression procedure and obtained lower N_o values and higher k values than were reported in the original paper.

The basic equation of Stanford and Smith (80) can be written in the form $N_m = N_o (1 - e^{kt})$, from which it is clear that N_m approaches N_o as t becomes large, implying eventual exhaustion of the mineralizable pool of soil N, which in fact does not occur. Thus, although the first-order model is conceptually attractive, it may not be superior to other methods of estimating N-supplying capacity of soils. It has long been known that a parabolic function of the form $y = at^b$ describes soil decomposition processes quite well, where y is a measure of the extent of decomposition, t is time and a and b are constants (21). The observation by Stanford and Smith (80) and others (74) that N_m plotted against the square root of time gives a straight line function for many soils is a special case of the general parabolic function where b = 0.5. As a matter of fact, the best fitting parabolas to their data have values which differ somewhat from 0.5.

Of probably greater importance than goodness of fit to mathematical models are the conditions of measurement of mineralized N. Smith *et al.* (76) observed that leaching the soil with reagents such as 0.01 M $CaCl_2$ to remove mineralized N also removes significant amounts of soluble organic N. They found estimates of N_o and k to be influenced considerably by the inclusion of soluble organic N in the leachate. Table 2.1 presents correlation coefficients among total soil N, measured values of N_m and N_o estimated by a hyperbolic model and by the first-order model for the 39 soils with which Stanford and Smith worked (80).

Table 2.1. Correlation of measured values of N mineralized with organic C, organic N and estimates of N_o by two methods for 39 soils*

	N_m	Organic C	Organic N	N_o hyperbolic model	N_o first-order model
N_m	––	.728	.817	.986	.974
Organic C		––	.959	.724	.768
Organic N			––	.819	.841
N_o (hyperbolic model)				––	.976

*From data of Stanford and Smith (80) and Talpaz *et al.* (83)

Measured N_m values and estimates of N_o by both models are very highly correlated. For practical purposes simple measurement of N_m appears to be as useful as any calculated values derived from it.

Recognizing that mineral N is unlikely to be derived from a single organic pool, Richter *et al.* (69) developed reaction coefficients for three components of soil organic N on the basis of long-term incubation values. If resistant organic matter is assigned a reaction coefficient of 1, decomposable plant residues would have a value of approximately 28 and autolyzing microbial biomass a value of 60. In a comparison of zero-tillage systems with conventional tillage Carter and Rennie (17) reported that estimated N_o values increased from 35 to 132% in the surface layers of zero-tillage systems. Values of N_o expressed as a percentage of total organic N ranged from 2.5 to 15% in zero-tillage and 3.9 to 9% in conventional tillage systems. Reserve N fertility in soil layers below 2 cm depth was significantly lower in zero-tillage systems.

2.2.3.3. Field estimates of N availability

One method of measuring rate of N supply to crops is based on the decline of organic N which often occurs over a long period of time. Using a first-order model Bartholomew and Kirkham (7) estimated decomposition rates in several long-term experiments which ranged from 2.2 to 10% per year. Reinhorn and Avnimelech (68) observed that N release from newly cultivated soils was linearly related to total organic N, with an annual loss rate of 32%. These very high rates obviously could not continue beyond the first few years of cultivation. Among the 21 pairs of virgin and cultivated soils for which they reported data were one which lost

42% of the organic N in the 0-30 cm layer in only two years of cultivation and another wich contained 144% of the original organic N after 30 years.

2.2.3.4. Crop uptake of soil N

Where isotopically labeled fertilizer N is applied to a crop it is possible by plant analysis to determine the contribution of soil N to growth of the crop. Broadbent (12) has summarized results of a number of such experiments for a few major crops. Values range from 32 to 216 kg N ha^{-1} over the growing season. Clearly such estimates are affected a great deal by management practices and nature of the crop as well as by soil properties. Rates of uptake of soil N by a variety of crops were found to vary over a rather narrow range from 0.3 kg N ha^{-1} day $^{-1}$ for sudangrass late in the season to 2.3 for corn during the period of most rapid growth. Presumably these N-uptake rates reflected mineralization rates in the soil. The A-value of Fried and Dean (28) provides another index of N availability, but when comparing values for different soils it is important that the measurements be made under similar conditions, since they are affected by fertilizer placement, type of fertilizer and other variables. A-values varying from 3 to 15% of total N have been reported (10).

2.2.4. Environmental influences

Among environmental factors which influence the release of N from soil organic matter, temperature and moisture have received the most attention. Myers et al. (60) recently reported a quantitative relationship between net N mineralization and moisture content over the range between -0.3 and -4.0 MPa, with the optimum between -0.01 and -.03. The effect of flooding is usually to increase rates of N mineralization (8, 63). Hirose and Kumada (39) reported that N mineralization increased with increasing moisture content within the range 50 to 90% of water holding capacity. Subsequent drying and then flooding showed an inverse relationship between quantities of N formed during aerobic and waterlogged stages. Stanford et al. (81) reported that the temperature coefficient Q_{10} for N mineralization was approximately 2.0 between 5 and 35°. Temperature effects on N mineralization are important to consider in predicting N-supplying capacity of soils.

2.2.5. Forest soils

Powers (64) reported that soil N mineralized during 14-days anaerobic incubation correlated significantly with N mineralized anaerobically for 6 months in the field and with site index, yield potential and foliar concentration of N in *Pinus ponderosa.* Net mobilization of N as measured by uptake by Douglas fir seedlings from forest floor materials under nine different stand understory conditions ranged from 0.34 to 9.9% of the amount added as organic matter. Net N mineralization was only partially related to N content of the material. In another study (88) mineralizable N estimated by anaerobic or aerobic incubation or by three different chemical extracts were not significantly correlated to a 5-year volume growth response to N fertilization, to site index or foliar N.

2.3. Phosphorus

The influence of organic matter on P supply to plants is somewhat more difficult to evaluate than in the case of N, since plants derive a significant proportion of their P from inorganic sources. In addition, the release of inorganic P from organic forms may be followed by sorption and precipitation reactions which alter the availability of the P mineralized. Moreover, organic matter may indirectly influence the P supply to plants through promoting growth and activity of phosphate-dissolving organisms.

2.3.1. *Nature of soil organic P*

Phosphate esters of inositol comprise a significant part of the total organic P (31), and nucleic acids and phospholipids have also been identified. Dalal (23) who recently reviewed the subject of soil organic P, states that the chemical nature of about half the organic P is unknown. On the basis of published data from seven countries he reported values from 0.4 to 83% of organic P in the form of inositol derivatives, 0.5 to 7.8% as lipid P and 0.2 to 2.4% in the form of nucleic acids. Some reports suggest that part of the unaccounted-for P may be in the form of monophosphorylated carboxylic acids and as teichoic acids (3, 23). Phosphoprotein and sugar phosphates have also been reported (23). None of these forms is inherently difficult to decompose, and direct addition to soil of compunds representative of these groups shows them to be readily available to plants. The low availability of native organic P in soils must therefore be attributed to sorption, fixation and precipitation reactions with other soil constituents. The presence of considerable phosphatase activity in most soils, at least part of which is cell-free, indicates the importance of P mineralization in plant nutrition.

2.3.2. *P mineralization*

Under conditions favoring accumulation of organic matter in soils organic P makes up a large proportion of total P (41, 46, 87). Moreover, P content of the soil parent material exerts a strong influence on the degree to which organic matter accumulates (41). The incorporation of P into soil organic matter involves assimilation both by plants and microorganisms, the decomposition products of which eventually become part of the soil humus (19, 20). In the early work of Thompson *et al.* (85) it was observed that the quantity of organic P mineralized was correlated positively with release of organic C, organic N and with soil pH. Rates of mineralization of all three elements were greater in virgin than in cultivated soils. Nucleic acids are readily hydrolyzed in soils (31). Islam and Ahmed (40) found that the maximum mineralization of inositol phosphates was obtained within the first 30 days of incubation. However, formation of insoluble salts retard the breakdown of inositol phosphates.

2.3.2.1. Phosphatases

Both free and combined phosphatases appear to be ubiquitous in soils (4, 25), and are intimately associated with other components of organic matter. Separation of phosphatase activity from humic substances is difficult. In an examination of 10 soils ranging in C content from 2.3 to 46% Sarathchandra and Perott (73) found that phosphatase activities generally increased with increasing C content, but two peat soils were exceptions to this trend.

2.3.2.2. Effect of liming

The frequently observed beneficial effect of liming on phosphate availability has been attributed in part to enhanced microbial activity (37). Halstead *et al.* (35) reported that liming produced an average decline of 3.6% of the total organic P in seven acid soils of eastern Canada. Marked increases in microbial activity were associated with lower values for extractable organic P. In a study of 50 woodland soils in England the rate of P mineralization was found to be closely related to phosphatase activity and to soil respiration in all soils except those overlying limestone (36). The mineralization rate of ^{32}P- labeled RNA applied to these soils was not significantly correlated with their organic matter content. Gerritse and van Dijk (29) found a correlation between V_{max} values for phosphatase and level of organic P.

2.3.3. Soil organisms in relation to P availability

Recently a great deal ·of attention has been focused on the role of mycorrhizae in uptake of nutrient elements, especially P, and enhanced P nutrition of mycorrhizal plants has frequently been reported. For many years it has been recognized that non-mycorrhizal organisms may also effect phosphate availability in soils. Inoculation of soils with bacterial cultures was common, particularly in eastern Europe where a culture called Phosphobacterin was used. The value of such inoculants for increasing crop yield was not confirmed in greenhouse and field trials in the U.S. (75). Louw and Webley (48) obtained more than 100 isolates of phosphate dissolving organisms from the root zone of oats. Some of these were found to produce lactic acid and 2- ketogluconic acid from glucose in pure culture. Presumably these acids would dissolve insoluble phosphates if produced *in situ*. In a study of factors affecting the solubility of phosphate during the microbial decomposition of plant material, Bromfield (14) observed that ferric phosphate and di- and tricalcium phosphates became more soluble during decompositon of subterranean clover in a submerged soil, whereas rock phosphate and aluminum phosphate were not dissolved under these conditions. Ferric phosphate was not dissolved during incubation in a well-aerated soil, but it became more soluble as aeration decreased or as the amount of clover increased. A survey of microorganisms isolated from the root surface, rhizosphere soil and non-rhizosphere soil of three pasture grasses showed that total numbers of organisms capable of attacking phenolphthalein diphosphate, sodium glycerophosphate, sodium phytate, lecithin, ribonucleic and deoxyribonucleic acids were higher on the root surface and in rhizosphere soil than in non-rhizosphere soil (30). Moghimi *et al.* (57, 58) attributed the ability of wheat plants to absorb P more efficiently from hydroxyapatite than did maize or peas to rhizosphere organisms and their products, of which 2-ketogluconic acid represented about 20% of the total. Banik and Dey (6) reported that two *Aspergillus* species, two *Bacillus* strains and one actinomycete were efficient in solubilizing phosphates of calcium, aluminum and iron. Isolates which produced oxalic and tartaric acids showed higher dissolving ability, whether or not citric acid was also produced. However, soils inoculated with these organisms showed no change in available P status.

2.4. Organic amendments

In recent years soils have been increasingly utilized as a repository for a wide variety of organic wastes. Although the diversity of such materials makes a quantitative treatment of mineralization processes very difficult, information on these materials is increasing and a substantial number of papers dealing with manures, composts and sludges is appearing in current scientific literature. An exhaustive review of this literature is beyond the scope of this paper, but a brief treatment will be included here.

2.4.1. Animal manures

Rates of mineralization of N and P in animal manures are usually somewhat higher than in soil organic matter, but vary a great deal depending on the nature of the material, condition of storage prior to application and climatic conditions. Herron and Erhart (38) reported that a metric ton of feedlot manure applied to sorghum was equivalent ot 11 kg N from NH_4NO_3, and that it increased the Bray and Kurtz-available P by 1 ppm. Mathers and Stewart (53) reported that in their experiments 2.4 kg manure N was equivalent to 1 kg N as NH_4NO_3. Numerous studies have shown that heavy manure applications have increased nitrate concentrations in the soil and subsoil, but quantitative estimates of the degree of mineralization associated with nitrate formation are rare (49, 51, 59). Pratt et al. (65) conducted a four-year field trial with animal manures and reported that manure containing 1.6 to 2.2% N appeared to mineralize at the rate of 40-50% during the first year following application, 10-20% the second year and 5% the third year in irrigated soils of Southern California. Mathers and Goss (52) developed equations of the form $R = At^B$ to describe the N supplying capacity of 9 manure sources with differing N contents. In these equations R is the amount of manure required to supply 100 kg available N per year, t is the time and A and B are functions of the total N content of the manure. Castellanos and Pratt (18) found that available N in several manures was well correlated with N released by a pepsin. The correlation with N availability was better than with pepsin digestion. They found that composting of manures drastically reduces their value as N fertilizers.

Manure effects on available P levels in soils appear to be dramatic and long lasting. Application of manure over a period of 117 years at the Rothamsted Experiment Station in England showed that manure produced large increases in soluble P and that superphosphate was much less effective than manure in raising soluble P levels when both were applied at approximately equivalent P levels (61). Meek et al. (54) reported that application of 392 kg P/ha as triple superphosphate to a calcareous soil in each of two years increased $NaHCO_3$- extractable P by 11 ppm, whereas 334 kg P/ha from applied manure resulted in an increase of 100 ppm in bicarbonate-soluble P. In a long term study (55) it was found that after a single 180 ton application of manure, $NaHCO_3$- extractable P decreased about 9% per year after manure application ceased.

2.4.2. Other organic amendments

A wide variety of other kinds of organic wastes are applied to soils, including crop residues, composted municipal wastes and sewage sludge. It is very difficult to make generally applicable statements about mineralization of N and P in such diverse materials. The familiar C/N ratio is a useful guide to the prediction of N mineralization in crop residues. The threshold ratio for net mineralization is about 30:1. However, substantially lower ratios are required before sufficient release of N to be of importance in crop production can be expected.

Sludge compositions vary a great deal according to the nature of the treatment process. As noted by Sommers (78) extreme variation exists in sludge composition and the parameters which are most useful in controlling application to agricultural land such as total N, NH_4^+ -N and metals, are those which exhibit the greatest variability. Anderson (2) found up to 55% of the N in aerobically digested sludge to be mineralized in 8 weeks, compared with 20% for anaerobic sludge. Peterson *et al.* (62) found that after an application of 1.25 cm of anaerobically digested sewage sludge recovery of N by two years of corn averaged about 15% and recovery of P 6%. At a 20 cm level of sludge application N recovery was 4% and P 2%. Amundson and Jarrell (1) reported that the value of the N in metric ton of aerobically digested sludge was equivalent to 24-37 kg N in the form of $(NH_4)_2SO_4$, and the corresponding value for anaerobically digested sludge was 9-13 kg N. In sludge applications the rates of mineralization of N and P are not usually limiting considerations; rather, the principal concern is to avoid excessive amounts.

2.5. Conclusions

Considerable progress has been made in attempts to express the conversion of soil organic nitrogen to mineral forms in quantitative terms, but like other biological processes it is influenced by environmental variables and by soil properties. The uncertainty in extrapolating measured data obtained either by chemical extraction or by incubation makes a precise prediction of the N-supplying capacity of soils an unattainable goal at the present time. However, existing methods of estimating potentially mineralizable N still have considerable value in practical N management in spite of their limitations.

Soil organic matter and microbial activities related to amount and nature of soil organic matter influence the supply of P available to plants, but interactions of mineralized P with inorganic soil constituents make it much more difficult to assess the contribution of organic P to the total. A considerable fraction of both organic N and organic P has not yet been adequately characterized, much of it probably incorporated into complex substances of high molecular weight. Fundamental research on the nature of soil organic matter has in the past usually produced gradual progress rather than dramatic breakthroughs, but much of this type of work is needed in relation to the processes which control mineralization of nitrogen and phosphorus.

2.6. References

1 Amundson R.G. and Jarrell W.M. 1983 A comparative study of bermudagrass grown on soils amended with aerobic or anaerobically digested sludge. J. Environ. Qual. 12, 508-513.

2 Anderson M.S. 1956 Comparative analyses of sewage sludges. Sewage Ind. Wastes 28, 132-135

3 Anderson G. 1980 Assessing organic phosphorus in soils. *In* The Role of Phosphorus in Agriculture, Eds. E.C. Sample and E.J. Kamprath, pp. 411-431 Amer. Soc. Agron., Madison, Wisconsin.

4 Appiah M.R. and Thomas R.L. 1982 Inositol phosphate and organic phosphorus contents and phosphatase activity of some Canadian and Ghanian soils. Can. J. Soil Sci. 62, 31-38.

5 Avnimelech Y. 1971 Nitrate transformation in peat. Soil Sci. 111, 113-118.

6 Banik S. and Dey B.K. 1982 Available phosphate content of an alluvial soil as influenced by inoculation of some isolated phosphate solubilizing microorganisms. Plant and Soil 69, 353-364.

7 Bartholomew W.V. and Kirkham D. 1960 Mathematical descriptions and interpretations of culture induced soil nitrogen changes. Trans. 7th Intern. Cong. Soil Sci. 2, 471-477.

8 Borthakur H.P. and Mazumder N.N. 1968 Effect of lime on nitrogen availability in paddy soil. J. Indian Soil Sci. Soc. 16, 143-147.

9 Bremner J.M. 1949 Studies on soil organic matter. I. The chemical nature of soil organic nitrogen. J. Agric. Sci. 39, 183-193.

10 Boadbent F.E. 1970 Variables affecting A-values as a measure of soil nitrogen availability. Soil Sci. 110, 19-23.

11 Broadbent F.E. 1979 Mineralization of organic nitrogen in paddy soils. *In* Nitrogen and Rice, International Rice Research Institute, Los Banos, Philippines, pp. 105-116.

12 Broadbent F.E. 1984 Plant use of soil nitrogen. *In* Nitrogen in Crop Production. Eds. R.D. Hauck and D.A. Russell American Society of Agronomy, Madison, Wisconsin.

13 Broadbent F.E. and Nakashima T. 1967 Reversion of fertilizer nitrogen in soils. Soil Sci. Soc. Am. Proc. 31, 648-652.

14 Bromfield S.M. 1960 Some factors affecting the solubility of phosphate during the microbial decomposition of plant material. Aust. J. Agric. Res. 11, 304-316.

15 Cameron R.S. and Posner A.M. 1979 Mineralisable organic nitrogen in soil fractionated according to particle size. J. Soil Sci. 30, 565-577.

16 Campbell C.A. and Souster W. 1982 Loss of organic matter and potentially mineralizable nitrogen from Saskatchewan soils due to cropping. Can. J. Soil Sci. 62, 651-656.

17 Carter M.R. and Rennie D.A. 1982 Changes in soil quality under zero tillage farming systems: Distribution of microbial biomass and mineralizable carbon and nitrogen potentials. Can. J. Soil Sci. 62, 587-597.

18 Castellanos J.Z. and Pratt P.F. 1981 Mineralization of manure nitrogen-correlation with laboratory indexes. Soil Sci. Soc. Am. J. 45, 354-357.

19 Chauban B.S., Stewart J.W.B. and Paul E.A. 1979 Effect of carbon additions on soil labile inorganic and organic and microbially held phosphate. Can. J. Soil Sci. 59, 387-396.

20 Chauban B.S., Stewart J.W.B. and Paul E.A. 1981 Effect of labile inorganic phosphate status and organic carbon additions on the microbial uptake of phosphorus in soils. Can. J. Soil Sci. 61, 373-385.

21 Corbet A.S. 1934 Studies on tropical soil microbiology: I. The evolution of carbon dioxide from the soil and the bacterial growth curve. Soil Sci. 37, 109-115.

22 Dahnke W.C. and Vasey E.H. 1973 Testing soils for nitrogen. *In* Soil Testing and Plant Analysis. Eds. L.M. Walsh and J.D. Beaton. Soil Science Society of America, Madison, Wisconsin, pp. 97-114.

23 Dalal R.C. 1977 Soil organic phosphorus. Adv. Agron. 29, 83-117.

24 Dubach P. and Mehta N.C. 1963 The chemistry of soil humic substances. Soils Fertil 26, 293-300.

25 Eivazi F. and Tabatabai M.A. 1977 Phosphatases in soils. Soil Biol. Biochem. 9, 167-172.

26 Fox R.H. and Piekielek W.P. 1978 Field testing of several nitrogen availability indexes. Soil

Sci. Soc. Am. J. 42, 747-750.

27 Freney J.R. and Simpson J.R. 1969 The mineralization of nitrogen from some organic fractions in soil. Soil Biol. Biochem. 1, 244-251.

28 Fried M. and Dean L.A. 1952 A concept concerning the measurement of available soil nutrients. Soil Sci. 73, 263-271.

29 Gerritse R.G. and van Dijk H. 1978 Determination of phosphatase activiteis of soils and animal wastes. Soil Biol. Biochem. 10, 545-551.

30 Greaves M.P. and Webley D.M. 1965 A study of the breakdown of organic phosphates by micro-organisms from the root region of certain pasture grasses. J. Applied Bacteriol. 28, 454-465.

31 Greaves M.P. and Wilson M.J. 1970 The degradation of nucleic acids and montmorillonite-nucleic acid complexes by soil micro-organisms. Soil Biol. Biochem. 2, 257-268.

32 Greenfield L.G. 1972 The nature of the organic nitrogen of soils. Plant and Soil 36, 191-198.

33 Guthrie T.F. and Duxbury J.M. 1978 Nitrogen mineralization and denitrification in organic soils. Soil Sci. Soc. Am. J. 42, 908-912.

34 Halm B.J., Ahenkorah Y. and Appiah M.R. 1982 Inositol penta- and hexaphosphate content of some cocoa soils in Ghana. Zeitschrift Pflanzenernaehr. Bodenkd. 145, 586-592.

35 Halstead R.L., Lapensee J.M. and Ivarson K.C. 1963 Mineralization of soil organic phosphorus with particular reference to the effect of lime. Can. J. Soil Sci. 43, 97-106.

36 Harrison A.F. 1982 Labile organic phosphorus mineralization in relationship to soil properties Soil Biol. Biochem. 14, 343-351.

37 Haynes R.J. 1982 Effects of liming on phosphate availability in acid soils: A critical review. Plant and Soil 68, 289-308.

38 Herron G.M. and Erhart A.B. 1965 Value of manure on an irrigated calcareous soil. Soil Sci. Soc. Am. Proc. 29, 278-281.

39 Hirose S. and Kumada K. 1963 Mineralization of native organic nitrogen. J. Soil Sci. Tokyo 34, 339-344.

40 Islam A. and Ahmed B. 1973 Distribution of inositol phosphates, phospholipids and nucleic acids and mineralization of inositol phosphates in some Bangladesh soils. J. Soil Sci. 24, 193-198.

41 Jackman R.H. 1964 Accumulation of organic matter in some New Zealand soils under permanent pasture. I. Patterns of change of organic carbon, nitrogen, sulphur and phosphorus. N.Z.J. Agric. Res. 7, 445-471.

42 Jansson S.L. 1958 Tracer studies on nitrogen transformations in soil with special attention to mineralization and immobilization relationships. Ann. Royal Agric. Coll. Sweden 24, 101-361.

43 Jorgensen J.R., Wells C.G. and Metz L.J. 1980 Nutrient changes in decomposing Loblolly pine forest floor. Soil Sci. Soc. Am. J. 44, 1307-1314.

44 Kawaguchi K. and Kyuma K. 1968 Fertility characteristics of the lowland rice soils in some southern Asian countries. Trans. 9th Intern. Cong. Soil Sci. 4, 19-31.

45 Ladd J.N., Oades J.M. and Amato N. 1981 Microbial biomass formed from ^{14}C, 15N-labelled plant material decomposing in soils in the field. Soil Biol. Biochem. 13, 119-126.

46 Larson W.E., Clapp C.E., Pierre W.H. and Morachan Y.B. 1972 Effects of increasing amounts of organic residues on continuous corn: II. Organic carbon, nitrogen, phosphorus and sulfur. Agron. J. 64, 204-208.

47 Lea R. and Ballard R. 1982 Predicting Loblolly pine growth response from N fertilizer, using soil-N availability indices. Soil Sci. Soc. Am. J. 46, 1096-1099.

48 Louw H.A. and Webley D.M. 1959 A study of soil bacteria dissolving certain mineral phosphate fertilizers and related compounds. J. Applied Bacteriol. 22, 227-233.

49 MacMillan K., Scott T.W. and Bateman T.W. 1972 A study of corn response and soil nitrogen

transformations upon application of different rates and sources and chicken manure. Proc. 1972 Cornell Agricultural Waste Management Conference, 481-494.

50 Marion G.M., Kummerow J. and Miller P.C. 1981 Predicting nitrogen mineralization in chaparral soils. Soil Sci. Soc. Am. J. 45, 956-961.

51 Marriott L.F. and Bartlett H.D. 1972 Contribution of animal waste to nitrate nitrogen in the soil. Proc. 1972 Cornell Agricultural Waste Management Conference, pp. 435-440.

52 Mathers A.C. and Goss D.W. 1979 Estimating animal waste applications to supply crop nitrogen requirements. Soil Sci. Soc. Am. J. 43, 364-366.

53 Mathers A.C. and Stewart B.A. 1970 Nitrogen transformations and plant growth as affected by applying large amounts of cattle feedlot wastes to soil. *In* Relationship of Agriculture to Soil and Water Pollution, Cornell Univ. Conference on Agricultural Waste Management, pp. 207-214.

54 Meek B.D., Graham L.E., Donovan T.J. and Mayberry K.S. 1979 Phosphorus availability in a calcareous soil after high loading rates of animal manure. Soil Sci. Soc. Am. J. 43, 741-744.

55 Meek B.D., Graham L.E. and Donovan T.J. 1982 Long term effects of manure on soil nitrogen, phosphorus, potassium, sodium, organic matter and water infiltration rate. Soil Sci. Soc. Am. J. 46, 1014-1019.

56 Michrina B.P., Fox R.H. and Piekielek W.P. 1982 Chemical characterization of two extracts used in the determination of available soil nitrogen. Plant and Soil 64, 331-341.

57 Moghimi A., Lewis D.G. and Oades J.M. 1978 Release of phosphate from calcium phosphates by rhizosphere products. Soil Biol. Biochem. 10, 277-281.

58 Moghimi A.M., Tate M.E. and Oades J.M. 1978 Characterization of rhizosphere products especially 2-ketogluconic acid. Soil Biol. Biochem. 10, 283-287.

59 Murphy L.A., Wallingford G.W., Powers W.L. and Manges H.L. 1972 Effects of solid beef feedlot wastes on soil conditions and plant growth. Proc. 1972 Cornell Agricultural Waste Management Conference, 449-464.

60 Myers R.J.K., Campbell C.A. and Weier K.L. 1982 Quantitative relationship between net nitrogen mineralization and moisture content of soils. Can. J. Soil Sci. 62, 111-124.

61 Olsen S.R. and Barber S.A. 1977 Effect of waste application on soil phosphorus and potassium. *In* Soils for Management of Organic Wastes and Waste Waters, American Society of Agronomy, Madison, Wisconsin, pp. 197-215.

62 Peterson A.E., Kelling K.A. and Walsh L.M. 1974 Crop yield and recovery of N and P as affected by liquid sewage sludge. Agron. Abst. 1974, p. 36.

63 Ponnamperuma F.N. 1964 Dynamic aspects of flooded soils and the nutrition of the rice plant. *In* The Mineral Nutrition of the Rice Plant, Johns Hopkins Press, Baltimore, Md., pp. 295-328.

64 Powders R.F. 1980 Mineralizable soil nitrogen as an index of nitrogen availability to forest trees. Soil Sci. Soc. Am. J. 44, 1314-1320.

65 Pratt P.F. Davis S. and Sharpless R.G. 1976 A four-year field trial with animal manures. I. Nitrogen balances and yields. II. Mineralization of nitrogen. Hilgardia 44, 99-125.

66 Preston C.M. 1982 The availability of residual fertilizer nitrogen immobilized as clay-fixed ammonium and organic nitrogen. Can. J. Soil Sci. 62, 479-486.

67 Reddy K.R. 1982 Mineralization of nitrogen in organic soils. Soil Sci. Soc. Am. J. 46, 561-566.

68 Reinhorn T. and Avnimelech Y 1974 Nitrogen release associated with the decrease in soil organic matter in newly cultivated soils. J. Environ. Qual. 3, 118-121.

69 Richter J., Nuske A., Habenicht W. and Bauer J. 1982 Optimized N-mineralization parameters of loess soils from incubation experiments. Plant and Soil 68, 379-388.

70 Robinson J.B.D. 1967 Soil particle size fractions and nitrogen mineralization. J. Soil Sci. 18, 109-117.

71 Sahrawat K.L. 1982 Assay of nitrogen supplying capacity of tropical rice soils. Plant and Soil 65, 111-121.

72 Sahrawat K.L. 1982 Simple modification of the Walkley-Black method for simultaneous determination of organic carbon and potentially mineralizable nitrogen in tropical rice soils. Plant and Soil 69, 73-77.

73 Sarathchandra S.W. and Perott K.W. 1981 Determination of phosphatase and arylsulfatase activities in soils. Soil Biol. Biochem. 13, 543-545.

74 Shimano K. and Osaki I. 1981 Investigations of nutrients and water supplying powers of typical soils in the Abashiri district. V. Estimation and evaluation of nitrogen mineralization potentials in soils. Bull. Hokkaido Prefect. Agric. Exp. Sta. 45, 27-36.

75 Smith J.H. and Allison F.E. 1962 Phosphobacterin as a soil inoculant. Laboratory, greenhouse and field evaluation. USDA Tech. Bull. 1263, 22 pp.

76 Smith J.L., Schnabel R.R., McNeal B.L. and Campbell G.S. 1980 Potential errors in the first-order model for estimating soil nitrogen mineralization potentials. Soil Sci. Soc. Am. J. 44, 996-1000.

77 Smith S.J. and Stanford G. 1971 Evaluation of a chemical index of soil nitrogen availability. Soil Sci. 111, 228-232.

78 Sommers L.E. 1977 Chemical composition of sewage sludges and analysis of their potential use as fertilizers. J. Environ. Qual. 6, 225-232.

79 Sowden F.J., Morita H. and Levesque M. 1978 Organic nitrogen distribution in selected peats and peat fractions. Can. J. Soil Sci. 58, 237-249.

80 Stanford G. and Smith S.J. 1972 Nitrogen mineralization potential of soils. Soil Sci. Soc. Am. Proc. 36, 465-472.

81 Stanford G., Fere M.H. and Schwaninger D.H. 1973 Temperature coefficient of soil nitrogen mineralization. Soil Sci. 115, 321-323.

82 Swift R.S. and Posner A.M. 1972 Nitrogen, phosphorus and sulphur contents of humic acids fractionated with respect to molecular weight. J. Soil Sci. 23, 50-57.

83 Talpaz H., Fine P. and Bar-Yosef B. 1981 On the estimation of N-mineralization parameters from incubation experiments. Soil Sci. Soc. Am. J. 45, 993-996.

84 Terry R.E. 1980 Nitrogen mineralization in Florida Histosols. Soil Sci. Soc. Am. J. 44, 747-750.

85 Thompson L.M., Black C.A. and Zoellner J.A. 1954 Occurrence and mineralization of organic phosphorus in soils with particular reference to associations with nitrogen, carbon and pH. Soil Sci. 77, 185-196.

86 Williams B.L. 1974 Effect of water-table level on nitrogen mineralization in peat. Forestry 47, 195-202.

87 Williams C.H., Williams E.G. and Scott N.M. 1960 Carbon, nitrogen, sulphur and phosphorus in some Scottish soils. J. Soil Sci. 11, 334-346.

88 Youngberg C.T. 1978 Nitrogen mineralization and uptake from Douglas-fir forest floors. Soil Sci. Soc. Am. J. 42, 499-502.

3. The role of organic matter and ammonium in producing high corn yields

S.R. OLSEN

3.1. Introduction

Soil organic matter has been recognized for centuries as the key to soil fertility and productivity. Organic matter plays a major role in the chemical, microbiological and physical aspects of soil fertility. Organic matter as a source of nutrients, primarily nitrogen, has received most attention and research. Soil organic matter is increased by organic manures or by growing herbage crops. All organic manures supply plant nutrients. For a century the classical experiments at Rothamsted indicated that yields of crops with farmyard manure could be achieved equally by applications of NPK fertilizers (5, 19). However since 1962 varieties with a high yield potential have been grown and evidence is accumulating from long-term tests that farmyard and other organic manures can give larger yields than can be obtained with fertilizers only (4, 19). It has been postulated that these effects are the result of organically bound nitrogen behaving in ways not easily imitated by fertilizer-nitrogen. This tentative explanation needs to be explored in more detailed experiments (4). Investigation of these effects may identify some nutritional needs of crops that prevent achievement of maximum yields and may lead to better utilization of nitrogen by crops.

In this paper, the concept will be examined that the beneficial effect of organic matter on yields may be related in part to the nitrogen supply being a combination of ammonium and nitrate nitrogen. The main nitrogen sources under natural conditions are the ions, ammonium and nitrate. Results with a number of plant species subjected to NH_4^+-N and NO_3^--N nutrition have shown that each ion produces a different physiological response within the plant (6,17,44). Plants vary in their ability to absorb and utilize the ammonium and nitrate forms. Weight gains by adding ammonium to an all nitrate system have been observed in twelve crop species (17, 27). Why NH_4^+-N has this growth promoting effect is not known. However, the reduction of NO_3^- to NH_3 requires energy (6, 53) and it may be reasoned that by supplying NH_4^+, energy is conserved and diverted to other metabolic processes including ion uptake and growth.

It is suggested that the combination of these two N forms may be the major reason the crops gave higher yields with farmyard manure (FYM), *i.e.* the plant was able to utilize N more efficiently and conserve energy for growth when the N is supplied as NH_4^+ and NO_3^- . FYM as a source of N avoids excessive and sometimes toxic levels of NH_4^+, and it provides a steady supply of NH_4^+ during the growth period. Also, NO_3^- N will be available during the growth period, but any limitations in its utilization will be overcome by the steady supply of NH_4^+ from FYM during its mineralization. These limitations is utilizing essentially all N as NO_3^- may be related to 1) less than optimum levels of nitrate reductase (NR), especially in shaded leaves, 2) more carbohydrate energy needed to reduce NO_3^- to NH_3 especially if some NO_3^- is reduced in the roots, and 3) the consumption of organic acids produced to neutralize OH^- produced in cells as NO_3^- is reduced to NH_3. These limitations will be more important at high yield levels where more stress develops for nutrients and photosynthetic products in order to produce grain. Very few, if any, critical studies have been made on the relative effectiveness of NH_4^+ and NO_3^- for plant growth through the grain filling period or to maturity, with soil as the growth medium.

These concepts need to be tested under field conditions where 1) high yields can be obtained with good varieties and high plant populations, 2) the plants can be grown to maturity, 3) plant growth can be observed over the entire growth period and under various environmental conditions of temperature and water, and 4) under conditions where other limitations in obtaining high yields may be observed or identified.

When we consider nutrient supply limitations and adequate levels, is there a common characteristic of farmers' soil and management practices that lead to high corn yields. An ample supply of nutrients in a large portion of the root zone seems to be a common feature. Usually the fields are deep ploughed to incorporate a heavy application of farm manure, or, crop residues have been incorporated into the soil surface by minimum tillage over a period of years to increase soil organic matter. A well-adapted mid-to long-season hybrid is planted at higher populations (93898 to 98840 plants/ha) than average practice (61775 plants/ha). The weather is favorable and water stress minimal. Weeds and insects are controlled.

Very few researchers have obtained 18.83 Mg/ha (300 bushels/acre), whereas 10 or more farmers in the USA have reached and exceeded this yield. The record yield in 1977 was 22.16 Mg/ha by Roy Lynn, Jr., at Schoolcraft, Michigan. A possible reason may be related to a wider range of favorable conditions in farmers' fields, and the fact that there are a lot more farmers than researchers. One farmer, Herman Warsaw, of Saybrook, Illinois, has 14 years of data from contest yields. For the first 7 years his yields increased from 9.98 Mg/ha in 1967 to 16.76 Mg/ha in 1973 with an average yield of 13.12 Mg/ha. The next 6 years his yields ranged from 13.94 to 21.22 Mg/ha in 1975 with an average of 16.95 Mg/ha. He achieved 19.58 Mg/ha in 1979, 20.47 Mg/ha in 1981, and 19.34 Mg/ha in 1982. However, the statistical odds that he can top 21.72 Mg/ha are 1 year out of forty (55).

In 1980, Roy Flannery obtained 19.58 Mg/ha followed by 21.22 Mg/ha in

1982 in a fertility, variety, population experiment in New Jersey. Other researchers closely approached the 18.83 Mg/ha level in Florida. In 1969, a corn yield of
19.08 Mg/ha was achieved in a variety trial in western Colorado. In 1981, a corn
yield of 19.15 Mg/ha was obtained in western Colorado followed by yields of
19.96 and 19.02 Mg/ha in two separate experiments in 1982 (34, 35). These experiments indicate significant progress in achieving high corn yields.

When we consider soil-plant factors associated with high yields, in addition to
gains from organic manures, more experiments are needed to explore the genetics
of nutritional variation. For example, significant advances can be made to increase
the efficiency of N utilization. Research indicates that the apparent rather complicated inheritance of nitrogen-utilization efficiency can almost certainly be as-
cribed to at least four major variables: 1) uptake of nitrate (and/or ammonium to
a lesser extent); 2) the level and activity of nitrate reductase; 3) the size of the
storage pool of NO_3^-, which is in turn related to nitrate reductase activity; and 4)
the ability to mobilize and translocate nitrogen from leaves and other parts to the
developing grain. Nitrate reductase is only one factor in utilization of fertilizer nitrogen, and translocation of remobilized vegetative nitrogen to the developing
grain is important (54).

3.2. Review of literature

In this section literature will be reviewed connected with the concept that the
beneficial effect of organic matter on yields may be related in a major way to the
nitrogen supply being a combination of NH_4^+ - N and NO_3^- -N.

3.2.1. Beneficial effect on yields of organic manures

When P and K levels are more than adequate, plots at Rothamsted regularly
receiving FYM produced larger yields than fertilizer-treated plots irrespective of
the rate of N applied (4, 5, 19). Since 1962 yields at Barnfield of mangolds, sugar
beet (*Beta vulgaris* L.), potatoes (*Solanum tuberosum* L.), and wheat (*Triticum
aestivum* L.), have been larger on FYM-treated plots than on plots receiving NPK
+ Na + Mg fertilizers. The shapes of the reponse curves to N fertilizer suggested
that yields with PK + Na + Mg fertilizers would never equal those with FYM, however much N was applied (4). Wheat grown continuously, after fallow, and in rotation with beans (*Phaseolus vulgaris* L.) yielded more with FYM than with fertilizers at Broadbalk. Maximum yields of barley (*Hordeum vulgaris* L.) were about the
same from FYM and from NPK + Na + Mg fertilizers, but they were acieved by
using much less N-fertilizer when FYM was applied. The total yield (grain + straw)
was larger and the crop contained much more N when grown with FYM than with
fertilizers (19). Similar results have been obtained on a sandy soil at Woburn, or-

ganic manures giving larger yields of sugar beet, potatoes and wheat than were possible with fertilizers. Growing grass-clover leys, or leguminous green manure crops, has also given larger yields of wheat and potatoes at Woburn than could be obtained without FYM. Peat and straw as manures, and non-leguminous leys or green manures had similar, but smaller effects. The results were summarized by stating that the gains in yield appear to be associated with the provision of nitrogen by the organic treatments to crops in ways and at times that are not fully imitated by fertilizers applied in conventional ways. There is no evidence that some other property of the organic matter has caused larger yields (4).

In the USA, long-term studies show the value of FYM. In the cornproducing and humid wheat-producing states, many investigators concluded that a sound system of fertility maintenance could not be developed without the return to the land of animal wastes from feeding the crops produced. The results of long-term experiments from both the Morrow and Sanborn Field plots indicate that there is relatively littele difference in increases in yields of crops if the N is supplied by manure or chemical fertilizers (38). These studies, however, do not include testing a wide range of N applications, *i.e.*, amounts needed for maximum yields.

3.2.2. *Ammonium and nitrate as a nitrogen source*

Combinations of these two N sources have produced larger plant yields or higher N contents than either source alone in twelve plant species (6, 17, 27, 58, 60). The relative effectiveness for corn growth of NH_4^+ and NO_3^- sources has been studied mainly in solution cultures and short-term greenhouse trials (6, 45, 60). Very few critical studies have been made of the relative effectiveness of NH_4^+ and NO_3^- for growth through the grain-filling period with soil as the growth medium, especially at high yield potentials (12556 to 18834 kg/ha for corn). The importance and potential role of controlled N nutrition in the production of food and fiber should not be underestimated. It is the largest and most significant laboratory-proven potential for increased crop growth that has not been demonstrated on a field scale. Whereas plant utilization of NO_3^- has been investigated extensively, the conditions associated with beneficial assimilation of NH_4^+ are largely unknown and call for further investigation in both laboratory and field (43).

Growth of sunflower (*Helianthus annus* L.) leaves was greater with NH_4^+ plus NO_3^- than with either NH_4^+ or NO_3^- alone. Shoot protein level was also greater with the combined source. It was suggested that the simultaneous presence of NH_4^+ and NO_3^- in the culture solution may lead to the synthesis in the roots of a complex of nitrogenous compounds which when transported to the leaves permits the establishment of a high protein level (62). After a 3-day growth period of sunflower leaves, a greater utilization was found of available N in the synthesis of protein by plants supplied with NH_4^+ plus NO_3^- , *i.e.*, the protein constituted

42.8, 49.1 and 26.2% of the total N for NH_4^+, NH_4^+ plus NO_3^-, and NO_3^- treatments, respectively (62). Investigators have established the generally accepted principle that protein synthesis proceeds at maximal rates only in the presence of all of the amino acids (59). This condition appears to have been obtained in this study only when NH_4^+ and NO_3^- were simultaneously available to the plants. The inability of the root systems of NH_4^+ and NO_3^- fertilized plants to supply the shoots with adequate amounts of all the amino acids may have contributed to the lower level of protein in the leaves of these plants (62).

The most suitable inorganic N source for growing plant cells of soybean (*Glycine max* L.), wheat (*Triticum monococcum* L.) and flax (*Linum usitatissimum* L.) was a mixture of NH_4^+ and NO_3^- (11). Cell cultures grew with NH_4^+ alone if Kreb's cycle acids were added, *i.e.*, citrate, malate, fumarate or succinate (11).

Yield increases by adding NH_4^+ to an all NO_3^- system have been shown in carnation (*Dianthus caryophyllus* L.) (12), chrysanthemum (*Chrysanthemum morifolium)* (20), wheat (*Triticum aestivum* L.) (6), corn (*Zea mays* L.) (45, 58), lettuce (*Latuca sativa* L.) and radish (*Raphanus sativa* L.) (60), Douglas fir (*Pseudotsuga menziesii* (Mirb.)), Sitka spruce (*Picea sitchensis* (Bong.), white spruce (*Picea glauca* (Moench.)) (50, 51), and Scots Pine (*Pinus sylvestis* L.) (32).

In wheat the growth rate and yield of plants grown with NH_4+ added to the root culture already containing adequate NO_3^- (200 u*M*) exceeded that of plants grown in NH_4^+ or NO_3^- alone, and the yield increased 54% with added NH_4^+ up to 40 u*M*. This yield increase was suggested to result from reduced energy requirement in using NH_4^+ instead of NO_3^- in protein synthesis and from increased photosynthetic capacity (6). In other experiments, maximum protein synthesis occurred in shoots of wheat seedlings grown in both NH_4^+ and NO_3^- culture (61). Several experiments suggest that corn grows faster on NH_4^+ plus NO_3^- than on NO_3^- alone. These experiments indicate that corn does not obtain enough N for maximal growth and yield from normal soils in which the predominate form of N is NO_3^- (21, 45, 58). In Wisconsin, corn hybrids grown on sandy soil and provided with supplemental urea or NH_4NO_3 produced more grain and stover of higher N content than plants supplied with KNO_3 (21). Nitrate reductase was not decreased by combination NH_4^+ and NO_3^- treatments with corn (45). When both forms of N were absorbed, NH_4^+ was used preferentially for synthesis of amino acids and protein (45). In corn both the quantity and quality of leaf protein were highest in an NH_4^+ plus NO_3^- mixture. Nitrate alone decreased the nutritive value of the protein by decreasing the methionine content, while NH_4^+ alone increased aspartate content and decreased the content of other amino acids (17). However, in barley (*Hordeum vulgare* L., cv Aramir) the form of N had no effect on protein concentration and composition and was of little importance as a source of differences in the total amino acid composition of the plant (44).

Other studies showed that aster (*Callistephus chinensis* L.) plants would tolerate much higher levels of NH_4^+ when levels of NO_3^- in the growth media were also high (16). A level of protein production with NH_4^+ plus NO_3^- was superior in *Bryophyllum calycinum* to that achieved with either NH_4^+ or NO_3^- alone (39). There was a marked enrichment of the leaves in protein at the 50% NH_4^+ level in the culture solution. Other data suggest that corn plants can maintain their normal organic salt content in the tops only if a part of the nitrogen is transported upward as NO_3^- (63).

For many species it appears that a mixture of NH_4^+ and NO_3^- produces greatest growth and protein content, the optimum ratio probably differs for different species and may change with age of the plant, temperature, and pH of the growth medium.

The difference between the cation (C) and inorganic anion (A) content of plants, the (C-A) or organic salt content, is numerically equal to the organic anion content (52, 63). Plants appear to maintain their (C-A) content at a constant value (63). Normal growth is possible at a wide variation of the C content, the A content and the content of the individual ionic species, provided that the (C-A) content is normal. This indicates that the organic anions are essential for good growth. A reduced growth with a normal (C-A) content may occur, but a low (C-A) content is never accompanied by a normal growth rate. Hence, a normal (C-A) content is one of the conditions for good growth (63). If NO_3^- in a growth medium for sugar beet (*Beta vulgaris* L.) is replaced by NH_4^+ the organic salt content and the yield decrease considerably. With NH_4NO_3 nutrition the organic salt content is also lowered, mainly through an increased accumulation of NO_3^- in the foliage, but NH_4NO_3 gives only a slight depression in yield compared with NO_3^- alone (52). This data also suggests a combination of NH_4^+ and NO_3^- produces better growth than NH_4^+ alone.

From solution culture work, plant growth appears to be stimulated by NH_4^+ only in plants receiving optimum or near-optimum levels of NO_3^-. Other observations suggest that understanding ionic interactions in a system utilizing both NH_4^+ and NO_3^- is one of the most challenging problems of plant nutrition (18). Further, the observed 50% increase in growth rates from the addition of NH_4^+ to solutions containing optimum levels of NO_3^- is impressive and merits study to determine the physiological reasons (18).

Use of ammonia rather than nitrate as an N source led to a 30% reduction of the doubling time of cell matter for *Dunaliella tertiolecta* (a marine algae). Ammonia-grown cells possessed a greater capacity for photosynthetic O_2 evolution at light saturation than did nitrate-grown cells and their content of ribulosediphosphate carboxylase was likewise greater (the enzyme required in photosynthesis) (36). A stimulatory effect of NH_4^+ on photosynthesis in intact chloroplasts has been observed. The authors concluded that NH_4^+ activates the ribulosediphosphate carboxylase reaction. This stimulatory effect occurred at concentrations well below those required for uncoupling of photophosphorylation (7). Others have suggested that the activity of ribulosediphosphate carboxylase is rate-limiting for photosynthesis under many conditions (57). These observations may be connected with the improved growth reported for combinations of NH_4^+ and NO_3^- in the growth medium (6, 17, 45, 62).

3.2.3. Energy requirements for NH_4^+ grown and NO_3^- grown plants and for the combined sources

Results with a number of plant species subjected to NH_4^+-N and NO_3^--N nutrition have shown that each ion produces a different physiological response within the plant (6, 17, 44). The plant response to a combined source of these ions differs also from either ion alone (6, 17, 45, 62). For example, NH_4^+ uptake depletes carboxylates in the roots more than NO_3^- uptake. During grain fill in wheat, the grain was calling on other parts of the plant to supply carbohydrates in addition to the supply from current photosynthesis (31). Clearly the photosynthetic system was not supplying sufficient carbohydrates during the latter half of grain formation to assure maximum yields. Other experiments with wheat showed NO_3^- was a better source of N than NH_4^+ in wheat after heading (47). Possibly this effect is due to a shortage of carbohydrates in the roots which decrease NH_4^+ uptake.

Why does the photosynthetic system in wheat fail to supply enough carbohydrates during grain fill? A possible answer is that the breakdown of vegetative protein, also required for grain filling, may interfere with photosynthetic processes. Thus, if more N can be supplied from the roots during this period of grain filling, perhaps the photosynthetic system would function at a higher level. More N could be supplied from a combination of NH_4^+ and NO_3^- which produces higher protein levels in younger plants (6, 62).

One author states that the choice of N transport compounds appears to change with the N source supplied and with the energy status of the plant. When carbohydrate status is high, glutamine which has a N:C ratio of 2:5, will predominate, whereas under carbohydrate shortage, asparagine (2N:4C) or arginine (4N:6C) are more abundant (29).

Two authors observed that yield of wheat was increased 50% (19 days growth) by adding NH_4^+ to cultures supplying the maximum utilizable concentration of NO_3^- (6). They suggested this yield increase may be related to more efficient utilization of the plant's limited energy supply. Energy is conserved when NH_4^+, as compared to NO_3^- is supplied to the plant since no reduction is necessary during assimilation. The oxidation of conserved NADH could then be linked to O_2 via the cytochrome system with a resultant increase in O_2 consumption as has been observed in *Chlorella vulgaris* (48). This leads to increased phosphorylation which in turn, provides increased energy for growth and ion absorption. Another possibility is that the rate of reduction of NO_3^- by nitrate reductase within the plant limited growth in the NO_3^- only series.

Another author refers to data showing that the reduction of NO_3^- to NH_3 by *Chlorella pyrenoidsa* L. required the input of 162,000 cal/mole. This represents 23.4% of the energy available from the combustion of a mole of glucose (14). He indicated that if energy (photosynthate) is the limiting factor in plant productivity, then NH_4^+ grown plants should always be more productive than NO_3^- grown plants, based on the energy required for reduction of NO_3^- to NH_3. However, this is contingent upon the coordination of NH_4^+ uptake with environment induced variations in photosynthetic supply so that toxic levels of NH_4^+ are never accumulated throughout the plant's life cycle.

While these energy considerations are valid, the mechanism of NO_3^- reduction and assimilation that occurs in illuminated green leaves eliminates the need for extensive carbohydrate catabolism. The reactions may be indicated as shown:

Cytoplasm			Chloroplast	
NADH	NAD	6 Fd(reduced)		6 Fd (oxidized)

$$NO_3^- \xrightarrow[\text{reductase}]{\text{nitrate}} NO_2 \quad NO_2^- \xrightarrow[\text{reductase}]{\text{nitrate}} NH_3$$

The NADH for NO_3^- reduction to NO_2^- would be derived via carbohydrate oxidative pathways. Because ferredoxin (Fd) is reduced by light energy trapped by the chloroplast, 75% of the 162,000 cal/mole required for NO_3^- reduction would not be channeled through carbohydrate metabolism. Although NO_2^- reduction in the chloroplasts competes with carbon reduction for light energy (via Fd), other authors state that available light energy exceeds the energy needs of carbon reduction (40). Sulfate reduction would also compete for light energy *via* Fd. Also, do these statements hold true to the same degree, *i.e.*, non-limiting, for higher plant populations where more leaves are shaded?

In the major route of NH_3 assimilation in the chloroplast the following reactions occur:

$$NH_3 + \text{glutamate} + ATP \xrightarrow[\text{synthetase}]{\text{glutamine}} \text{glutamine} + ADP + H_2PO_4^-$$

$$\text{glutamine} + 2 \text{ Fd (red.)} + 2 \text{ -oxoglutarate} \xrightarrow[\text{synthase}]{\text{glutamate}} 2 \text{ glutamate} + \text{Fd(ox.)}$$

Fd (reduced) would be produced *via* light energy and bypasses carbohydrate metabolism (29). Glutamate synthase has been found to be widely distributed in the plant kingdom and to exist in two forms, 1) one utilizing NADH as an electron donor and present in roots and other nongreen tissue, and 2) one form capable of accepting electrons from reduced ferredoxin, and present in leaves, chloroplasts, and algae (25, 28, 29, 49). With glutamate synthase, a mechanism is provided whereby the superior assimilatory characteristics of glutamine synthetase can be linked to α-amino N production.

For plants grown on NH_4^+, it has been shown that most of it is assimilated in the root. Thus, ATP and equivalent energy (possibly NAD(P)H as a substitute for Fd (red.)) required for glutamate formation or NAD(P)H required in the glutamate dehydrogenase reaction would be derived from carbohydrate catabolism.

From these observations, Hageman concludes that reduction and assimilation of NO_3^- by the leaf requires no more carbohydrate catabolism than assimilation of NH_3 (14). Note that energy for nitrate reductase reaction does not appear to be taken into account in this statement, nor the organic acids required for maintaining pH as NO_2^- is reduced. In NH_4^+ grown plants, photosynthate for assimilation energy and carbohydrate skeletons for amino acids and amides must be transported to the root. Portions of the N compounds must then be transported to the leaf. The energy cost for transport and the efficiency of distribution of photosynthate and N compounds to the various organs is not known (14).

Reduction of NO_3^- in the root is dependent on carbohydrate metabolism and would require 162,000 cal/mole of energy to produce NH_3. With cereal species the amount of root NO_3^- reduction varied from 25 to 60% (37). Using a similar procedure, other data showed that corn roots reduced about one-third of the absorbed NO_3^- (range of 22 to 54% for 11 genotypes). Preliminary evidence indicated that the % of NO_3^- reduced by the roots decreases with age (41).

Other research indicates that the reduction of NO_3^- is the largest energy-requiring process carried out by all crop plants, after the reduction of CO_2 (29). Various attempts have been made to calculate the carbohydrate (CHO) cost of producing protein. Some authors have calculated that to produce 1 g of protein will require 1.62 g glucose if NH_4^+ is used as the N source, but 2.42 g if NO_3^- is the starting material. In contrast, 1 g starch requires only 1.21 g glucose (29). Thus, if these figures are a correct model of crop growth, they would predict that plants would yield much more on NH_4^+ than on NO_3^- (which in general they do not) and that there would be a negative relationship between protein content and yield, which is often observed particularly in comparisons between different varieties of the same species of crop plant (29). However, the calculations are based on dark (heterotrophic) metabolism and thus assume that all ATP and NAD (P)H required to power the process is derived from glucose. This is unlikely to represent the true situation for crop plants. If current ideas about the role of chloroplasts in N metabolism are correct, then a considerable proportion of the energy from N metabolism can be derived directly from light-dependent electron transport. Exactly how much is difficult to calculate, but certainly all of the steps from NO_2^- to glutamate can be powered by light energy (29).

The use of reducing power and ATP for N metabolism is only likely to detract from yield if yield is limited by the rate of CO_2 fixation and if CO_2 fixation is limited by light energy. These two conditions have not been clearly established, and many authorities believe that plants may have a surplus fixation capacity and that for most of the day CO_2 fixation is not limited by light energy but by CO_2 concentration. Note that these concepts and limiting conditions may not apply, however, for high yield levels of corn. It is then probable that most, if not all, of NO_3^- reduction and its further metabolism in leaves could occur at no energy cost to the plant. This statement applies, however, to the portion of NO_3^- reduced in leaves.

In the longer term increased N nutrition results in increased production of leaf protein (of which about 30-50% is ribulosediphosphate carboxylase) and chlorophyll, thereby boosting photosynthesis and CO_2 fixation, both on a leaf area and per plant basis. On the other hand, although NO_2^- reduction may be powered directly by light energy, the resultant charge imbalance may divert carbohydrates from sucrose and starch into organic acids.

Two authors have discussed the energetic aspects of pH regulation in plants (42). Nitrate assimilation produces almost 1 OH^- per NO_3^-. NH_4^+ assimilation produces at least 1 H^+ per NH_4^+. H^+ or OH^- produced in excess of that required to maintain cytoplasmic pH must be in some way neutralized. If these H^+ and OH^- ions were not removed physically or chemically from the cytoplasm, they would lead to enormous pH changes in the cytoplasm. Thus the pH would decrease to zero or increase to 14 in less than a cell generation time (42). In higher land plants, much of assimilated N occurs as shoot protein; the shoot cells have no direct access to the H^+ and OH^- sink of the soil solution.

When NH_4^+ is the N source, it is assimilated into organic-N in the roots. The shoot is supplied with a mixture of amino acids, amides, and organic acids which can be incorporated into cell material without damaging pH changes.

When NO_3^- is reduced in the roots, the organic compounds involved in N transport to the shoot are similar to those used when NH_4^+ or N_2 (legumes) is the N source, with similar implications for regulation of shoot pH. The excess OH^- generated in the roots in partly excreted to the soil solution, and partly neutralized by the 'biochemical pH stat' which produces strong acids from essentially neutral precursors.

When NO_3^- is assimilated solely in shoots, the excess OH^- is initially neutralized by the biochemical pH stat. Storage of the inorganic cation organate in shoot cell vacuoles could lead to turgor and volume regulation problems in these cells. These are avoided when an insoluble salt (Ca oxalate) is the product of the pH stat, or when the cation organate is translocated to the roots where organate breakdown regenerates OH^-, which is lost to the soil solution.

The use of the biochemical pH stat to counter OH^- generated in NO_3^- assimilation involves the commitment of relatively large quantities of reduced carbon as stored organic acid anion. Oxalate is the most economical acid to produce, because it can neutralize one OH^- per carbon used. Up to 100% of the organic carbon in the plant can be involved in organic anion storage consequent upon operation of the biochemical pH stat. This storage involves sequestration of organic carbon which could have otherwise been used to generate ATP. This ATP consumption by the biochemical pH stat amounts to 5-10 ATP molecules per N assimilated, depending on the nature of the anion stored (42).

With regards to the potential sources of energy for N assimilation and related reactions, the location of N assimilation reactions in the roots deprives the plant of the option of the direct use of photo-produced cofactors rather than respiration-derived cofactors, and is particularly important for the very energetically demanding processes of nitrate reduction and N_2 fixation (42).

Other authors have discussed ionic balance in relation to N nutrition of plants (23). Irrespective of the form of nutrition, a very close balance was found in the tissues investigated (leaves, petioles, stems, and roots) between total cations (Ca, Mg, K, Na), and the total anions (NO_3^-, $H_2PO_4^-$, $SO_4^=$, Cl^-), total non-volatile organic acids, oxalate, and uronic acids. In comparison with the tissues of NO_3^- - fed plants, the corresponding NH_4^+ tissues contained lower concentrations of inorganic cations and organic acids, and a correspondingly higher proportion of inorganic anions. In all tissues approximately equivalent amounts of diffusible cations (Ca^{++}, Mg^{++}, K^+, Na^+), and diffusible anions (NO_3^-, $H_2PO_4^-$, $SO_4^=$, Cl^-) and nonvolatile organic acids were found. An almost 1:1 ratio occurred between the levels of bound Ca and Mg, and oxalate and uronic acids. This points to the fact that in the tomato plant the indiffusible anions are mainly oxalate and pectate. Approximately equivalent values were found for the alkalinity of the ash and organic anions (total organic acids including oxalate, and uronic acids) (23).

The influence of N nutrition is shown very clearly, nitrate tissue having very much higher concentrations of organic acids than corresponding NH_4^+ tissues, especially malic acid, i.e., in leaves the concentration of malic acid is 22 times higher in NO_3^- - fed leaves than NH_4^+ - fed leaves (malic acid was 71.6% of the total organic acids). The authors indicate this result is connected with the assimilation of NO_3^- insofar as it provides a continuous source of negative charges which may be transferred to form organic anions in the plant or returned to the nutrient medium as HCO_3^- (23).

Form of N affected pH of macerated tissues of leaves, petioles, stems and roots of tomato, i.e., pH is lower NH_4^+ - fed tissues of leaves by 0.5 unit. The pH values of macerated plant tissue (tomato) give only a rough indication of the real pH decrease which may occur at metabolic sites. It is possible for example that at low pH, photosynthetic CO_2 fixation may be reduced.

Some important characteristics of nitrate reductase (NR) are listed below. After NO_3^- is inside the plant evidence indicates that NR is a rate limiting enzyme in the biochemical pathway for reduction of NO_3^-. The amount of NR is under strong regulatory control. The enzyme itself has a 4-hour half-life ($35°C$) with simultaneous synthesis and degradation. Its steady-state concentration is closely related to the energy status of the plant (1, 13, 53). The characteristics are:

1) NR is NADH-specific (reduced nicotinamide adenine dinucleotide).

2) The enzyme is a metaloprotein and contains both Fe and Mo.

3) Maximum activiteis of 600 to 700 n moles of NO_2^- produced/min/mg protein have been reported for the enzyme.

4) The enzyme is adaptive, *i.e.*, inducible by NO_3^- Bulk of NR is in the leaves (75-80%).

5) The enzyme is unstable both *in vivo* and *in vitro*.

6) Plants subjected to mild moisture and temperature stress exhibit reduced levels of NR activity.

7) NR exhibits both a seasonal and diurnal variation. NR is less in early morning than afternoon. NR is less in shaded leaves. As plant population increases, NR will decrease in shaded leaves. This population effect varies with inbreds and hybrids.

8) As NO_3^- disappears or is used by the tissues, NR disappears. Thus, synthesis must be maintained to balance against continual degradation. Higher levels of NR activity are associated with higher rates of NO_3^- reduction.

A report indicates that the assimilation of $^{15}NO_3^-$ into amino N was strictly dependent on light and ceased abruptly when light was extinguished. Incorporation of $^{15}NO_2^-$ was also markedly dependent on light and little reduction occurred in the dark (2). The mechanism by which NO_2^- reduction ceases in the dark is not understood, but NO_2^- never accumulates in healthy tissue in the dark (46). Incorporation of NH_4^+ into amino acids was 8 to 10-fold higher than NO_3^- or NO_2^- in the dark in bean leaves (2). In barley leaves in the light NH_4Cl (0.1 or 1 m*M*) increased incorporation of ^{15}N into soluble amino acids from Na $^{15}NO_3$ and Na $^{15}NO_2$ (2). Nitrate reductase in leaves of Perilla is an adaptive enzyme, the induction of which is shown to require CO_2, light and NO_3^-. A continuous breakdown of the enzyme takes place even during a period of net synthesis (22). Later studies showed that soybean leaves lost 95% of their initial NR activity after a 16-h dark period at 20°C. Upon re-exposure to light *in vivo* NR activity increased rapidly to maximum levels after 4-h light. The rate of increase was proportional to light intensity and independent of temperature (20°, 30° and 40° C) (33).

Another report shows that, in the dark, plant growth on NO_3^- requires more dark respiration than those on NH_4^+ , whereas in the light there was no difference in respiration between the two sources (29). These results indicates an advantage to the plant from a mixed source of NH_4^+ and NO_3^- forms such as the plant could obtain from fertilizer N and farmyard manure.

3.3. Results

Long term experiments with farmyard manure (FYM, 35 Mg/ha containing 224 kg of total N. One-half of this N is assumed to be available in application year) at Rothamsted were modified in 1962 because the large increases in yield from adding N fertilizer to soils treated with FYM had not been explained (19). Four amounts of N (0, 75, 150 and 225 kg/ha as 'Nitro-Chalk') were tested (1962, 1964) on potatoes and mangolds. Yields are shown in Figures 3.1a and 3.1b. Yields were always larger on FYM- than on PK- treated soils at all amounts of N for both crops. The shape of the response curves to N suggested that yields on PKtreated soils would not have equalled those on FYM- treated soils no matter how much N was applied (19).

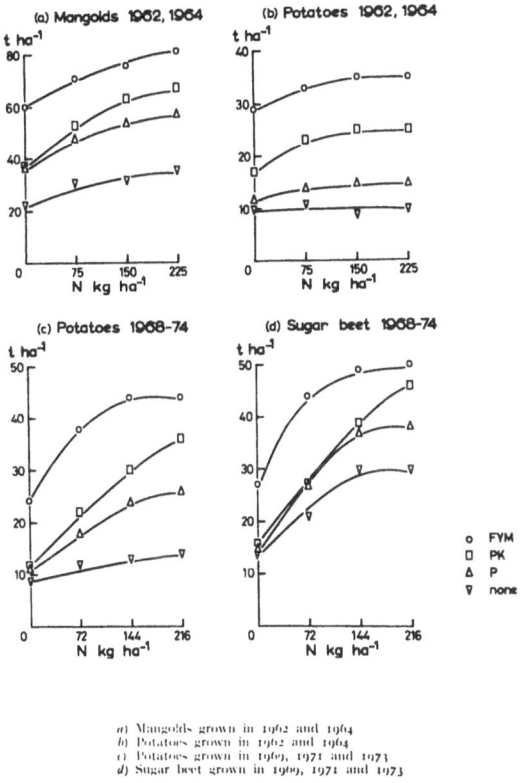

a) Mangolds grown in 1962 and 1964
b) Potatoes grown in 1962 and 1964
c) Potatoes grown in 1969, 1971 and 1973
d) Sugar beet grown in 1969, 1971 and 1973

Fig. 3.1. a, b,c, d. Relationship between yields of mangolds, potatoes and sugar beet and fertilizer N applied to soils treated with farmyard manure, O, PK fertilizers, ; P fertilizers ; or unmanured ; during 1962-73, Barnfield.

42

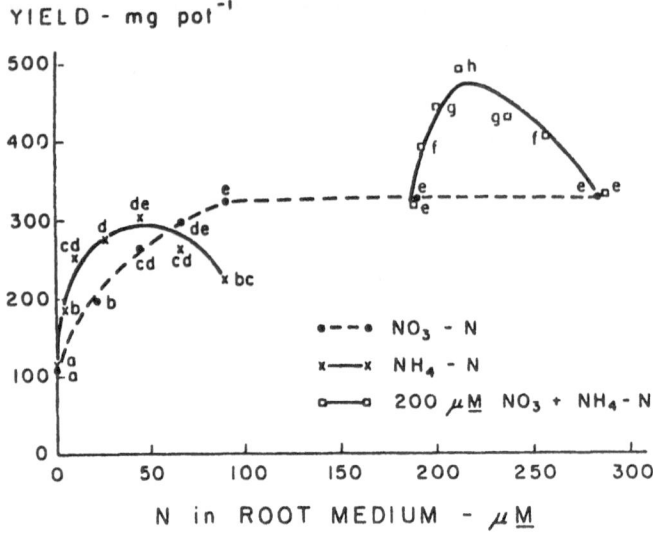

YIELD - mg pot^{-1}

N in ROOT MEDIUM - μM

Fig. 3.2. The effect of source and concentration of N on yield of wheat. (Data points are steady-state N concentrations of the NO_3^- and NH_4^+ sources and supplied levels for the NO_3^+ NH_4 source. Points labeled with the same letter do not differ significantly, P 0.05, by Duncan's multiple range test.)

Experiments were continued during 1968-1974. Four amounts of N were again tested (0, 72, 144 and 216 kg/ha for root crops; 0, 48, 96 and 144 kg/ha for cereals). Results for potatoes and sugar beets are shown in Figures 3.1c and 3.1d. Both crops yielded more on FYM than on PK-treated soil at all N rates. With optimum fertilizer N barley yielded a little more and spring wheat slightly less when grown on FYM-treated soil than on PK-treated soil. Winter wheat yielded higher with FYM than fertilizer N during 1970-75 (19).

A new organic manuring study was initiated at Woburn, England, in 1964, on sandy loam soils by applying FYM and other organic manures from 1964-71. Supplies of P and K were balanced in all contrasts made. Wheat and potatoes were grown (1972-74) with a full range of N fertilizer amounts to test the value of the organic residues. The shapes of the response curves showed that however much N was used, the fertilizer-treated soils could not achieve the yields from organically-treated soils. For potatoes most gains from the organic treatment were in the range of 5-10 Mg/ha, while gains in wheat yields ranged up to 1.5 Mg/ha (5, 26).

Similar experiments on corn have been conducted (1980-82) in Colorado where the effect of FYM at 40 Mg/ha was tested with three rates of N (246, 403 and 515 kg/ha). This crop yielded more on FYM than on PK-treated soil at all N rates (unpublished data). The yield increase with the best variety was 2700 kg/ha in 1982.

As indicated in the introduction various reasons have been suggested to explain why FYM or other organic manures produce larger yields than those usually observed with fertilizers only. In this section results will be cited to examine the concept that the beneficial effect of organic matter on yields may be related in part to the nitrogen supply to roots and tops is absorbed from a combination of NH_4^+ - N and NO_3^- - N. As organic matter is mineralized the first available form of N is assumed to be NH_4^+, then nitrification converts NH_4^+ to NO_3^-. Many studies have been made of the plant response to either ion alone but relatively few studies have measured the response to combinations of these two sources.

The response of corn to N sources is shown in Table 3.1. Yield, N content of grain and tissue increased with urea or NH_4NO_3 compared with KNO_3 (21).

In nutrient solutions, yield and total N assimilated into organic N with corn were larger in a mixed NH_4^+ plus NO_3^- than with either ion alone as shown in Tables 3.2 and 3.3 (45).

Table 3.1. Response of corn to N sources on an irrigated sandy soil, Wisconsin

Source of N	WN 273, 1969			WN 433, 1970		
	Grain yield	N in grain	N in tissue	Grain yield	N in grain	N in tissue
	kg/ha	%	%	kg/ha	%	%
Urea	4875 a	1.65 a	1.07 a	6775 a	1.38 a	1.10 a
NH_4NO_3	4972 a	1.66 a	1.10 a	6647 a	1.36 a	1.11 a
KNO_3	4475 b	1.59 b	0.99 b	5868 b	1.33 b	1.05 b

Letters in common indicate no significant difference at 5% level. Yields and N concentrations are means of four rates of N (56, 112, 168 and 224 kg/ha. From Jung *et al.*) (21)

Table 3.2. Influence of ammonium- and nitrate-N on the fresh weight of corn plants in nutrient solutions

| Part of plant | N source, | Fresh weight, g/plant | |
		ppmNO$_3^-$ - N/ppm NH$_4^+$ - N	
	0/100	100/0	50/50
Leaves	93.4 a	84.1 a	121.1 a
Stems	127.9 fg	104.8 g	202.6 f
Roots	70.4 kl	52.8 l	112.0 k
Total	291.7	241.7	435.7

Means within a plant part followed by the same letter do not differ at the $P = 0.05$ level. Plants sampled 47 days after planting. From Schrader *et al.* (45)

Table 3.3. Utilization of ammonium- and nitrate-N supplied to corn plants in nutrient solutions

| | | Content of N, mg | | | |
| | | N source, ppm NO$_3^-$ - N/ppm NH$_4^+$ - N | | | |
	0/50	0/100	50/0	100/0	50/50
Total uptake of N by plants					
NO$_3^-$ - N	0	0	864	1320	886
NH$_4^+$ - N	990	1338	0	0	940
NO$_3^-$ - N content of plants	0	0	147	340	386
Total N assimilated into organic N					
NO$_3^-$ - N	0	0	717	980	500
NH$_4^+$ - N	990	1338	0	0	940

From Schrader *et al.* (45)

Yield of wheat in nutrient solution increased 54% when NH_4^+ was added to NO_3^- solutions at optimum concentrations as shown in Figure 3.2 and Table 3.4. Total N uptake was greater from the mixed source (6).

Sunflower leaves produced more dry weight and protein content on a mixed N source than from NH_4^+ or NO_3^- alone as shown in Table 3.5 (62).

Table 3.4. Effect on yield and total N uptake by wheat of N sources in nutrient solutions

N supplied (uM)		Tops	Yield (mg/pot) Roots	Total	Tops	N uptake (mg/pot) Roots	Total
NH_4^+	60	203	102	305	9.42	5.09	14.51
NO_3^-	200	214	114	328	7.13	4.03	11.16
$NO_3^- + NH_4^+$	200+40	347	146	493	16.66	6.93	23.59

Concentrations, uM, for maximum yield and uptake from Cox and Reisenauer (6)

Table 3.5. Dry weight, total N and protein N of sunflower leaves as influenced by N source

Treatment	Dry wt mg/set	Total N mg/g	mg/set	Protein N mg/g	mg/set
Leaf set 1					
NH_4^+	15.2	59.5	0.90	29.9	0.45
$NH_4^+ + NO_3^-$	16.9	59.0	1.00	33.1	0.56
NO_3^-	13.2	57.4	0.76	29.1	0.38
Leaf set 2					
NH_4^+	0.9	89.8	0.08	37.1	0.03
$NH_4^+ + NO_3^-$	1.3	93.8	0.12	46.1	0.06
NO_3^-	0.7	149.0	0.10	39.0	0.03

From Weissman (62)

Table 3.6. Influence of the form of N nutrition on the various N fractions in the leaves of white mustard (meq/100 g dry weight)

N Source	Total N	Protein N	Soluble Org N	Amino N	Nitrate N
NO_3^-	302	186	83	3.9	1
$NO_3^- + NH_4^+$	412	300	111	6.9	1
$NH_4^+ + NO_3^-$	362	266	95	4.2	1
NH_4^+	486	346	139	8.2	1

From Kirkby (24)

White mustard grown on a mixed source contained more total N, protein N, soluble organic N, and amino N than on NO_3^- alone as shown in Table 3.6. The NH_4 - N source contained 25% of total N as NO_3^- . Concentrations of several amino acids in leaves of white mustard was higher with the mixed N source than with NO_3^- alone as shown in Table 3.7 (24).

Table 3.7. Influence of the form of N nutrition on the free amino acids (ug N/g fresh wt) in the leaves of white mustard

Amino acid	N source			
	$NO_3^- + NO_3^-$ *	$NO_3^- + NH_4^+$	$NH_4^+ + NO_3^-$	$NH_4^+ + NH_4^+$
Aspartic	12	12	16	17
Glutamic	10	24	11	28
Serine	3	8	5	16
Asparagine	Tr	Tr	Tr	Tr
q Alanine	33	44	29	45
Glutamine	13	14	12	18
Valine	4	5	5	5
Leucine	Tr	Tr	1	3
Arginine	Tr	Tr	Tr	9
Proline	12	52	25	68
Total	87	159	104	209

*First source - plants grown 2 weeks.
Second source - plants grown an additional 4 weeks.
From Kirkby (24)

Cell cultures from several plant species produced more dry weight on NO_3^- plus NH_4^+ or glutamine than with NO_3^- alone as shown in Table 3.8 (10, 11). These cells required NH_4^+ or glutamine for rapid growth.

Two studies show that NH_4^+ stimulates the activity of ribulosediphosphate carboxylase compared with NO_3^- in a marine alga, Table 3.9 (36), and in chloroplasts, Table 3.10 (7), compared with a control containing no NH_4^+ or NO_3^- .

Temperature and pH have different effects on uptake of NH_4^+ and NO_3^- as show in Table 3.11 (9), Table 3.12 (30) and Table 3.13 (3). Uptake of NH_4^+ increases with increases in pH whereas NO_3^- uptake increases as pH decreases. Uptake of NH_4^+ is relatively greater than NO_3^- as temperature decreases and in these experiments, NH_4^+ uptake was greater at low and high temperature (3, 30).

Table 3.8. Cell yield of cultures from several plant species (mg dry wt) grown in defined medium in various nitrogen sources

Culture	Inoculum	N source			Growing period (days)
		NO_3^-	$NO_3^- + NH_4^+$	NO_3^- + glutamine	
Reseda	16	173	386	415	11
Wheat	80	238	290	305	6
Flax Embryo	15	127	356	361	11
Horseradish	20	132	181	226	10
Soybean	23	74	388	301	6

The concentration of KNO_3 was 25 mM, $(NH_4)_2 SO_4$ was 1 mM, and L-glutamine was 4 mM.
From Gamborg (10)

Table 3.9. Ribulosediphosphate carboxylase activity in *D. tertiolecta* with NH_4^+ or NO_3^-

Activity	NH_4^+ cells	NO_3^- cells
cpm/10^6 cells	10,974 + 317 (152)	7,242 + 355 (100)
cpm/ug chlorophyll a	5,354 + 177 (133)	4,020 + 144 (100)
cpm/ug protein	778 + 27 (109)	711 + 35 (100)

Relative values in parentheses, From Paasche (36)

Table 3.10. Effect of NH_4^+ on $^{14}CO_2$ fixation and O_2 evolution (nmoles/mg chl)

Time in light, sec	Control			NH_4^+ 2 nM		
	^{14}C fixed	O_2 evolved	$^{14}C/O_2$	^{14}C fixed	O_2 evolved	$^{14}C/O_2$
10	3.8	––	—	4.1	––	––
20	13.5	––	—	20.7	––	––
40	38.4	36.3	1.05	67.4	41.4	1.62
60	68.4	81.9	0.84	200	106	1.89

From de Benedetti *et al.* (7)

Table 3.11. The effect of pH on the uptake of NH_4^+ and NO_3^- by excised roots of 2 weeks old rice seedlings from solutions of labeled NH_4NO_3 (ug N/g dry wt)

N source	pH		
	4.0	5.5	7.0
NO_3^-	72	56	37
NH_4^+	220	300	350

Uptake period was 90 minutes. From Fried *et al.* (9)

Table 3.12. Influence of source of nitrogen and root temperature on the levels of ammonium, nitrate and organic nitrogen in the roots and tops of maize inbred ND 203 at 28 days growth

N source	Root temp °C	Tops			Roots		
		NH_4^+ - N ppm	NO_3^- - N ppm	Org N %	NH_4^+ - N ppm	NO_3^- - N ppm	Org N %
NH_4^+	14.8	1290	0	3.15	500	0	3.34
	28.2	210	0	3.51	1050	0	4.29
NO_3^-	14.8	35	750	2.24	130	3410	2.93
	28.2	180	6370	2.57	280	6280	2.27
LSD (*P*=0.05)		640	950	0.52	290	1350	0.63

From Moraghan and Porter (30)

Table 3.13. Influence of temperature and pH of culture solution on absorption of labeled ammonium and nitrate (mg N/g root dry wt) by tillers of Italian ryegrass (*L.multiflorum*, S.22).

Root temperature °C	N absorbed (% translocation)			
	NH_4^+		NO_3^-	
	Buffered	Unbuffered	Buffered	Unbuffered
20	19.7+2.1(59)	21.4+2.8(82)	11.5+2.2(52)	15.8+1.05(55)
5	14.0+1.7(51)	13.8+1.0(63)	3.6+0.3(38)	4.2+1.1(48)

Buffered- A small quantity of solid $CaCO_3$ was added to each culture to give a constant pH of 7.9.

Unbuffered- The pH of the culture solution fell from an initial value of 5.5 to 3.9 (20°C) or 4.3 (5°C).

Culture solution contained 500 uM NH_4NO_3 and plants were treated for 22 h. From Clarkson and Warner (3)

3.4. Discussion

The achievement of high yields of corn and other crops with combinations of farmyard manure and organic residues with fertilizer NPK opens ways to identify factors that limit plant performance. Soil organic matter can be increased by applications of manure or organic residues; tillage proctices to conserve crop residues can maintain or improve organic matter levels. As soil organic matter increases, reserves of all nutrients in soils, and particularly in subsoils, will increase and these nutrients may be absorbed more effectively from deeper soil layers than fertilizers applied in mormal ways to surface soil. Cooke states "we cannot assume that 'fertilizer NPK' can necessarily be equated to 'soil NPK'. Similarly there are conditions where accumulating a reserve of organic matter results in larger yields than can be achieved in a 'fertilizer only' regime, perhaps because N from the organic source, is more effective than N from fertilizer applied to surface soil" (5). Our results with high yielding corn hybrids support this statement.

In addition to the effects of organic matter indicated above, data were cited in this paper to suggest a concept that the beneficial effect of organic matter on yields may be related also to the N supply being a combination of NH_4^+ and NO_3^- .

Plants vary greatly in their ability to absorb and utilize NH_4^+ and NO_3^- as their sources of N. The main limitations for maximum growth rate in utilizing all N as NO_3^- may be connected with 1) less than optimum levels of NR, especially in shaded leaves, 2) more carbohydrate energy needed to reduce NO_3^- to NH_3, especially when some NO_3^- is reduced in the roots, and 3) to the consumption of organic acids to neutralize OH^- produced in cells as NO_3^- is reduced to NH_3. Some plants become Fe deficient in all NO_3^- systems. Other plants grow better with NH_4^+ than with NO_3^-, but plants usually have narrow tolerance limits for NH_4^+ which becomes toxic at higher levels (6, 8, 17). Farmyard manure as a source of N avoids excessive and sometimes toxic levels of NH_4^+ and it provides a steady supply of NH_4^+ during the growth period as well as a supply of NO_3^-.

The main N sources under natural conditions are the ions NH_4^+ and NO_3^- but their levels or ratios to one another may be changed by fertilizer and organic residue management practices, including the use of nitrification inhibitors. Results with a number of plant species subjected to NH_4 - N and NO_3^- - N nutrition have shown that each ion produces a different physiological response within the plant (6, 17, 44). The plant response to a combined source of these ions differs also from either ion alone (6, 17, 45, 62). Weight gains by adding NH_4^+ to an all NO_3^- system were observed in twelve crop species (17, 27). Ammonium uptake depletes carboxylates in the roots more than NO_3^- uptake, but a combined source leads to higher total N, organic and protein N in some plants (6, 24, 45, 62). Other observations indicate a greater utilization of available N in the synthesis of protein by plants supplied with NH_4^+ plus NO_3^- (24, 62). Other data indicate a close correlation between the total protein content of *Xanthium* leaves and their capacity for nitrate reductase synthesis (10). *In vitro* nitrate reductase activity was higher in leaf blades of corn from those plants grown on a mixture of NH_4^+ and NO_3^- (45).

Studies with wheat show that the photosynthetic system does not supply enough carbohydrate during the latter half of grain fill to assure maximum yields. A possible reason is that the breakdown of vegetative protein, also required for grain filling, may interfere with the photosynthetic process. Thus, if more N can be supplied from the roots during this period and N can be utilized to form leaf or grain proteins, perhaps the photosynthetic system would function at a higher level. Data cited in this paper indicate that more N could be supplied from a combination of NH_4^+ and NO_3^- which also produces higher protein levels in younger plants (24, 62, 62). Two papers also indicate that NH_4^+ stimulates the activity of ribulosediphosphate carboxylase, the enzyme that reacts with CO_2 in photosynthesis.

In wheat, NH_4^+ promotes more vegetative growth, which would produce a higher leaf area index, than does NO_3^- (47). The effect of temperatue on uptake of NH_4^+ and NO_3^- may explain this early growth response (3, 30). With Italian ryegrass, the amount of labeled N reaching the shoot can be 4 to 12 times as great from labeled NH_4^+ as from labeled NO_3^- in a solution of NH_4NO_3 at 8° C. In U.K. cereal crops the vegetative phase of growth, and thus the major fraction of total N uptake, is usually completed before the soil has reached a temperature of 12°C (3).

The importance of early growth response to N forms is indicated in another study showing that the potential yield of corn grain, which is produced late in the season, is determined by the leaf area, which is always produced early in the season. However, less than this potential yield of grain will actually be obtained if 1) the net assimilation rate is decreased by any factor such as moisture stress, or 2) the leaf area is prematurely reduced by some factor such as a nutrient deficiency, insects, disease, or hail damage. Fertilization practices are important in determining the leaf area/plant and also in preventing premature death of leaves due to nutrient deficiencies (15).

3.5. References

1 Beevers L. and Hageman R.H. 1969 Nitrate reduction in higher plants. Ann. Rev. Pl. Physiol. 20, 495-522.
2 Canvin D.T. and Atkins C.A. 1974 Nitrate, nitrite and ammonia assimilation by leaves: Effect of light, carbon dioxide and oxygen. Planta 116, 207-214.
3 Clarkson D.T. and Warner A.J. 1979 Relationships between root temperature and the transport of ammonium and nitrate ions by Italian and Perennial ryegrass (*Lolium multiforum* and *Lolium perenne*). Pl. Physiol. 557-561.
4 Cooke G.W. 1977 The roles of organic manures and organic matter in managing soils for higher crop yields- a review of the experimental evidence- Proc. Int. Seminar on Soil Environment and Fertility Management in Intensive Agriculture. Tokyo-Japan.
5 Cooke G.W. 1978 Some priorities for British soil science. J. Soil Sci. 30, 187-213.
6 Cox W.J. and Reisenauer H.M. 1973 Growth and ion uptake by wheat supplied nitrogen as nitrate, or ammonium, or both. Plant and Soil 38, 363-380.
7 De Benedetti E., Forti G., Garlaschi F.M. and Rosa L. 1976 On the mechanism of ammonium stimulation of photosynthesis in isolated chloroplasts. Plant Sci. Letters 7, 85-90.
8 Dibb D.W. and Welch L.F. 1976 Corn growth as affected by ammonium vs nitrate absorbed from soil. Agron. J. 68, 89-94.
9 Fried M., Zsoldos F., Vose P.B. and Shatokhin I.L. 1965 Characterizing the NO_3 and NH_4 uptake process of rice roots by use of ^{15}N labelled NH_4NO_3. Physiol. Plant 18. 313-320.
10 Gamborg O.L. 1970 The effects of amino acids and ammonium on the growth of plant cells in suspension culture. Pl. Physiol. 45, 372-375.
11 Gamborg O.L. and Shyluk J.P. 1970 The culture of plant cells with ammonium salts as the sole nitrogen source. Pl. Physiol. 45, 598-600.
12 Green J.L., Holley W.D. and Thaden B. 1973 Effects of the NH_4^+ : NO_3^- ratio, chloride, N-serve, and simazine on carnation flower production and plant growth. Fla. St. Hort Soc. Proc. 86, 383-388.
13 Hageman R.H., Lambert R.J., Loussaert D., Dalling M. and Klepper L.A. 1974 Nitrate and nitrate reductase as factors limiting protein synthesis. Workshop Proceedings National Academy of Sciences (U.S.)pp. 103-134.
14 Hageman R.H. 1980 Effect of form of nitrogen on plant growth. *In* Nitrification inhibitors-potentials and limitations. Eds. J.J. Meisinger, G.W. Randall and M.L. Vitosh ASA Special Publication No 38, 47-62. Madison, WN.

15 Hanway J.J. 1962 Corn growth and composition in relation to soil fertility: I. Growth of different parts and relation between leaf weight and grain yield. Agron. J. 54, 145-148.

16 Haynes R.J. and Goh K.M. 1977 Evaluation of potting media for commercial nursery production of container-grown plants. II. Effects of media fertilizer nitrogen, and a nitrification inhibitor on yield and nitrogen uptake of *Callistephus chinensis* (L.) Nees 'Pink Princess'. N.Z.J. Agric. Res. 20, 371-381.

17 Haynes R.J. and Goh K.M. 1978 Ammonium and nitrate nutrition of plants. Biol. Rev. 53, 465-510.

18 Hiatt A.J. 1978 Critique- of absorption and utilization of ammonium nitrogen by plants. *In* Nitrogen in the Environment, Vol 2. Eds. D.R. Nielsen and J.G. MacDonald pp. 191-199. Academic Press, NY.

19 Johnston A.E. and Mattingly G.E.G. 1976 Experiments on the continuous growth of arable crops at Rothamsted and Woburn experimental stations: Effects of treatments on crop yields and soil analyses and recent modifications in purpose and design. Ann. Agron. 27, 927-956.

20 Joiner J.N. and Knoop W.E. 1969 Effect of ratios of NH_4^+ to NO_3^- and levels of N and K on chemical content of *Chrysanthemum morifolium* 'Bright Golden Ann'. Fla. St. Hort. Soc. Proc. 82, 403-407.

21 Jung Jr. P.E., Peterson L.A. and Schrader L.E. 1972 Response of irrigated corn to time, rate, and source of applied N on sandy soils. Agron. J. 64, 668-670.

22 Kannangara C.G. and Woolhouse H.W. 1967 The role of carbon dioxide, light and nitrate in the synthesis and degradation of nitrate reductase in leaves of *Perilla frutescens*. New Phytol. 66, 553-561.

23 Kirkby E.A. and Mengel K. 1967 Ionic balance in different tissues of tomato plant in relation to nitrate, urea or ammonium nutrition. Pl. Physiol. 42, 6-14.

24 Kirkby E.A. 1968 Influence of ammonium and nitrate nutrition on the cation-anion balance and nitrogen and carbohydrate metabolism of white mustard plants grown in dilute nutrient solutions. Soil Sci. 105, 133-141.

25 Lea P.J. and Miflin B.J. 1975 The occurrence of glutamate synthase in algae. Biochem. Biophys. Res. Commun. 64, 856-862.

26 Mattingly G.E.G. 1974 The Woburn organic manuring experiment. 1. Design, crop yields and nutrient balance 1964-72. Report of Rothamsted Expt. Sta. for 1973, Part 2, p. 98-133.

27 Mengel K. and Kirkby E.A. 1982 Principles of Plant Nutrition. 3rd Ed. Int. Potash Inst, Bern Switzerland, 655 p.

28 Miflin B.J. and Lea P.J. 1975 Glutamine and asparagine as nitrogen donors for reductant-dependent glutamate synthesis in pea roots. Biochem. J. 149, 403-409.

29 Miflin B.J. 1980 Nitrogen, metabolism and amino acid biosynthesis in crop plants. *In* The Biology of Crop Productivity. Ed. P.S. Carlson p. 255-296. Academic Press, NY.

30 Moraghan J.T. and Porter O.A. 1975 Maize growth as affected by root temperature and form of nitrogen. Plant and Soil 43, 479-487.

31 Moss D.N. 1976 Studies on increasing photosynthesis in crop plants. *In* Burris R.H. and Black C.C. (eds.) CO_2-Metabolism and Plant Productivity, p. 31-41. Univ. Park Press, Baltimore.

32 Nelson L.E. and Selby R. 1974 The effect of nitrogen sources and iron levels on the growth and composition of sitka spruce and scots pine. Plant and Soil 41, 573-588.

33 Nicholas J.C., Harper J.E. and Hageman R.H. 1976 Nitrate reductase activity in soybeans (*Glycine max* [L. Merr.) I. Effects of light and temperature. Pl. Physiol. 58, 736-739.

34 Olsen S.R. 1981 Removing barriers to crop productivity. Agron. J. 74, 1-4.

35 Olsen S.R., Champion D.F., Young D., Keenan J., Golas H., Schweissing F. and Christensen D. 1983 Soil and plant properties associated with high corn yields. Agron. Abs, p. 177.

36 Paasche E. 1971 Effect of ammonia and nitrate on growth, photosynthesis, and ribulosedi-

phosphate carboxylase content of *Dunaliella terriolecta*. Physiol. Plant. 25, 294-299.

37 Pate J.S. 1973 Uptake assimilation, and transport of N compounds by plants. Soil Biol. Biochem. 5, 109-119.

38 Peterson J.R., McCalla T.M. and Smith G.E. 1971 Human and animal wastes as fertilizers. Fertilizer Technology and Use. *In* R.A. Olson, T.J. Army, J.J. Hanway and V.J. Kilmer. pp. 557-596. Soil Sci. Soc. of Am, Madison, WN.

39 Pucher G.W., Leavenworth C.S., Ginter W.D. and Vickery H.B. 1947 Studies in the metabolism of crassulacean plants: The effect upon the composition of *Bryophyllum calycinum* of the form in which nitrogen is supplied. Pl. Physiol. 22, 205-227.

40 Radner R. and Kok B. 1977 Photosynthesis: limited yields, unlimited dreams. BioSci. 27, 599-605.

41 Raghuveer P. 1977 Characteristics of nitrate uptake from nutrient solution and root nitrate reductase activity among corn genotypes. Ph. D. Thesis, Univ. of IL, Urbana.

42 Raven J.A. and Smith F.A. 1976 Nitrogen assimilation and transport in vascular land plants in relation to intracellular pH regulation. New Phytol. 76. 415-431.

43 Reisenauer H.M. 1978 Absorption and utilization of ammonium nitrogen by plants. *In* Nitrogen in the Environment. Eds. D.R. Nielsen and J.G. MacDonald. Vol 2, p. 157-170. Academic Press, NY.

44 Richter R., Dijkshoorn W. and Vonk C.R. 1975 Amino acids of barley plants in relation to nitrate, urea of ammonium nutrition. Plant and Soil 42, 601-618.

45 Schrader L.E., Domska D., Jung Jr. P.E. and Peterson L.A. 1972 Uptake and assimilation of ammonium-N and nitrate-N and their influence on the growth of corn (*Zea mays* L.). Agron. J. 64, 690-695.

46 Schrader L.E. 1978 Uptake, accumulation, assimilation and transport of nitrogen in higher plants. *In* Nitrogen in the Environment. Eds. D.R. Nielsen and J.G. MacDonald. Vol 2. Soil-Plant-Nitrogen Relationships. Academic Press, NY., p. 101-141.

47 Spratt E.D. 1974 Effect of ammonium and nitrate forms of fertilizer-N and their time of application on utilization of N by wheat. Agron. J. 66, 57-61.

48 Syrett P.J. 1956 The assimilation of ammonia and nitrate by nitrogenstarved cells of *Chlorella vulgaris:* III. Differences of metabolism dependent on the nature of the nitrogen source. Physiol. Plant. 9, 28-37.

49 Tempest D.W., Meers J.L. and Brown C.M. 1970 Synthesis of glutamate in *Aerobacter aerogenes* by a hitherto unknown route. Biochem. J. 117, 405-407.

50 Van den Driessche R. 1971 Response of conifer seedlings to nitrate and ammonium sources of nitrogen. Plant and Soil 34, 421-439.

51 Van den Driessche R. and Dangerfield J. 1975 Response of douglas-fir seedlings to nitrate and ammonium nitrogen sources under various environmental conditions. Plant and Soil 42, 685-702.

52 Van Tuil H.D.W. 1965 Organic salts in plants in relation to nutrition and growth. Agricultural Research Reports No. 657, 83 p. Centre for Agricultural Publications and Documentation, Wageningen.

53 Viets F.F. Jr. and Hageman R.H. 1971 Factors affecting the accumulation of nitrate in soil, water, and plants. Agric. Handbook No. 413, USDA-ARS, Wash. D.C.

54 Vose P.B. 1981 Effects of genetic factors on nutritional requirements of plants. *In* Handbook of Nutrition and Food. Ed. M. Rechcigl. p. 1-64. CRC Press, Inc. Cleveland, OH.

55 Walker W.M. and Welch L.F. 1980 "Super yield" odds. Crops Soils Mag. 32(8), 12-14.

56 Wallace W. and Pate J.S. 1967 Nitrate assimilation in higher plants with special reference to cocklebur (*Xanthium pennsylvanicum* Wallr.). Ann. Bot. 31, 213-228.

57 Wareing P.F., Khalifr M.M. and Treharne K.J. 1968 Rate-limiting processes in photosynthesis at saturating light intensities. Nature 220, 453-457.

54

58 Warncke D.D. and Barber S.A. 1973 Ammonium and nitrate uptake by corn (*Zea mays* L.) as influenced by nitrogen concentration and NH_4^+ /NO_3^- ratio. Agron. J. 65, 950-953.

59 Webster G.C. 1959 Nitrogen Metabolism in Plants. Row-Peterson, Evanston, IL.

60 Weir B.L., Paulson K.N. and Lorenz O.A. 1973 The effect of ammoniacal nitrogen on lettuce (*Latuca sativa*) and radish (*Raphanus sativus*) plants. Soil Sci. Soc. Am. Proc. 36, 462-465.

61 Weissman G.S. 1959 Influence of ammonium and nitrate on the protein and free amino acids in shoots of wheat seedlings. Am. J. Bot. 46. 339-346.

62 Weissman G.S. 1964 Effect of ammonium and nitrate nutrition on protein level and exudate composition. Pl. Physiol. 39, 947-952.

63 Wit C.T. de, Dijkshoorn W. and Noggle J.C. 1963 Ionic balance and growth of plants. Versl. Landbouwk. Onderz. 69.15, 68 p.

4. Nitrogen transformations in Histosols

R.E. TERRY

4.1. Introduction

Deposits of organic soils (Histosols) occur mainly in cool, humid regions such as the northeastern United States and Canada. Histosols are also found in the San Joaquin Delta of California and the Hula Valley of Israel where the climate is hot and dry. Large deposits of organic soils have formed in the warm humid climate of the southern coastal plains of the United States. The Florida Everglades contains more than 8×10^5 ha of peat and muck soils and is the largest known tract of organic soils in the world (20).

Histosols contain more than 20 to 30% organic matter (the minimum percentage depends on the clay content of the mineral fraction), and are at least 30 cm thick (8). The organic nitrogen content of these soils generally ranges from 0.5 to 2.5%, thus accumulation of organic soils represents a geological sink for fixed nitrogen and carbon.

The requirements for formation of organic soils are excess water, organic material, and low oxygen or anaerobic conditions. Under these conditions soil organic matter remains protected against decomposition. However, when the soil is drained the organic matter is subject to microbial decomposition and the reserve of organic carbon and nitrogen begins to be converted to inorganic forms. The processes of decomposition and soil subsidence occur in all drained organic soils and spell the ultimate demise of agricultural use of Histosols.

Although large portions of organic soil deposits throughout the world remain undrained and undeveloped, some of these deposits have been drained for purposes of mining fuel and horticultural peat and for agricultural development. Drained Histosols are very productive soils and are used for the production of truck crops, sugarcane, and minor crops such as mint and cranberries.

4.2. Histosol subsidence

Drainage of organic soil deposits results in a decrease in surface elevation

Y. Chen and Y. Avnimelech (eds.), The Role of Organic Matter in Modern Agriculture.
ISBN 90-247-3360-X.
© *1986, Martinus Nijhoff Publishers, Dordrecht.*

(subsidence). Currently, the organic soils of the Everglades are subsiding at the rate of 3 cm year^{-1}. Similar rates have been recorded elsewhere: 1.75 cm year^{-1} in the Netherlands (19), 2.07 cm year^{-1} in Quebec (15), 7.6 cm year^{-1} in the San Joaquin Delta (34) and 10 cm year^{-1} in the Hula Valley of Israel (2).

Among the reasons for subsidence are 1) shrinkage due to drying, 2) loss of the bouyant force of groundwater, 3) compaction, 4) wind erosion, 5) burning, and 6) microbial oxidation. Microbial oxidation of soil organic matter is the predominant cause of Histosol subsidence (20). A comparison of CO_2 evolution rates from Terra Ceia muck from the Everglades with subsidence rates established by field measurement indicated that microbial oxidation accounted for approximately 73% of the loss of surface elevation in Everglades Histosols (33). The role of microbial oxidation in subsidence of Histosols was recently reviewed by Tate (23).

Among the products of microbial oxidation of organic soils are carbon dioxide, water, and inorganic nitrogen. The nitrogen cycle of organic soils is of great interest due to the large quantities of nitrogen contained in these soils. Nitrogen plays a vital role in the biochemical processes of living organisms and is the fertilizer element used in largest quantities by field crops. Nitrogen undergoes a variety of transformations in soil, most of which are mediated by microorganisms. Biological N_2 fixation adds to the quantities of fixed nitrogen which may be used by microorganisms and plants. Organic nitrogen which is the major form of nitrogen in soils may be released as inorganic nitrogen through mineralization. The major inorganic forms, ammonium and nitrate, are available for plant uptake by crops or immobilization by microorganisms. Ammonium may be oxidized to nitrate by nitrifying bacteria. This nitrate may in turn be lost from the soil through leaching or denitrification. The purpose of this paper is to examine the various transformations of nitrogen in drained Histosols.

4.3. Nitrogen mineralization

Nitrogen mineralization occurs simultaneously with microbial oxidation of the soil organic matter. Plant-available inorganic N is one of the products of Histosol subsidence.

There have been a few studies of nitrogen mineralization in Histosols. Avnimelech (1) studied the accumulation of nitrates in samples of Histosols from the Hula Valley of Israel and found a linear relationship between time and nitrate accumulation in organic soil samples. Nitrate accumulated slowly in samples incubated with 60% (wt/wt) moisture. The rate of accumulation increased as soil moisture increased to field capacity (120%), but nitrate accumulation decreased in soil samples incubated near saturation (150%). The average nitrate accumulation rate in the Hula peat at field capacity was calculated to be 2.7 μg NO_3^- - N g^{-1} day^{-1}. Avnimelech *et al.* (2) estimated that the rate of nitrate accumulation under field conditions in muck soil of the Hula Valley was about 1,000 to 2,000 kg NO_3-N ha^{-1} $year^{-1}$.

Isirimah and Keeney (13) examined the accumulation of inorganic N in samples of Wisconsin Histosols incubated aerobically in the laboratory. On the average, approximately 30 kg N per hectare-meter of organic deposit was mineralized each month. These researchers also studied changes in the distribution of organic nitrogen forms following six months of aerobic incubation. Much of the mineralizable nitrogen was derived from the acid soluble organic N, mostly from hexosamine N, amino acid N, and unidentified N. They found considerable microbial turnover of hydrolyzable N to refractory N during aerobic incubation.

The methods used in the above studies of nitrogen mineralization did not account for mineralized nitrogen which was subsequently immobilized or denitrified. Terry (26) examined the effects of soil moisture tension and organic carbon additions on nitrogen mineralization rates in Florida Histosols. He employed a ^{15}N-tracer technique in laboratory incubation studies to correct for immobilization and denitrification of mineralized N. Microbial oxidation of the soil organic matter was measured by trapping CO_2 released from 100-g samples of soil incubated for six months. Soil moisture tension over the range of 0.1 to 3.0 bar had no significant effect on nitrogen mineralization or microbial oxidation in surface samples of Pahokee muck.

Microbial oxidation and nitrogen mineralization were as much as 50% greater in soil samples collected from cropped fields when compared to those collected from a fallow (bare) field (26). Nitrogen mineralization in the cropped soil was accelerated during the first 28 days of incubation. The mineralization rates of samples from cropped and uncropped fields were similar during the final five months of incubation. The enhanced microbial oxidation and nitrogen mineralization in the cropped soils during the first month of incubation was attributed to decomposition of crop residues.

Additions of St. Augustinegrass (*Stenataphrum secundatum*) (Walt) Kuntz) clippings or glucose to Histosol samples increased the decomposition of native soil organic matter (26). The enhanced CO_2 evolution from samples amended with available organic carbon exceeded the quentities of carbon contained in the amendments. A change in the decomposition rate of native soil organic matter when organic materials are added to soil is described as a "priming effect". This finding points to the possibility that subsidence of organic soils is enhanced by the addition of available carbon in the form of root exudates and crop residues.

Samples of Pahokee muck collected from various depths of the soil profile were incubated in the laboratory to allow estimation of microbial oxidation and nitrogen mineralization rates (26). Annual soil subsidence and nitrogen mineralization rates of the profile were determined for each 25 cm interval to a depth of

Table 4.1. Soil loss and nitrogen mineralization due to microbial oxidation of Pahokee muck

Soil depth (cm)	Bulk density (g cm^{-3})	CO_2-C evolution rate (kg ha^{-1} year^{-1})	Soil loss (%)	Subsidence rate (cm year^{-1})	N mineralization rate (kg ha^{-1} year^{-1})
0-25	0.34	4,475	1.23	0.31	266
25-50	0.19	2,829	1.34	0.33	269
50-75	0.17	3,154	1.60	0.40	178

75 cm, the typical depth of the water table in this soil (Table 4.1). Microbial oxidation throughout the profile produced the equivalent of 10 metric tons of CO_2-C and 713 kg inorganic N per hectare on an annual basis. This loss of soil represented a subsidence rate of 1.-4 cm year^{-1}. Approximately 686 kg N ha^{-1} were mineralized for each centimeter of Pahokee muck lost due to microbial oxidation. The observed subsidence rate for Pahokee muck measured in the field was 3 cm year^{-1} (20). Volk (33) estimated that approximately 73% of subsidence in these soils was due to microbial oxidation. Thus, it is likely that about 1,500 kg N ha^{-1} are mineralized each year as the organic soil subsides.

Reddy (18) reported the results of a study of nitrogen mineralization in Florida Histosols incubated as natural soil columns (7.5 cm i.d. x 70 cm long). These columns were subjected to fluctuating seasonal temperatures and flooding events. Nitrogen mineralization was calculated by summing the NH_4^+ - N, NO_3^- - N, and soluble organic N leached periodically from the columns. The cultivated and virgin organic soils included in that study were from the Zellwood area of central Florida and the Everglades agricultural area of south Florida. Samples of the Lauderhill muck from the Zellwood area were collected from a cultivated field with freshly incorporated sweet corn (*Zea mays* L. Rugosa) residues. The N mineralization rate of this soil was enhanced due to the decomposition of the crop residues.

Mineralization rates were significantly higher in the samples of cultivated and virgin Pahokee muck from the Everglades when compared to the mineralization rates of the soils from central Florida. From the soil column data, Reddy (18) estimated annual nitrogen mineralization rates ranging from 410 to 938 kg N ha^{-1} for cultivated organic soils and 874 to 1250 kg N ha^{-1} for virgin soils. Of the total nitrogen leached from the soil columns, 48 to 84% was in the nitrate form.

As nitrogen in Histosols is mineralized by microbial oxidation there are sever-

al transformations which alter the concentration and the form of the inorganic nitrogen in the soil. Inorganic N may accumulate in the surface layers of soil, be lost
to the ground water through leaching, or be lost to the atmosphere *via* denitrification. Inorganic N may also be returned to the organic form through crop uptake
or microbial immobilization.

In early studies of crop growth in Everglades peat, Neller (16) reported a lack
of crop response to nitrogen fertilizer. The accumulations of nitrate in cropped
and uncropped Everglades peat were studied from October 1934 to June 1936.
There was no indication that pasture grass, corn or sugarcane (*Saccharum* spp.)
would have responded to nitrogen fertilizers. Nitrate-N concentrations as high as
455 μg g^{-1} were found in samples of the surface 15 cm of the organic soil.

Variations in the inorganic nitrogen levels in surface soils of fallow, St. Augustinegrass, and sugarcane fields in the Everglades were measured at monthly intervals from July 1977 through August 1978 (28). Soil samples from the 60-70
cm depth of the fallow field were also analyzed. The monthly variations in the
$(NO_3^- + NO_2^-)$ - N concentrations in the soil and precipitation from rainfall events
are plotted in Fig. 4.1. The maximum NH_4^+ - N concentration measured in these
soils was 20 μg g^{-1} of soil. The $(NO_3^- + NO_2^-)$ - N concentrations in the surface

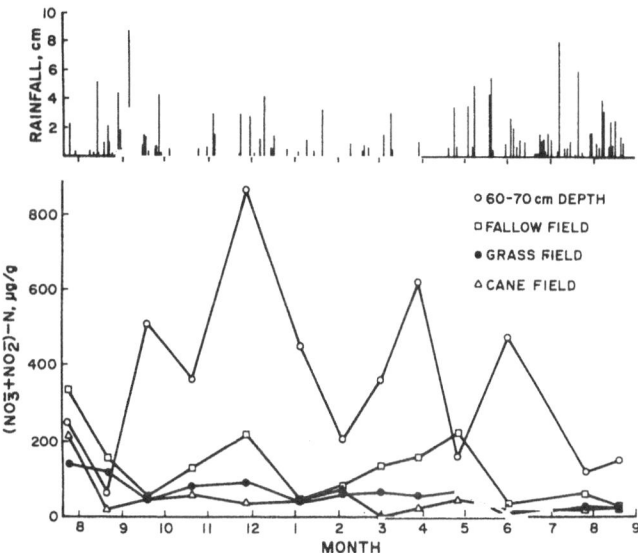

Fig. 4.1. Variation of $(NO_3^- + NO_2^-)$ -N contents of surface cropped, surface fallow, and subsurface fallow Pahokee muck from July 1977 through August 1978.

soils ranged from near O to 335 $\mu g\,g^{-1}$ indicating that the ammonium was rapidly nitrified as it was mineralized. Nitrate is highly mobile in soils; therefore, the accumulation of nitrate was closely linked to rainfall events (Fig. 4.1.). For example, a total of 46 cm of precipitation occurred during August and September 1977. During this period, the $(NO_3^- + NO_2^-)$ - N concentration of the fallow surface soil declined from 335 to 58 $\mu g\,g^{-1}$. Much of this nitrate was leached to the lower horizons as indicated by the increase in $(NO_3^- + NO_2^-)$ - N at the 60-70 cm depth during the same period. Nitrate levels were higher at each sampling period in the fallow soil than in the cropped soils due, presumably, to the lack of crop uptake and the lower availability of organic carbon as an energy source for denitrification.

4.4. Nitrification

Nitrification is the biological oxidation of reduced nitrogen forms to a more oxidized state. In soils this aerobic process is conducted mainly by autotrophic nitrifiers which gain their energy from the oxidation of ammonium and nitrite while using CO_2 as their carbon source. Tate (21) reported the presence of large populations of heterotrophic nitrifiers in Pahokee muck the Everglades. In laboratory studies these organisms have been shown to produce hydroxylamine, nitrosoethanol, nitrite, and nitrate when grown in the presence of ammonium and a carbon source (32).

Tate (24) found no production of nitrite or nitrate in ammonium amended soil samples that had been heated to kill the autotrophic nitrifiers, while preserving the heterotrophic nitrifiers. In unheated control soils the added ammonium was rapidly oxidized to nitrite and nitrate. These findings suggested that the autotrophic nitrifiers were the sole population responsible for nitrification in Pahokee muck.

The rapid accumulation of nitrate rather than ammonium in a variety of aerated Histosols has been reported (1, 16, 18, 28). It has been shown, however, that under the anaerobic contitions of flooded organic soils, ammonium accumulates in the soil profile. In the soil column leaching experiment conducted by Reddy (18), the annual output of NH_4^+ - N ranged from 19 to 30 kg ha^{-1} for a variety of Florida Histosols. About 40 to 70% of the total NH_4^+ - N released from the soil columns occurred during two 25-day flooding periods.

Changes in the oxygen status of Pahokee muck from a flooded vegetable field in the Everglades are reflected in the variations in the NH_4^+ - N and $(NO_3^- + NO_2^-)$ - N contents of the soil shown in Fig. 4.2. (29). Nitrate plus nitrite nitrogen concentrations of Pahokee muck decreased from 50.5 $\mu g\,g^{-1}$ soil prior to flooding to 5 $\mu g\,g^{-1}$ after three days of flooding. This decreased concentration of oxidized nitrogen was maintained until the field was drained, when again the

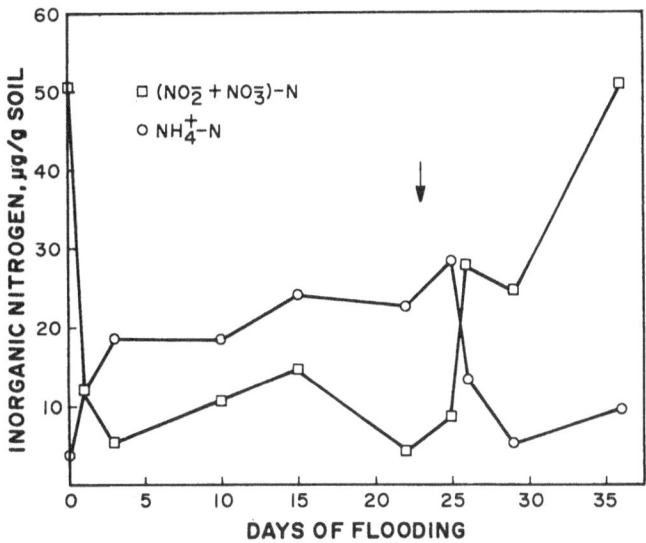

Fig. 4.2. Variation in nitrate and ammonium concentrations in Pahokee muck during flooding, July 1978. The arrow indicates the end of the flood period.

$(NO_3^- + NO_2^-)$ - N concentrations reached the preflood levels. Ammonium-N concentrations increased during the period of flooding. Preflood concentrations of 4 μg NH_4^+ -N g^{-1} were augmented to about 20 μg g^{-1} during the flooding. Upon drainage of the field, the NH_4^+ -N concentration returned to 4 μg g^{-1}.

4.5. Nitrogen in drainage water of organic soils

In view of the large amounts of nitrogen mineralized and nitrified in organic soils, it is not surprising that organic soils contribute to dissolved nitrogen in agricultural drainage water. In a study of the drainage water from three organic soils in southern Ontario, Miller (14) found average annual NO_3^- - N concentrations ranging from 14.8 to 42.7 mg l^{-1}. The average annual export of NO_3^- - N from these three sites ranged from 37 to 245 kg ha^{-1}. Amounts of NH_4^+ - N and dissolved organic N in the drainage water of these soils was reported to be quite low compared to NO_3^- - N.

Duxbury and Peverly (6) monitored the nutrient content of drainage water from New York Histosols and calculated annual losses of 39.2 to 87.5 kg ha^{-1} for NO_3^- - N and ⟨ 1.0 to 1.9 kg ha^{-1} for NH_4^+ - N. Nutrient concentrations were found to increase as drainage water flow from the soils increased, so that the

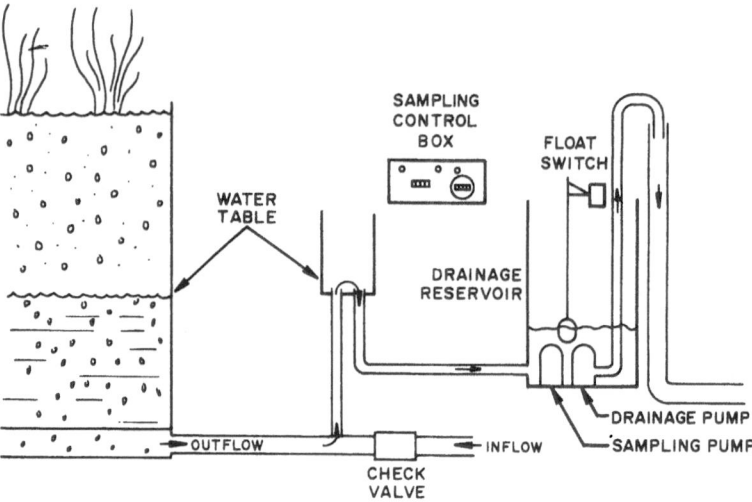

Fig. 4.3. Diagram of the lysimeter drainage water outflow system.

greatest output of nitrogen was during high-flow events of the late winter and spring.

Nitrate concentrations of the drainage water from the 2,000 hectares of drained Histosols of the Hula Valley of Israel were monitored from 1967 to 1970 (2). The average annual output of nitrate during that period was 1,100 metric tons per year with a high output of 2,200 metric tons during the rainy year of 1968-1969. As much as 50 percent of the nitrogen input into Lake Kinneret (Sea of Galilee) came from the drained organic soils of the Hula Valley during that period.

An extensive study of nutrient loadings of drainage water in the Everglades agricultural area was conducted by the Florida Sugar Cane League (10), in cooperation with the South Florida Water Management District. The quality of drainage water from eight farms was monitored from October 1976 through September 1977. The annual nitrogen loading rates of these drainage waters ranged from low as 12 kg N ha^{-1} for a cattle ranch to as high as 40 kg N ha^{-1} for a vegetable farm.

Terry *et al.* (27) conducted a study on the effects of water-table depth on the quality of drainage water from large lysimeters filled with Pahokee muck. The water-table lysimeters (5.5 m diameter and 1.2 m deep) were constructed from vinyl-lined, above-ground swimming pools. Corrugated plastic drainage tiles (10 cm diameter) placed at the bottom of the lysimeters allowed for drainage of excess water from rainfall and for addition of irrigation water. Fig. 4.3. is a diagram of the lysimeter water-table control and drainage water outflow system. This study was a portion of a lysimeter project designed to evaluate the response of five sugercane varieties to water-table depths of 30, 60, and 90 cm.

Table 4.2. Effect of water-table depth on annual nutrient loading of the lysimeter drainage water

Year	Water-table depth (cm)	Drainage outflow (cm)[+]	TDKN# (kg/ha)	NO_3^- - N (kg/ha)	NH_4^+ -N (kg/ha)
1978	30	113a‡	68.1a	1.1b	37.5a
	60	70a	44.4b	1.9b	27.9ab
	90	54c	33.4b	80.2a	8.1b
1979	30	96a	57.7a	6.6b	29.6a
	60	59b	43.8ab	3.2b	18.0b
	90	41c	34.7b	90.9a	3.2c

+ One cm of drainage outflow represented 236 liters.

‡ Values within the same column for each year followed by the same letter are not significantly different at the 95% level (Duncan's multiple-range test).

Total dissolved Kjeldahl nitrogen.

Total dissolved Kjeldahl nitrogen (TDKN), NH_4^+ -N, and NO_3^- -N in the drainage water samples were monitored from January 1978 to December 1979. The concentrations of nitrogen in lysimeter drainage water increased with increasing drainage water outflow (27). The annual drainagewater output and nitrogen loading rates for lysimeter drainage water for 1978 and 1979 are given in Table 4.2. During both years of the study approximately twice as much water flowed from the 30 -cm lysimeters as from the 90 -cm lysimeters while an intermediate a-mount flowed from the 60 -cm lysimeters. Gascho and Shin (11) reported greater sugarcane yields for the 60 - and 90 -cm lysimeters than for the 30 - cm watertable lysimeters. The smaller crop canopies of the 30 - cm lysimeters likely intercepted less rainfall than the canopies of the 60 - and 90 -cm lysimeters causing greater outflow. There was also a greater volume of drained soil in the 60- and 90 -cm ly-simeters to provide water storage after rainfall events. The annual TDKN load of drainage water from the lysimeters was proportional to drainage water outflow. Less than 8 kg NO_3^- - N ha^{-1} were lost in the drainage water of the 30-60 cm water-table lysimeters over the 2-year study. Approximately 170 kg NO_3^- - N ha^{-1} were lost in drainage water of the 90 -cm lysimeters for mineralization of organic N to take place, and there was no drainage after small rainfall events

(< 50 mm), therefore it is possible that nitrate accumulated in the lysimeters until the larger rainfall events leached the soil. Concentrations as great as 49 mg l^{-1} were found in drainage water from the 90 cm lysimeters. The larger quantity of NH_4^+ -N from the 30- and 60 -cm lysimeters may have been caused by the greater quantity of drainage output and the smaller volume of drained soil for nitrification to take place. The average annual nitrogen output of the 30- and 90 -cm lysimeters were 65 and 120 kg ha^{-1}, respectively. The nitrogen output of the 60 -cm lysimeters was 45 kg ha^{-1}, near the range reported by the Florida Sugar Cane League (10).

The quantities of nitrogen found in the drainage water of Histosols are very large in comparison with N loading of drainage water from mineral soils (6, 14), but the quantities reported are much lower than the quantities of nitrogen known to be mineralized from Histosols. Only about 10% of the mineralized N (600 to 2,000 kg ha^{-1} $year^{-1}$) can be accounted for in drainage water (12 to 200 kg N ha^{-1}) plus harvested crop (*e.g.* 80-100 kg N ha^{-1} for sugarcane 4). It is apparent that most of the mineralized nitrogen must be lost from the soil environment *via* denitrification.

4.6. Denitrification

Denitrification, which is mediated by soil bacteria, is the process by which nitrate is reduced to dinitrogen by the following reduction steps:

$$NO_3^- \rightarrow NO_2^- \rightarrow NO \rightarrow N_2O \rightarrow N_2$$

Raveh and Avnimelech (17) have reported that denitrification can be used as an effective means of lowering nitrate levels in organic soils of the Hula Valley. Flooding of the soil by subsurface irrigation led to the reduction of nitrate levels in the soil from 2,650 to 800 kg NO_3^- - N ha^{-1}. They reported that most of the decrease occurred in the top 30 cm of soil. Nitrate concentrations remained unchanged in the 30- to 90-cm layer of the soil during flooding. It was apparent that the difference in denitrification losses was related to differences in microbial activity and the availability of organic carbon within the soil profile. Sprinkler irrigation of the soil was also found to be an effective means of creating temporary anaerobic conditions in the surface soil, thus promoting dentrification losses.

The dependence of denitrification in organic soils on available organic carbon was demonstrated by Terry and Tate (28). Potential denitrification rates of soils collected from fallow sugarcane, and St. Augustine grass fields were determined in the laboratory. Freshly collected soil samples were amended with a nitrate solution and incubated anaerobically with 0.1 atm acetylene in the headspace gases. In the presence of acetylene gas the sole gaseous product of denitrification is N_2O

Table 4.3. Potential denitrification rates (μg N cm^{-3} day^{-1}) in Pahokee muck.

Soil use	Average rate	Range
Fallow (0-10 cm)	16.2c[+]	10.2 - 21.1
Fallow (60-70 cm)	6.1c	1.4 - 9.6
Sugarcane (0-10 cm)	38.1b	29.6 - 52.0
St. Augustinegrass (0-10 cm)	65.6a	52.4 - 78.5

[+]Values followed by the same letter are not significantly different at the 95% level (Duncan's multiple-range test).

due to the inhibition of nitrous oxde reductase by acetylene (35). Denitrification loses were determined by measuring the accumulation of N_2O in the samples by gas chromatography. The mean denitrification rates for four sampling times from March through August 1978 are listed in Table 4.3. Denitrification rates were significantly lower in the surface layer of the fallow soil than in the cropped soils. The fallow field had been kept clear of all vegetation for a period of one year prior to this experiment. It was apparent that the enhanced denitrification rates in the cropped soils were due to increased available carbon provided by root exudates and decaying plant residues. This relationship of denitrification rates with microbial activity is also illustrated by the decrease in denitrification rate at the 60- 70 -cm depth of the fallow soil. Experiments have shown that microbial numbers and microbial activities are significantly higher in the cropped organic soils than in fallow soils, and are also higher in surface organic soils than in subsurface soils (22, 25). Avinmelech and Raveh (3) reported that denitrification activities were very low in layers of organic soils below 60 -cm, even though the soils were rich in organic matter and anaerobic conditions existed. The decreased denitrification activity below 60 -cm was attributed to the lack of readily decomposable organic substrates.

By comparison of N_2O production in the presence of and in the absence of acetylene, both the total nitrogen dinitrified and that portion of the reduced NO_3^- which remains as N_2O can be determined (30). The relative portions of gaseous products of denitrification can be described in terms of a N_2/N_2O -N ratio. Terry and Tate (30) studied the effect of cropping practice on denitrification and the N_2/N_2O -N ratio. They collected surface samples of cropped and fallow Pahokee muck in July 1979 for laboratory incubation experiments. The soil conditions for denitrification should have been optimum at this sampling time due to the high rainfall and rapid crop growth of the rainy season in south Florida

Table 4.4. Effect of soil use and flooding on denitrification rates (μg N cm^{-3} day^{-1}) and N_2/N_2O-N ratios in Pahokee muck after 4 hours of incubation.

Soil use	Denitrification rate	N_2/N_2O-N
Fallow	7.2	0.17
Sugarcane	18.0	0.36
St. Augustinegrass	50.4	0.68
Vegetable (flooded 22 days)	18.0	36.00

The denitrification rates were greater in the soils of sugarcane and St. Augustine-grass fields than in the soil of a fallow field (Table 4.4.). This increased denitrification activity was attributed to greater availability of organic carbon in the cropped soils. In these freshly collected samples of Pahokee muck, not only were the denitrification rates greater in the cropped soils, but the N_2/N_2O-N ratios were increased. After 4 hours of incubation 17, 27, and 40% of the gaseous denitrification products were in the form of N_2 in fallow, sugarcane, and St. Augustinegrass field samples, respectively.

To determine the effects of flooding on denitrification rates and the N_2/N_2O-N ratio in Pahokee muck, soil samples were collected from a vegetable field which had been flooded for 22 days (30). The denitrification rate was similar to that of samples from the sugarcane field; however, the N_2/N_2O-N ratio was much greater (Table 4.4.). This indicated that N_2O was reduced much more rapidly in the flooded soil than in the drained soil. It has been reported that the presence of nitrate in soil samples inhibits the reduction of N_2O to N_2 by denitrifiers (5). To test this hypothesis Terry and Tate (30) added NO_3^- - N to the fallow, cropped, and flooded samples of Pahokee muck to an initial concentration of approximately 100 mg l^{-1} in the soil solution. The nitrate levels may have inhibited N_2O reduction in the drained soil samples, however N_2O was readily reduced in the flooded soil (Table 4.4.). Rapid N_2O reduction in the presence of nitrate also occurred in sediment samples dredged from a small drainage ditch and from the Hillsboro Canal in the Everglades (30). The enzymes required for N_2O reduction were apparently present in the bacteria of samples collected from the anaerobic environments, and nitrate apparently did not inhibit this enzyme. However, nitrate may inhibit the synthesis of nitrous oxide reductase in fresh soil samples collected from well-drained fields (9, 29, 30).

Unfortunately, there are not many reports on field measurements of denitrification from organic soils. Emissions of N_2O from Histosols of Florida and New

York have been reported, however (6, 31). Considerable interest in nitrous oxide emissions from soils has been sparked because of the proposed role of N_2O in regulating stratospheric ozone levels, and in contributing to the atmospheric greenhouse effect. Nitrous oxide emissions from Histosols are of particular interest due to the large quantities of N denitrified in these soils.

Nitrous oxide fluxes from cropped and uncropped soils in Florida and New York were measured by the static chamber technique described by Terry et al. (31). Study sites on Pahokee muck in Florida were established in a fallow field and in fields cropped to St. Augustinegrass and sugarcane. Nitrous oxide emissions from an undrained Histosol in Conservation Area 3 of the Everglades were also determined. In New York, N_2O emissions were measured from organic soils cropped to onions (Allium) and sweet corn. Fluxes of nitrous oxide were measured from May 1979 through May 1981. Nitrous oxide emissions were found to be closely related to increased soil moisture from rainfall events (7, 31) and maximum fluxes of 4.5 kg N_2O-N ha^{-1} day^{-1} occurred following rainfall events. High N_2O fluxes were sustained for substantial periods of time; for example, the daily flux of N_2O from the fallow field in Florida was above 1 kg N ha^{-1} for 27 days in July 1979. Peak emissions of N_2O occurred during the wet summer months in Florida and in the wet spring and fall periods in New York. A striking feature of the N_2O emission pattern in New York was a daily flux commonly in the range of 0.2-0.4 kg N ha^{-1} during January and February, despite the fact that the surface

Table 4.5. Emissions of nitrous oxide (kg N ha^{-1}) from Histosols of Florida and New York.

Crop	Maximum observed daily N_2O flux	Annual N_2O emission	
		May 1979 -May 1980	May 1980 -May 1981
Onions (NY)	4.5	85	72
Sweet Corn (NY)	2.9	76	152
Sugarcane (FL)	3.1	48	7
St. Augustinegrass (FL)	4.6	97	16
Mone (fallow) (FL)	4.5	165	59
Sawgrass (FL) (undrained)	⟨ 0.1	–	1

soil was generally frozen. Soil atmosphere measurements indicated that N_2O was generated in the section of the soil profile that was unfrozen and above the water table.

The annual N_2O emissions from these organic soils of New York and Florida are reported in Table 4.5. Annual N_2O emissions from the Florida sites differed substantially in the two study years. The dramatic reduction in the second year was attributed to an unusually low amount of annual precipitation (92 cm compared with 157 cm in the first year and a 54 yr average of 145 cm). Annual emissions varied considerably with crop and showed the same relative pattern in both years increasing in the order sugarcane ⟨ grass ⟨ fallow. At the 5% confidence level, the differences between the sites were all significant in the first study year. In the second year, N_2O loss from the fallow field was significantly higher than from the other sites, which were not different. The large loss from the fallow field compared with the cropped fields was probably the combined result of wetter soil, higher soil nitrate levels in the absence of plants, and a lower N_2/N_2O-N ratio.

Emissions of N_2O from the drained Histosols were as much as two orders of magnitude greater than N_2O emissions from the undrained organic soil of Conservation Area 3 in south Florida (7). It was concluded that although undrained organic soils represent a small portion of earth's land area, these soils may produce a significant portion of the N_2O found in the atmosphere.

4.7. Conclusions

Drained Histosols are an important agricultural resource as many of these soils are used in the production of high-value crops such as winter vegetables, onions, and mint. The useful productive life span of many organic soil deposits is limited because of soil subsidence. The natural process of microbial oxidation of organic soils occurs when the water saturated conditions which led to the formation of the soil are reversed to allow agricultural utilization.

The foregoing presentation has revealed some unique properties of organic soils. The quantities of inorganic nitrogen released from and transformed in organic soils are much greater than those in mineral soils. This situation provides an advantage to the farmer, in that, expensive nitrogen fertilizers are rarely needed for crop production on Histosols. At the same time, the vast quantities of nitrogen mineralized from organic soils pose a threat to the quality of surface and ground waters which may be polluted by nitrogen contained in drainage water.

The magnitude of the nitrogen cycle in organic soils provides the scientist with a unique opportunity to study nitrogen transformations. Continued study of nitrogen transformations in organic soils is needed not only to provide information for those who manage organic soils, but to provide a better understanding of the many microbial transformations of nitrogen in various ecosystems. A better understanding of microbial oxidation of soil organic matter and the microbial nitrogen transformations in soils will aid those who manage organic soils to extend the productive live span of these soils and to preserve the quality of our environment.

4.8. References

1 Avnimelech Y 1971 Nitrate transformation in peat. Soil Sci. 111, 113-118.
2 Avnimelech Y, Dasberg S., Harpaz A., and Levin I. 1978 Prevention of nitrate leakage from the Hula Basin, Israel: A case study in watershed management. Soil Sci. 125, 233-239.
3 Avnimelech Y. and Raveh A., 1974 The control of nitrate accumulation in soils in induced denitrification. Water Research 8, 553-555.
4 Barnes A.C. 1974 The Sugar Cane, John Wiley and Sons, New York 572 p.
5 Blackmer A.M. and Bremner J.M. 1978 Inhibitory effect of nitrate on reduction of N_2O to N_2 by soil microorganisms. Soil Biol. Biochem. 10, 187-191.
6 Duxbury J. M. and Peverly J.H. 1978 Nitrogen and phosphorus losses from organic soils. J. Environ. Qual. 7, 566-570.
7 Duxbury J.M., Bouldin D.R., Terry R.E. and Tate R.L. 1982 Emissions of nitrous oxide from soils. Nature 298, 462-464.
8 Farnham R.S. and Finney H.R. 1965 Classification and properties of organic soils. Adv. Agron. 17, 115-162.
9 Firestone M.K., Smith M.S. Firestone R.B. and Tiedje J.M. 1979 The influence of nitrate, nitrite, and oxygen on the composition of gaseous products of denitrification in soil. Soil Sci. Soc. Am. J. 43, 1140-1144.
10 Florida Sugar Cane League 1978 Water Quality Studies in the Everglades Agricultural Area of Florida. The Florida Sugar Cane League. Clewiston, Florida.
11 Gascho G.J. and Shih S.F. 1979 Varietal response of sugarcane to water table depth: 1. Lysimeter performance and plant response. Soil Crop Sci. Soc. Fla. Proc. 38, 23-27.
12 Guthrie T.F. and Duxbury J.M. 1978 Nitrogen mineralization and denitrification in organic soils. Soil Sci. Soc. Am. J. 42, 908-912.
13 Isirimah N.O. and Keeney D.R. 1973 Nitrogen transformations in aerobic and waterlogged Histosols. Soil Sci. 115, 123-129.
14 Miller M.H. 1979 Contribution of nitrogen and phosphorus to subsurface drainage water from intensively cropped mineral and organic soils in Ontario. J. Environ. Qual. 8, 42-48.
15 Millette J.A. 1976 Subsidence of an organic soil in southwestern Quebec. Can. J. Soil. Sci. 56, 499-500.
16 Neller J.R. 1944 Influence of cropping, rainfall, and water table on nitrates in Everglades peat. Soil Sci. 57, 275-280.
17 Raveh A. and Avnimelech Y. 1973 Minimizing nitrate seepage from the Hula Valley into Lake Kinneret (Sea of Galilee): I. Enhancement of nitrate reduction by sprinkling and flooding. J. Environ. Qual. 2, 455-458.
18 Reddy K.R. 1982 Mineralization of nitrogen in organic soils. Soil Sci. Soc. Am. J. 46, 561-566.
19 Schothorst C.J. 1977 Subsidence of low moor peat soils in the western Netherlands. Geoderma 17, 265-291.
20 Stephens J.C. 1969 Peat and muck drainage problems. J. Irrig. and Drainage Div. Proc. Am. Soc. Civil Eng. 95, 285-305.

21 Tate R.L. 1977 Nitrification in Histosols: A potential role for the heterotrophic nitrifier. Appl. Environ. Microbiol. 33, 911-914.

22 Tate R.L. 1979 Microbial activity in organic soils as affected by soil depth and crop. Appl. Environ. Microbiol. 37, 1085-1090.

23 Tate R.L. 1980 Microbial oxidation of soil organic matter in Histosols. Adv. Microbiol. Ecol. 4, 169-201.

24 Tate R.L. 1980 Variation in heterotrophic and autotrophic nitrifier populations in relation to their role in nitrification in organic soils. Appl. Environ. Microbiol. 40, 75-79.

25 Tate R.L. 1980 Effect of several environmental parameters on carbon metabolism in Histosols. Microbial Ecol. 5, 329-336.

26 Terry R.E. 1980 Nitrogen mineralization in Florida Histosols. Soil Sci. Soc. Am. J. 44, 747-750.

27 Terry R.E., Gascho G.J. and Shih S.F. 1980 Effect of depth to water table on the quality of water in the Everglades agricultural area. pp. 700-704, Proc. 6th Int. Peat. Cong., Duluth, Minn.

28 Terry R.E. and Tate R.L. 1980 Denitrification as a pathway for nitrate removal from organic soils. Soil Sci. 129, 162-166.

29 Terry R.E. and Tate R.L. 1980 Effect of flooding on microbial activiteis in organic soils: Nitrogen transformations. Soil Sci. 129, 88-91.

30 Terry R.E. and Tate R.L. 1980 The effect of nitrate on nitrous oxide reduction in drained, organic soils and sediments. Soil Sci. Soc. Am. J. 44, 744-746.

31 Terry R.E., Tate R.L. and Duxbury J.M. 1981 Nitrous oxide emissions from drained, cultivated organic soils of south Florida. J. Air Poll. Cont. Assoc. 31, 1173-1176.

32 Verstraete W. and Alexander M. 1973 Heterotrophic nitrification in samples of natural ecosystems. Environ. Sci. Technol. 7, 30-42.

33 Volk B.G. 1972 Everglades Histosol subsidence: 1. CO_2 evolution as affected by soil type, temperature and moisture. Soil and Crop Sci. Soc. of Fla. 32, 132-135.

34 Weir W.W. 1950 Subsidence of peatlands of the Sacramento;San Joaquin Delta, California. Hilgardia. 20, 37-56.

35 Yoshinari T. and Knowles R. 1977 Acetylene inhibition of nitrous oxide reduction and measurement of denitrification and nitrogen fixation in soil. Soil Biol. Biochem. 9, 177-183.

Section II

**Effects of soil organic matter and redox on micronutrients availability
to plants**

5. Soil organic matter interactions with trace elements

Y. CHEN and F.J. STEVENSON

5.1. Introduction

The availability of trace elements to plants is governed by a variety of reactions that include complexation with organic and inorganic ligands, ion exchange and adsorption, precipitation and dissolution of solids, and acid-base equilibria (90). A key role is often played by organic matter, both in enhancing availability to plants (65, 66, 113, 114) and in reducing toxicity effects of the free cation (20, 31, 62). Trace metals that would ordinarily convert to insoluble precipitates (as carbonates, sulfides, or hydroxides) at the pH's found in many soils are undoubtedly maintained in solution through chelation.

Consideration is given in this chapter to the role of soil organic matter in promoting the uptake of Fe, Mn, Zn, and Cu by higher plants, the nature of organic complexing agents in soil, the use of micronutrient-enriched organic wastes and naturally occurring metal-organic complexes as soil amendments, and stability constants for the binding of select micronutrient cations to humic and fulvic acids.

5.2. Importance of complexes of Fe, Mn, Zn and Cu with humic substances to agriculture.

5.2.1. Physiological and biochemical functions

The functions of Fe, Mn, Zn and Cu in higher plants are numerous and variable. Their over-all effects on biochemical processes are listed in Table 5.1. A deficiency of any given micronutrient can seriously affect plant growth, and subsequently yields. Deficiencies are especially common in calcareous soils or those to which lime has been applied.

Transition metals (*e.g.*, Fe, Mn, Cu and Zn) typically occur in plant tissues in chelated forms, where they are strongly and specifically attracted to N and S con-

Table 5.1. Over-all effects of Fe, Mn, Zn, and Cu on biochemical processes in plants (adapted from Mengel and Kirby (93)).

Nutrient element and uptake	Functions	
	General	Specific
Mn in the form of ions from the soil solution	The ion brings about optimum conformation of an enzyme protein (enzyme activation). Bridging of reaction partners.	Structural influences by binding to organic molecules, particularly enzymes, thereby altering their conformation Formation of a Lewis acid, whereby the ion accepts an electron pair and catalyzes or polarizes reactive groups. Essential component of prosthetic groups which bring about electron transfer.
Fe, Cu, Zn in the form of ions or chelates from the soil solution	Present predominantly in chelated forms as prosthetic groups of enzymes. Enable electron transport by valency changes	Formation of Lewis acids whereby the ions accept an electron pair and thus catalyze or polarize reactive groups. Participate in redox reactions (Fe and Cu). Essential components of prosthetic groups which bring about electron transfer.

taining functional groups. However complexes with carboxyl groups and phosphates also occur. The most important naturally occurring chelates are those containing the heam group, and chlorophyll. The former are Fe-porphyrins in which the heam group forms the prosthetic group for a number of enzymes (catalase, peroxidase, cytochromes, cytochrome oxidase). The Fe present in the haem moiety can undergo valency changes (*e.g.* from Fe^{2+} to Fe^{3+} and *visa versa*). Copper functions in enzyme systems in an analogous way to that described for Fe. For Mn, an active Mn-ATP complex is formed.

It is likely that micronutrients are predominantly transported to plant roots as soluble chelate complexes. Therefore, chelation in the vicinity of the root plays an important role in their availability.

5.2.2. Transport of micronutrients to plant roots

The transport of micronutrients in general, and Fe in particular, from the solid phase of the soil to plant roots was clearly demonstrated by Lindsay (78) and O' Connor et al (98). Since Fe deficiencies are more wide-spread than those for Mn, Zn, and Cu, this micronutrient will be used to demonstrate both the transport problem and the significance of humic substances in facilitating the movement of micronutrients to plant roots.

In comparison to the macronutrients, the Fe content of green plant tissues is low, generally of the order of 100 μg/g on a dry-weight basis. Somewhat lower a-mounts are found in cereal grains, tubers, and roots. A soil containing 0.5 μg Fe/g throughout the plough layer will contain sufficient Fe to meet the requirements of most agricultural plants (78). Total Fe levels in soil are substantially higher, usually of the order of 2%, or 20.000 μg/g. Thus, the total amount of Fe in most soils greatly exceeds crop requirements. The main problem of Fe supply in one of availability and transport.

The solubility of inorganic Fe in soil is highly pH dependent. In order for sufficient Fe to be transported to roots through mass-flow, the total solubility of Fe must be at least (10^{-7} M (79), a level that is only achieved at pH 3. By raising the pH to just over 4, only 1% of the Fe demand can be met. At normal soil pH levels, even with allowance for the contribution of diffusion, the concentration of inorganic Fe in the soil solution is far below that required to meet the Fe requirements of plants. It appears, therefore, that soluble organic complexes of Fe must play an important role in supplying Fe to plants. These soluble organic compounds consist of root exudates, humic substances from the soil organic matter, metabolic products of microorganisms, or applied Fe-chelate fertilizers (93).

Two basic mechanism--diffusion and mass-flow (convection)--may be operational in the movement of micronutrients to plant roots or groundwater. Seldom has there been a clear division between the two processes; both occur to some degree whenever micronutrients are transported within the soil system. The dominant mechanism will depend on the rate and direction of water movement, the micronutrient involved, plant species, and environment conditions surrounding the plant root.

The movement of micronutrients to plant roots through diffusion and mass-flow has recently been reviewed by Ellis et al. (48). Therefore, we shall only briefly discuss some major points related to the two mechanisms.

Diffusion: the activity gradient of ions is the driving force for the net transfer of ions and molecules from regions of high to low activity. This transfer is called diffusion and is described by Fick's law for steady state diffusion in pure liquid (concentration terms):

$$\partial Q/\partial t = -DA(\partial C/\partial X)$$

where Q is the quantity of nutrient diffusing across a unit cross-sectional area in time t, D is the diffusion coefficient, A is the activity or concentration gradient, X is the distance in the direction of net movement, and C is the concentration of ions in the bulk solution. As the system $\partial C/\partial X$ approaches zero, the rate of movement of ions approaches equality in all directions and net diffusion approaches zero. Values of D for the various micronutrients in water have been measured or calculated and are summarized by Ellis et al. (48). In general, D_o (the diffusion coefficient in water) for Fe, Mn, Zn, and Cu varies from 10^{-6} to 10^{-5}. Apparent diffusion coefficients (D_e) in soils are much lower.

Most work on diffusion rates (D_e) of micronutrients in soils has focused on Zn. As one might expect, several factors affect diffusion, the most important being the moisture content of the soil. An increase in soil moisture content tends to increase D_e by making the paths less tortuous. Bulk density of a soil is related to diffusion *via* its effect on soil tension and moisture content. Warnacke and Barber (151) have shown that D_e, in general, increases with increasing bulk density. Soil pH also alters the D_e of Zn (38, 98). Specifically, D_e increases with a decrease in pH below 7.0. The effects of pH may be ion specific and may also be influenced by soil type. Values of D_e for Zn are also concentration dependent (97, 152). Diffusion of cations such as Fe, Mn, Zn and Cu is also affected by the presence of other ions. To maintain electroneutrality, either co-diffusion (of ions of opposite charge) or counter-diffusion (of ions of the same charge but in the opposite direction) must take place (84).

Factors that increase the concentration of a micronutrient in the soil solution should increase its diffusion rate. Accordingly, the presence of chelates and naturally occurring or synthetic organic complexes should increase the D_e of the micronutrient. Lindsay (78) concluded that diffusion as a chelate complex was a major mechanism for the transport and uptake of Fe by plants.

Mass-flow (convection, miscible displacement): The fundamental principle of the movement of micronutrients (and other soluble ions, molecules and colloidally dispersed particles) through mass-flow is simple and obvious: whenever water moves in the soil due to potential gradients it will carry the micronutrients that are in solution with it. Thus, micronutrients as ions or in the form of complexes or chelates may be transported to plant roots by mass-flow.

In general, Fe, Mn, Zn, and Cu are considered immobile in soil because they readily form precipitates and are strongly bound to clay surfaces. Autoradiography (11, 75, 77, 156, 157) and thin slicing techniques (71) have been used to differentiate between transport mechanisms of ions to plant roots. If the autoradiograph or thin section shows depletion of a nutrient around the root, absorption is assumed to have exceeded the rate at which the nutrient has been transported through mass-flow; thus, diffusion is the major mechanism of ion movement. An accumulation of the micronutrient at the root surface indicates that movement to the root has occurred at a faster rate than absorption; in this case, massflow is assumed to be the major transport mechanism. Some reports on the movement of Fe, Mn, Zn, and Cu will be briefly summarized.

1. Iron

O'Connor *et al.* (98) found that Fe uptake by plants increased linearly with the concentration of Fe in the soil solution. The greater uptake was attributed to increased diffusion caused by chelates increasing the level of Fe in solution. They suggested that mass-flow was a significant mechanism for the movement of Fe to plant root, but only at pH's below 4.5, or when the Fe in solution is in complexed forms. Oliver and Barber (99) reported that diffusion, mass-flow, and root interception were all involved in the uptake of Fe by plant roots. At low rates of water transpiration from the plant, diffusion accounted for about 70% of the Fe uptake.

2. Manganese

According to several investigators (10, 61, 99), at low Mn concentrations in the soil solution, diffusion accounts for most of the movement of Mn to plant roots; at high concentrations, mass-flow and root interception occur.

3. Zinc

Oliver and Barber (99) and Wilkinson *et al.* (156, 157) regarded diffusion as the major mechanism of Zn transport to plant roots. Wilkinson *et al.* (156, 157) based their conclusion on the fact that a three-fold increase in transpiration rate did not effect Zn absorption by the plant. Halstead *et al.* (61) also concluded that diffusion was the main mechanism for Zn uptake; mass-flow could not account for Zn absorption in a number of plants which were tested.

4. Copper

The status of Cu uptake seems to differ from that of Fe, Mn, and Zn. Although its concentration in the soil solution is similar to that of Zn (65, 66), uptake by plants is an order of magnitude lower. Therefore, mass-flow can account for adequate movement of Cu to plant roots. This was confirmed by Oliver and Barber (99), who found that diffusion of Cu accounted for only 5% of its uptake.

It is likely, therefore, that complexing agents (*e.g.,* defined biochemical compounds and humic substances) play a prominent role in the dissolution of micronutrients and their transport to plant roots. The effects of humic substances on plant growth will be discussed in greater detail in the next section.

5.2.3. *Supply of micronutrients to higher plants*

Beneficial effects of humic substance on plant growth have been recognized by many workers (95, 124, 125, 138). Applications of humic substances to soils low in clay (and organic matter), or to nutrient solutions, have produced significant growth responses. Among other mechanisms, it has been suggested that growth is enhanced by increasing the uptake of micronutrients by the plant, thereby favorably affecting metabolic relationships.

De Kock and Strmecki (44) investigated the growth promoting effects of lignite on mustard plants grown in nutrient solution containing additional Fe. They concluded that the growth promoting effect resulted from humic substances that solubilized Fe, even at high phosphate levels where insoluble Fe-phosphates were expected to be formed. Chlorotic plants were found to contain high amounts of Fe-phosphates in their roots while healthy plants did not, indicating that humic substance from the lignite not only provided Fe to the roots, but also imporved Fe translocation in the plant. Lee and Bartlett (76) reported that the addition of Na-humate at a 5 mg/l level to a nutrient solution resulted in an enhanced growth response along with an increase in root and shoot Fe. When $FeCl_3$ or Fe-citrate were added to the solution, the Fe content of the shoots increased further while that of the roots decreased. This is another indication of the influence of humates on the mobility of Fe in plant tissues.

Linehan (80, 81) grew wheat plants in a Hogland solution (pH 5) to which humic substances at various rates were added. The solution also contained M Fe - EDTA and Fe - citrate at 2 M concentrations. Addition of up to 200 mg/l humic substances resulted in a decrease in the Fe content of the roots. However, this decrease was accompanied by an increase in plant growth and development, due apparently to better translocation of Fe in the plant. Similar results were obtained when fulvic acid (FA) and polymaleic acid (PMA) were added to the nutrient solutions (82). Iron in shoots of the wheat seedlings increased for those treatments that contained up to 5 mg/l of HA in solution, as well as up to 25 mg/l FA and PMA. The diverse effects of the chelating agents on the Fe content of roots was attributed to sorption of large positively charged colloidal iron particles on the root surface at low or zero chelate concentration. Lindsay and Schwab (79) state that in the absence of a solid reservoir of Fe in contact with a nutrient solution, an Fe chelate will not be effective in providing Fe to plant roots. A number of investigators (35, 36, 37, 44, 122, 123) have used humic substances or humic-like substances enriched with Fe as Fe amendments to deficient soils. Many of these amendments have been found to be efficient in providing Fe to the test plants, an approach that will be discussed separately in Section 5.4.

In sumary, it may be concluded that humic substance can imporve both the Fe supply to plant roots and its translocation in the plant. As noted below, humic substances also have the potential for improving the uptake of other micronutrients.

Jalai and Takkar (67) reported that the uptake of Fe, Zn and Cu by rice was directly correlated to the organic matter content of the soil. In contrast, tests performed on beet disks by Vaughan and McDonald (146) showed that additions of HA to the nutrient media led to slight decreases in Zn uptake; after 3 days, a decrease was observed at HA concentrations greater than 25 mg/l. At lower HA concentrations, no significant effects on uptake were observed. White and Chaney (153) followed the uptake of Mn, Zn, and Cd in two soils containing variable amounts of organic matter (Sassafras, 1.2% organic matter; Pocomoke, 3,8% organic matter). Various amounts of Zn and Cd were added and chlorosis and yield were recorded. Damage to plants from toxic levels of the applied microelements was inhibited in the soil containing the high amount of organic matter (Pocomoke soil). Strickland et al. (135) grew soybean plants in pots of quartz sand to which variable amounts of peat were applied (0.5% to 8% by weight). Addition of the peat led to increased yields. When Cd was added at a rate of 20 μg/g of soil weight, plant growth was inhibited. The decrease in growth was less prominent at higher additions of peat. Increases in Cd concentration were observed in roots, stems and leaves at the low rate of peat additions (0.5%); at higher rates, adsorption of Cd to the peat apparently inhibited its uptake by the plant. Tyler and McBride (142) found that the addition of HA to a nutrient solution containing Cd resulted in a decrease in Cd activity due to complex formation and absorption of Cd by corn roots.

Increases in the uptake of N, P, K, Ca, Mg, Fe, Zn, and Cu has been reported by Schnitzer and Rauthan (119) for cucumber plants grown in Hoagland solutions containing 100 to 800 mg/l of FA. For this range of FA concentration, a significant increase in plant growth was observed. Additions of FA at concentrations of 500 mg/l or more were less beneficial. This study differs from those of Lee and Bartlett (76) and Linehan (80, 81, 82) in the range of concentration that was beneficial. A plausible explanation for the divergent findings is that cucumber plants in the study of Schnitzer and Rauthan (119) were grown to maturity; in most other studies, seedlings only were tested. The same explanation may apply to the positive results obtained when Fe-enriched amendments containing humic substances have been applied to soils (see Section 5.4).

Although enhancement of micronutrient uptake by plants in the presence of humic substances is the common case, a number of reports have indicated decreases in micronutrient uptake. Some of these reports were briefly reviewed earlier in this chapter. The differences can be attributed to one or more of the following factors: (1) plant uptake mechanism may vary with the plant and its physiological age; (2) uptake is affected when a solid phase of the micronutrient comes in contact with the nutrient solution; (3) uptake is affected by the concentration of soluble organic chelating agents (humic acid, fulvic acid, defined biochemicals), by the concentration of the micronutrient and by the stability constant of the metal-organic matter complex; and (4) uptake is influenced by the presence of insoluble organic matter, such as humin, which may contain active functional groups that "fix" the micronutrient and thus prevent its uptake by the plant. The last mechanism would act in an opposite direction to soluble organic complexing agents.

The beneficial effect of organic materials containing humic substances will be further clarified in Section 5.4. where emphasis will be given to micronutrient-enriched organic wastes and naturally occurring metal organic complexes and their use as soil amendments.

5.3. Nature of organic complexing agents in soil

Two groups of compounds are responsible for the binding of trace elments in soil, namely: (i) biochemicals of the type known to occur in living organism such as simple aliphatic acids, amino acids, sugar acids, and polyphenols, and (ii) humic substances, represented by the so-called humic and fulvic acids. The latter represent end products of the humification process whereas the former are synthesized by microorganisms during the decay of plant and animal residues. A second source of biochemical chelating agents is excretion products of plant roots.

5.3.1. Defined Biochemical Compounds

Biochemical compounds having chelating characteristics, such as simple aliphatic acids, amino acids, and polyphenols, are periodically produced in soil through microbial activity. These constituents normally have only a transitory existence; accordingly, the amounts present in the aqueous phase will vary over time and will represent a balance between synthesis and destruction by microorganisms. High amounts would be expected to be produced during periods of intense biological activity, such as following additions of crop residues to the soil. Their production in the rhizosphere may also enhance the availability of insoluble forms of micronutrients to plants. Natural chelating substances of various types have been observed in soil leachates (15, 28, 41).

The assumption has often been made that simple biochemicals are of little importance as complexing agents because of their rapid destruction by microorganisms. Absolute elimination of metabolites from biogenic habitats is seldom achieved, however, and it is more reasonable to conclude that small but measurable quantities of biochemical compounds will normally be present in the aqueous phase. It should be noted, in this respect, that the molar concentrations of micronutrients in the soil solution will normally be very low, usually $< 10^{-6} M$ (90).

The approximate molar concentrations of individual biochemical species in the soil solution, as deduced by Stevenson and Ardakani (133) from published literature values, are as follows:

Simple aliphatic acids	1×10^{-3} to $4 \times 10^{-3}\,M$
Amino acids	2×10^{-5} to $6 \times 10^{-4}\,M$
Aromatic acids	5×10^{-5} to $3 \times 10^{-4}\,M$

Simple aliphatic acids are of special interest as natural chelators because of their ubiquitous nature and because many of the hydroxy derivatives are effective solubilizers of mineral matter. The organic acids most effective in forming stable chelate complexes with metal ions are those of the di- and tricarboxylic hydroxy types, a typical example being citric acid.

A variety of organic acids have been identified in soil but in very small amounts (94, 132, 150). Anaerobic conditions are particularly suitable for their production. Accordingly, high amounts may be found in the saturated phase during warm periods when conditions are optimum for microbial activity. The importance of oxalic acid as a chelator of Fe in forest soils has been emphasized by Graustein et al. (57).

Excretion products of roots include a variety of aliphatic organic acids, many of which (e.g., citric, oxalic, and tartaric) are capable of forming complexes with metal ions (107, 126). Organic acids were found by Smith (126) to be the most abundant compounds in tree root exudates; those identified were citric, fumaric, malic, malonic, and succinic. Differences in the susceptibilities of plant species to trace element deficiencies have often been attributed to variations in organic acid production (see review of Hodgson (64)).

Low-molecular-weight phenolic acids occur widely in soils (68, 154, 155) but their effect on trace element chemistry is unknown. Many of the compounds identified, such as vanillic, p-hydroxybenzoic, p-coumaric, and ferulic acids, would not be expected to form chelate-type complexes with metal ions.

Recent evidence indicates that hydroxamate and catecholate siderophores produced by microorganisms play an important role in the iron nutrition of plants (4, 40, 104). These substances, which contain the hydroxamate (I) or catecholate (II) anionic reactive group, represent a class of microbially produced, Fe^{3+} transport moities with log K values ranging from 25 to 52. Greater amounts appear to be produced when the organism is under Fe stress.

Biologically significant levels of hydroxamic siderophores have been observed in soils (10^{-8} to 10^{-7} M at 10% moisture) (40). The amounts contained in the rhizosphere of plants appear to be particularly high (104). Hydroxamate siderophores have been shown to be produced by soil bacteria and fungi, including ectomycorrhizal fungi.

Other potential chelating agents in soil include amino acids, chlorophyll and chlorophyll-degradation products, and polyphenols. Neutral and basic amino acids are adsorbed by most soils, and, for this reason, they may play a subservient role to organic acids as chelators of metal ions. Polyphenols may be especially important as chelating agents in forest soils as these compounds occur in leaf drip, stem flow, and forest litter and can be leached directly into the mineral soil.

5.3.2. Humic substances

Humic substances are best described as a series of acidic, yellow- to black-colored, moderately high-molecular-weight polyelectrolytes that are formed by secondary synthesis reactions and which have characteristics dissimilar to any of the compounds occurring in living organisms. The modern view is that polyphenols derived from lignin, or synthesized by microorganisms, are enzymatically converted to quinones, which then polymerize in the presence or absence of amino compounds to form brown colored polymers of variable composition and molecular weight (63, 69, 118, 132).

5.3.2.1. Extraction and fractionation

Alkali, usually 0.1 to 0.5 M NaOH at a soil to extractant ratio of about 1:5 (w/v), has been the most popular extractant of soil organic matter. Repeated extraction is required to obtain maximum recovery. The solubility of humic substances in alkali is due in part to conversion of carboxyl (COOH) groups to their Na-salts (RCOOH → RCOONa), in which form the humates are soluble; salts of diand tri-valent cations are insoluble. Leaching the soil with dilute HCl, which removes Ca and other polyvalent cations, increases the efficiency of extraction of organic matter with alkaline reagents. However, a certain amount of organic matter, normally ⟨ 5% of the total for surface soils, is removed in the process. As a general rule, about two-thirds of the soil organic matter is solubilized by extraction with alkali.

In recent years, several milder and more selective extractants have been used, including salts of complexing agents (e.g. $Na_4P_2O_7$ and EDTA), organic chelating agents (e.g. acetylacetone), dilute acid mixtures containing HF, and organic solvent mixtures of various types. Of the mild extractants, $Na_4P_2O_7$ (usually 0.1 or 0.15 M aqueous solution) has been the most widely used. The amount of organic

matter removed (⟨ 30%) is considerably less that with caustic alkali but less alteration occurs. To minimize chemical changes in the humic material, extraction at
high pH's should be carried out in the absence of oxygen.

Reactions leading to extraction of organic matter by $Na_4P_2O_7$ have been postulated to be as follows (5):

$$R(COO)_4Ca_2 + Na_4P_2O_7 \longrightarrow R(COONa)_4 + Ca_2P_2O_7 \downarrow \qquad [1]$$

$$2[RCOOX(OH)_2] (COO)_2 Ca + Na_4P_2O_7 \longrightarrow$$
$$2[RCOOX(OH)_2](COONa)_2 + Ca_2P_2O_7 \downarrow \qquad [2]$$

where X is a trivalent cation. Humic acids recovered from soil by $Na_4P_2O_7$ usually contain Fe and Al as contaminants.

The organic matter extracted from soils is usually fractionated on the basis of
solubility characteristics (Fig. 5.1.). The fractions commonly obtained include:
humic acid, soluble in alkali, insoluble in acid; *fulvic acid,* soluble in alkali, soluble
in acid; *hymatomelanic acid,* alcohol-soluble part of humic acid; *humin,* insoluble
in alkali. Humic acids are sometimes further divided into two groups by partial
precipitation with electrolyte (salt solution) under alkaline conditions. The first
group, the brown humic acids (*Braunhuminsäure)*, are not coagulated by an electrolyte and are characteristic of humic acids in Histosols and Alfisols. The second
group, the gray humic acids (*Grauhuminsäure*), are easily coagulated and are characteristic of humic acids in Altoll and Rendoll soils. In the older German literature, considerable attention was given to the so-called *"apocrenic"* and *"crenic"*
acids, which were light-yellow fulvic acid-type substances.

In the natural state, humic and fulvic acids are intimately bound to one another, and to other organic consituents. Even after extraction, it is difficult to free
humic substances from organic impurities (carbohydrate, protein) (47, 132). Separation of "true fulvic acids" from the acid extract following removal of humic
acids may be possible by selective adsorption-desorption on a macroporous resin,
such as XAD-8 (3).

One useful concept that has evolved over the years, and popularized by Kononova (69), is that the various humic fractions represent a *system of polymers*
which vary in an orderly manner in elemental content, acidity, degree of polymerization, and molecular weight. The proposed relationships are shown in Fig. 5.2
where it can be seen that degree of polymerization, carbon and oxygen contents,
and exchange acidity all change systematically with molecular weight. No sharp
differences are shown to exist between the two main fractions (humic and fulvic
acids) or their subgroups. The humin fraction (material not extracted with alkali)
is not represented but this component may consist of:(1) humic acids so intimately bound to mineral matter that the two cannot be easily separated, and (2) highly

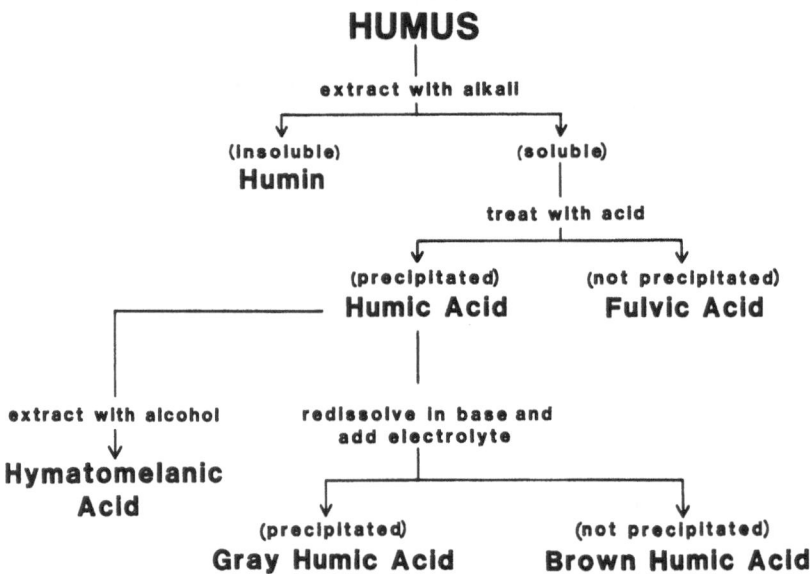

Fig. 5.1. Scheme for the fractionation of humic substances.

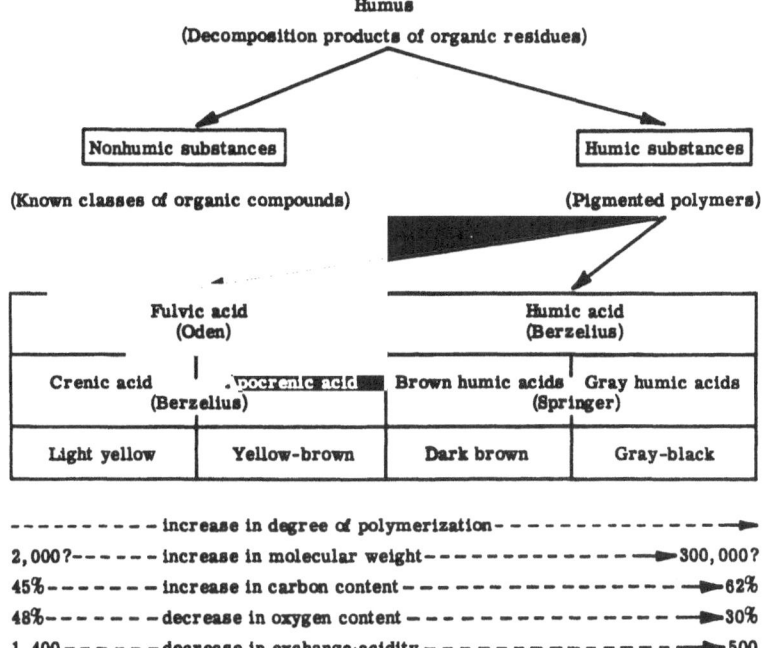

Fig. 5.2. Classification and chemical properties of humic substances

condensed humic matter having a high carbon content (⟩ 60%) and thereby insoluble in alkali.

All soils would be expected to contain a broad spectrum of humic substances, as depicted in Fig. 5.2. However, distribution patterns will vary considerably from one soil type to another. The humus of forest soils (Alfisols, Spodosols, and Ultisols) is characterized by a high content of fulvic acids; that of peat and grassland soils (Mollisols) contains high amounts of humic acid. As noted earlier, the humic acids of forest soils are mostly of the brown humic acid type; those of grassland soils are of the gray humic acid type.

Several attempts have been made to divise structural formulas for humic and fulvic acids, but as Dubach and Mehta (47) first pointed out, few molecules may have the precise identical structure. A "type" molecule for humic acid is believed to consist of micelles of a polymeric nature, the basic structure of which is an aromatic ring of the di- or tri-hydroxyphenol type bridged by -O-, -NH-, -N=, -O-, and other groups and containing both free OH groups and the double linkages of quinones (63, 69, 132).

Table 5.2. Oxygen-containing functional groups in humic and fulvic acids.*

Material	Total acidity	COOH	‡Acidic OH	Weakly acidic + alcoholic OH	C=O
		recorded range, meq/100 g			
Humic acids	560-890	150-570	210-570	20-496	10-560
Fulvic acids	640-1420	520-1120	30-570	260-950	120-420

* Adapted from Schnitzer (115) and Schnitzer and Khan (118)

‡ Usually reported as phenolic OH

5.3.2.2. Mechanisms of metal ion binding by humic and fulvic acids

The ability of humic substances to form stable complexes with metal ions can be attributed to their high content of oxygen containing functional groups, including COOH, phenolic-, alcoholic- and enolic-OH, and C=O structures of various types. Results presented in Table 5.2. show that the total acidities of fulvic acids (640 to 1420 meq/100 g) are considerably higher than for humic acids (560 to 890 meq/100 g). Both COOH and acidic OH groups (generally presumed to be phenolic OH) contribute to the acidic nature of these substances, with COOH being the most important.

Schnitzer (116) and Gamble *et al.* (53) concluded that two types of reactions are involved in metal-fulvic acid interactions, the most important one involving both phenolic OH and COOH groups. A reaction of lesser importance involved COOH groups only. The two reactions are:

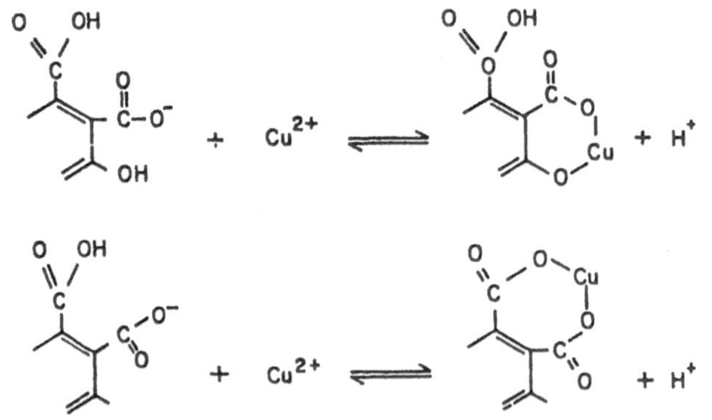

The formation of phthalate-type complexes (bottom reaction) is likely because humic acids have been shown to contain COOH groups that are located on adjacent positions of aromatic rings (132). Positive proof for the formation of salicylate-like ring structures (top reaction) has yet to be achieved. Other structures considered to be present in humic substances, and that have the potential for binding with metal ions, include the following:

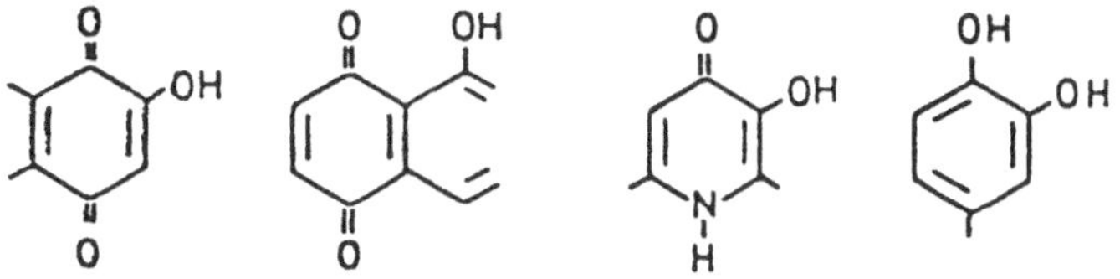

Results of infrared spectroscopy studies have confirmed that COOH groups, or more precisely carboxylate (COO^-), play a prominent role in the complexing of metal ions by humic and fulvic acids (23, 103, 147). Some evidence indicates that OH, C=O, and NH groups are also involved (22, 103, 147). Complexes may be formed with conjugated ketonic structures, according to the following reactions (103).

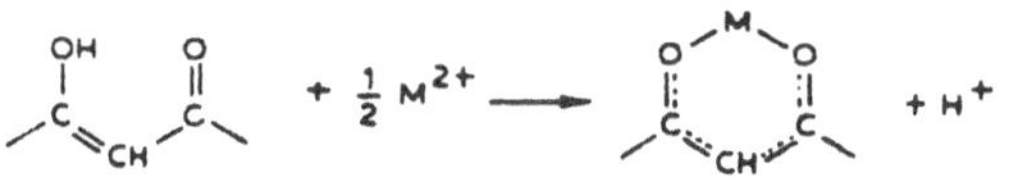

Considerable controversy exists as to the extent to which COO^- linkages are covalent or ionic. The asymmetric stretching vibration of COO^- in ionic bonds occurs in the 1630-1575 cm^{-1} region; when coordinate linkages are formed, the frequency shifts to between 1650-1620 cm^{-1}. Frequency shifts with metal-humate complexes have been variable and slight, a result that may be due to the formation of mixed complexes. Interpretations in the 1620 cm^{-1} region are further complicated because of interference from covalent bonding with other groups (103).

Results of electron spin resonance spectroscopy (ESR) studies have also been inconclusive. Lakatos *et al.* (73) reported that Cu^{2+} was bound to humic acid by a nitrogen donor atom and two carboxylates. On the other hand, McBride (91) concluded that only oxygen donors (COO^-) were involved; furthermore, a single bond was formed between Cu^{2+} and humic acid. Boyd *et al.* (24, 25) obtained evidence indicating that Cu^{2+} formed two equatorial bonds with oxygen donor atoms of humic acids, such as would be formed by the reactions shown above.

Goodman and Cheshire (55, 56) obtained evidence suggesting that Cu retained by a peat humic acid after acid washing was coordinated to porphyrin groups, from which they concluded that a small fraction of the Cu in peat was strongly fixed in the form of porphyrin-type complexes. Confirmation for this finding has been given by Abdul-Halim *et al.* (2). In contrast, spectra obtained by Bloom and McBride (19) for acid-washed peat failed to show the participation of groups other than COO^- in the binding of Cu^{2+}.

The observation that the strength of binding of Cu by humic acids decreases with an increase in the amount of Cu applied (43, 55, 56) is in agreement with other observations suggesting that Cu, when present in low amounts in peat soils, is so tightly complexed that it cannot be taken up by crop plants. Ennis and Brogan (49) prepared a Cu-humic acid complex by saturation with $CuSO_4$ and found that the Cu became increasingly unavailable to oat plants as more and more of the Cu was removed by chemical extraction (increasing concentrations of HCl were used).

5.3.2.3. Solubility characteristics

Humic substances form both soluble and insoluble complexes with polyvalent cations, depending on degree of saturation. Because of their lower molecular weights and higher contents of acidic functional groups, metal complexes of fulvic acids are more soluble than those of humic acids.

A number of processes affect the solubility characteristic of metal-humate and metal-fulvate complexes. A major factor is the extent to which the complex is saturated with metal ions. Other factors affecting solubility include pH, adsorption of the complex to mineral matter, and biodegradation. Under proper pH conditions, trivalent cations, and to some extent divalent cations, are effective in precipitating humic substances from very dilute solutions; monovalent cations are generally effective only at relatively high particle concentrations.

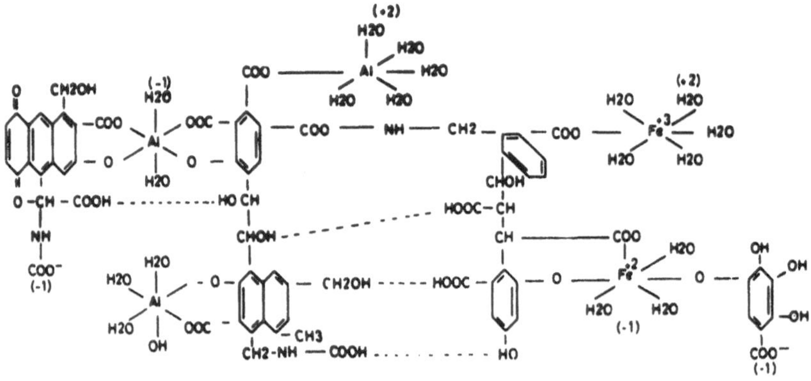

Fig. 5.3. Hypothetical polymerized compound with Al^{3+}, Fe^{3+}, and Fe^{2+} in 6-coordination with protonic bridges (dotted lines). The net charge is zero. From De Coninck (45).

The insolubility of humic substances in mineral soils is due to some extent to interactions with clay minerals, primarily through linkages with polyvalent cations (formation of a clay-metal-organic matter complex). Humic substances are also insoluble in organic-rich horizons of mineral soils, as well as peat, which can be attributed to intermolecular associations involving H-bonding and to polymerization through bridging by polyvalent cations. A hypothetical structure for a polymerized agglomerate in the B horizon of a Spodosol is shown in Fig. 5.3. (45).

5.3.2.4. Metal ion binding capacity

Approaches used to determine the binding capacities of humic substances for metal ions include coagulation (105), proton release (129, 130, 145), metal ion retention as determined by competition with a cation-exchange resin (42, 159), dialysis (160), anodic stripping voltammetry (60, 100), and ion-selective electrode measurements (26, 29). The maximum amount of any given metal ion that can be bound has been found to be approximately equal to the content of COOH groups. The COOH content of humic substances generally fall within the range of 1.5 to 5.0 meq/g. For Cu, this corresponds to retention of from 48 to 160 mg per g of humic acid. Assuming a carbon content of 56% for humic acids, one Cu atom would be bound per 20 to 60 carbon atoms in the saturated complex.

Factors influencing the quantity of metal ions bound by humic substances include pH, ionic strength, molecular weight, and functional group content (51, 133 134). For any given pH and ionic strength, trivalent cations are bound in greater amounts than divalent cations; for the latter, those forming strong coordination complexes (*e.g.*, Cu) will be bound to a greater extent than weakly coordinated ones (*e.g.*, Ca and Mg).

5.3.2.5. Reduction properties

Humic substances have the ability to reduce oxidized forms of certain metal ions, a typical case being the reduction of ferric Fe (72). Reduction of ionic species is of considerable importance because the solubility characteristics of the metal ions (and hence mobilities) are modified. The ESR approach has been used in conjunction with Mössbauer spectroscopy to obtain information on oxidation states and site symmetrics of Fe bound by humic and fulvic acids (59, 120).

5.4. Use of micronutrient-enriched organic wastes and naturally occuring metal-organic complexes as soil amendments

Most publications on soil amendments, enriched or unenriched with micronutrients, have focused on Fe-organo complexes as sources of Fe for sensitive crops growing on deficient soils. A smaller number of investigations have been concerned with Zn and Mn. Therefore, we shall deal primarily with Fe; other micronutrients will be discussed briefly.

5.4.1. Iron-organo complexes

Since soluble soil Fe exists mainly as Fe-organo complexes, efforts have been made to use solid organic wastes, either enriched with Fe or unenriched, as Fe sources for plants. Among the organic wastes found effective in increasing Fe uptake are animal manures of various types, composts of plant refuse, and extracts of forest by-products. Growth promoting effects of powdered charcoal, brown coal, and lignite have been found in retrospect to also be due to Fe-organo complexes.

The effectiveness of Fe-organo complexes is usually attributed to their similarity to soil organic matter, and in particular to the humic substances contained therein. The following discussion shall cover such raw organic materials as peat, coal, and lignite, along with lignosulfonates and polyflavenoids. The latter represent somewhat well-defined organic fractions of wood processing wastes.

5.4.1.1. Polyflavenoids and lignosulfonates

Polyflavenoids and lignosulfonates are by-products of forest-product manufacturing. Attention has been drawn to these materials as potentially economically Fe chelates. Although polyflavenoids and lignosulfonates have been tested as both foliar sprays and soil amendments, only the latter will be considered here.

"Rayplex", a polyflavenoid product marketed by Rayonier Inc., is a chemical extract of western hemlock bark enriched to 11% Fe. Chesnin (37) found that maximum dry matter yield and total Fe uptake were achieved with 2.5 mg Fe as Fe-polyflavenoid/kg soil for sorghum in pots of a 10% $CaCO_3$ sandy loam. Richardson (106) used 34 kg/ha of Rayplex by placing the granules 10 cm deep and 10 cm to one side of chlorotic sorghum plants in Arizona; some effect was apparently achieved but the plants failed to yield harvestable amounts of mature grain. Walker and Smither (148) obtained positive response by application of 2.3 kg Rayplex (granular) per peach or pear tree. In other work, Salardini and Murphy (112) measured DTPA-extractable Fe in a normal and Fe-deficient calcareous soil incubated at field capacity with 0, 20, 40 and 80 μg Fe/g as Fe-polyflavenoid. Extractable Fe levels fell quickly during the first week and by the end of 8 weeks ranged from 8 to 13 μg Fe/g for the normal soil, regardless of treatment, and from 3 to 6 μg/g for the Fe-deficient soil.

Lignosulfonates have been marketed as "Orzan" (Crown Zellerbach Corp.) and as "Greenz 26" (Little (83)). In a study where soybeans were grown in pots of a calcareous loam soil, Wallace and Ashcroft (149) found that the relative effectiveness of Fe-lignosulfonate (4.6% Fe), FeEDTA and FeEDDHA was 1:3:15 at a rate of 260 kg Fe/ha (about 100 μg Fe/g soil). Chesnin (37) compared a lignosulfonate with Kraft processed lignin and a polyflavenoid. The Fe-enriched lignosulfonate and polyflavenoid were comparable to each other; the Kraft lignin seemed to have an advantage over the other treatments at 2.5 μg Fe/g, presumably due to a lower molecular weight. After an 8 week incubation period, soil from the lignosulfonate treatments contained 9 to 13 and 3 to 6 μg/g of DTPA-extractable Fe for a normal and an Fe-deficient soil, respectively. Overall, polyflavenoids and lignosulfonates seemed to evoke similar plant responses and to have similar persistence in soil.

5.4.1.2. Manure and composts

A number of reports on the effectiveness of enriched- or unenriched manure in solving problems of Fe chlorosis have been published. For example, in an Fe-deficient calcareous soil in Senegal, Blondel (18) increased dry matter yields of sorghum from 520 to 970 kg/ha by adding 20 tons of farmyard manure (FYM) per ha. Poultry manure has been recommended for correction of both Zn and Fe deficiencies (12). Parsa and Wallace (101) have shown that dog manure (1,850 μg Fe/g) added at a rate of 1.5% to a calcareous loam soil (pot experiment) significantly increased dry matter yield and Fe uptake by sorghum; ashed dog manure produced a lesser increase in dry matter and no increase in Fe uptake. An infrared spectrum of the dog manure was found to be similar to that of soil fulvic acids, from which it was concluded that complexing by COOH and phenolic OH groups was the mechanism by which Fe nutrition was improved by the manure.

Enrichment of manure with Fe (as $FeSO_4$) has also been of interest. Thomas and Mathers (139) grew three crops of sorghum in pots of calcareous sandy loam treated with combinations of up to 4% manure and up to 40 μg Fe/g (as $FeSO_4$). For the first crop, dry matter yields were substantially increased when manure only was applied. Also, Fe-deficiency symptoms were effectively reduced. Ferrous sulfate at rates up to 40 μg Fe/g had no effect on yields and failed to correct chlorosis. For the second and third crops, yields were significantly higher for the manure plus Fe treatment than for manure or Fe alone. Maximum yields were obtained with as little as 0.5% manure when Fe was added at a rate of 40 μg/g. These observations indicate that organic compounds in the manure were effective in keeping Fe in plant available forms. In a field experiment on a calcareous soil, Mann et al. (86) investigated the interaction between the addition of 0 and 25 μg Fe/g (as $FeSO_4$) and 0 and 10 ton manure/ha on DTPA-extractable Fe and maize yield. The yield increase for the Fe-manure treatment was greater than for the sum of Fe and manure alone. Furthermore, the DTPA-extractable Fe after maize harvest showed the same type of Fe-manure interaction; values were 5.5, 5.9, 7.0 and 10.8 μg Fe/g for the control, + manure, + Fe, and + manure and Fe treatments, respectively.

Composts have been shown to be of use in alleviating Fe deficiencies. Francis *et al.* (52) reported results of a pot experiment where sorghum was grown on a calcareous soil to which $FeSO_4$ and compost were added. Application rates ranged from 0 to 45 μg Fe/g + levels of cotton leaf compost equivalent to from 0 to 20 tons/ha. Significant increases in dry matter yield and Fe uptake were observed; the Fe + compost interaction was also statistically significant. Takker (137) found that additions of sugar trash (up to 8%) to a calcareous soil initially caused a decline in ammonium acetate (pH 3) extractable Fe but after 35 days of incubation at field capacity extractable Fe exceeded that present initially and rose to significantly higher values.

5.4.1.3. Sewage sludge

Sewage sludges of industrial cities often pose a disposal problem because of toxic levels of heavy metals (*i.e.* Pb and Cd); otherwise, sewage sludge may find use as an Fe source. Parsa and Wallace (101) reported that activated sewage sludge (4,200 μg Fe/g) from Los Angeles, when mixed with a calcareous loam soil (pot experiment) at a rate of 1.5%, increased both dry matter yield and Fe uptake by sorghum. Neither sewage sludge ash nor 5 μg Fe/g as $FeSO_4$ or FeEDDHA were able to match the yield and Fe uptake of 1.5% sludge. At an application rate of 2% sludge, yields and Fe uptake were lower than for the 1.5% rate, indicating that a toxic level had been attained--perhaps with another microelement in the sludge. Baham *et al.* (9) experimented with gel filtration of two fulvic acid fractions of sewage sludge (50 and 105 μg Fe/g). Elution data showed multiple maxima in organic carbon concentration, a single maximum in Cl concentration, and multiple maxima of Fe concentration which coincided with both the organic carbon and Cl maxima. Analytical data showed that from 13 to 24% to the Fe was eluted at the Cl maximum, indicating an inorganic complex, possibly hydroxides. From 61 to 85% of the Fe was distributed in the fractions collected near the carbon maxima, indicating Fe-organo complexes in sewage sludge. Abdou and El-Nennah (1) have shown the cumulative effect of domestic sewage sludge when disposed on a calcareous loamy sand for up to 45 years; total Fe in the soil was increased six-fold, ammonium acetate extractable Fe (pH 2.8) was increased ten-fold, and water-soluble Fe was increased six-fold. Availability of Mn and Zn were increased simultaneously. Although there is little doubt that urban sewage sludge may serve as an Fe source, problems of toxicity and heavy metal contamination must be solved first.

5.4.1.4. Coal, lignite, and peat

The growth promoting effect of charcoal was recognized at the turn of the century and generated considerable interest on the fertilizer value of brown coals and lignite. Some brown coals have been found to promote plant growth; others are growth inhibiting (44). Theories based on hormonal or antibiotic characteristics of coals and lignite were advanced in the 1930s; at the same time, evidence accumulated linking the stimulatory effect of coals to the HA's contained therein, and linking the positive effect of HA's to maintenance of Fe in soluble forms available to microorganisms and plants. De Kock and Strmecki (44) investigated the growth promoting effect of a lignite dust from Yugoslavia by comparing various size fractions of the lignite with a HCl extract, a 2% sodium oxalate extract (HA) and the ash of both whole lignite and extracted residues; mustard plants were grown in nutrient solution without added Fe. Of the various lignite size fractions, the finest fraction (diametter ⟨ 0.15 mm) gave the greatest growth response. Only treatments that combined HA with either the HCl extract (which contained Fe) or 13 μg Fe/g as $FeCl_3$ gave dry matter yields equal to or surpassing those of the whole lignite. Ether extracts of the lignite (to recover possible hormonal stimulants) had no effect on growth. Although the growth promoting effects were clearly due to Fe--humates, a separate experiment showed that the greening ability of the lignite was lower than for EDTA, though greater than citrate, on a per Fe basis. From a commercial point of view, this cannot be considered a disadvantage due to price differences between the two types of compound, which are highly in favor of the natural cheiate.

From time to time, rather extravagent claims have been made for the use of lignites and commercial humates to imporve the physical, chemical, and biological properties of normally productive agricultural soils. Whereas such products show promise as carriers of trace elements for sensitive crops growing on deficient soils (noted above), their general use for field crops grown on noncalcareous silt loam and loam soils cannot be recommended (131). For these soils, correction of macronutrient deficiencies (N, P, K) can best be achieved using conventional inorganic fertilizers.

Results of tests on the fertilizer value of Fe-enriched HA's extracted by KOH or ammonia from a moderately to well-decomposed reed sedge/sphagnum moss peat, a decomposed woody peat, and a partly decomposed reed sedge peat have been reported by Bureau et al. (30). Soybeans were grown in pots of a silty clay soil containing 9% organic matter and 6% $CaCO_3$. Liquid HA was Fe-enriched by contact with fresh $Fe(OH)_3$ for 100 hours, or by additions of $Fe_2(SO_4)_3$ or ammonium ferrous sulfate and titration to neutral pH; precipitates were removed by centrifugation. Iron humates were applied at rates of 0, 10, 25, and 40 μg Fe/g of soil; FeEDDHA and FeHEEDTA were applied at rates of up to 5 μg Fe/g of soil. In addition, $FeCl_3$ at rates equal to those of the Fe-humates was applied in 150 ml of solution brought to pH 2 with HCl. After 30 days growth, FeEDDHA at the 1 μg Fe/g rate gave plant Fe concentrations equivalent to or slightly superior to the Fe- humate at the 10 μg Fe/g rate, whereas application rates of 0.04 and 5 μg Fe/g resulted in much lower and much higher uptake, respectively. No significant differences were observed for the Fe-humate due to peat type, extractant, or method of Fe enrichment. The Fe-humates were in fact equivalent to or slightly inferior to acidified $FeCl_3$. The conclusion of Bureau et al. (30) was that "further development of alkali extracts from peats as amendments to calcareous, chlorosis-producing soils for the purpose of increasing availability of Fe does not appear warranted". Little (83) reported that $FeSO_4$, applied alone or mixed with peat (14 g per bushel of peat, type unspecified) at rates up to 8.2 kg $FeSO_4$ per tree (type unspecified), was only partially successful in aleviating Fe deficiencies. However, the Fe-peat mixture was reported to be of value as an amendment for greenhouse tomatoes, roses, and pot plants.

Similar research has been carried out using an ammonium nitro HA (NH_4-NHA), which was produced by treating brown coal with HNO_3 and neutralizing it with NH_3. Aso and Sakai (8) showed that NH_4-NHA formed a complex with Fe in nutrient solution and that the complex led to regreening of chlorotic barley. Overall, NH_4-NHA seemed equivalent or slightly less effective than EDTA when both were used in nutrient solution at a rate of 2.5 mg/l in the presence of 1 mg/l $FeCl_3$.

Chen et al. (34, 35) reported the remedy of lime-induced chlorosis by Fe-enriched peat in a pot experiment with peanuts grown on a Rendzina soil containing 63% $CaCO_3$. Chlorophyll levels reached those of FeEDDHA-treated plants. In a separate field experiment, the application of Fe-enriched peat to peanuts grown on a soil containing 42% $CaCO_3$ resulted in yields amounting to 130% and 110% of those of control and FeEDDHA treatments, respectively. Iron-enriched peat has been found to be superior to FeEDDHA and $FeSO_4$ in a field experiment with gladioli (36). In the series of experiments carried out by Chen et al. (34-36), application rates of Fe as Fe-enriched peat were considerably higher than those of Fe as FeEDDHA. The basic concept was that application rates of the Fe-organo complex need not equal those of FeEDDHA because of the large differences in solubility, duration of effect, and price. Except for solubility, the Fe-organo complexes were superior to the commercial preparations.

Summarizing the research on the use of coal and peat as Fe fertilizers for Fe deficient soils, it can be said that whereas some positive effects have been noted, Fe-humates are relatively ineffective as compared to synthetic chelates (i.e., FeEDDHA) on a per Fe and per weight basis. The bulk of the research using lignites has been performed in eastern Europe, much of which has not been translated into English. Abstracts given in various bibliographies indicate that the research has proceeded primarily in the direction of the use of humates to improve the nitrogen status of the soil with limited application to problems of Fe deficiencies.

5.4.2. Zinc-, copper-, and manganese-organo complexes

Use of Zn-chelates to overcome Zn deficiencies has largely involved soil application of these materials. Zinc-EDTA has been the most frequently utilized compound although other Zn chelates are available. Several by-products containing Zn have been formulated from wood processing materials, such as lignosulfonates and polyflavenoids. Manure enriched with $ZnSO_4$ has also been used.

Anderson (6) compared the effectiveness of synthetic chelates as sources of Zn for plants grown in calcareous soils. The conclusion was reached that chelated forms were generally more effective at low rates than was $ZnSO_4$; the latter was superior to a Zn-polyflavenoid. Boehl and Lindsay (21) state that lignosulfonates and polyflavenoids are acceptable sources of Zn and have low phytotoxicity. However, the natural products are somewhat less stable in soil as compared to synthetic chelates. More promising results have been obtained with Zn-amended poultry manure. In a study by Singh et al. (123), various amounts of Zn-amended poultry manure and $ZnSO_4$ were applied to corn grown in pots in the greenhouse. Zinc-amended poultry manure at all rates of Zn application was more effective than $ZnSO_4$ alone, both in enhancing dry matter yield and Zn uptake. Using a radio assay technique, Singh *et al.* (123) concluded that both percent of Zn in the plant that was derived from the fertilizer and percent of the applied Zn that was assimilated were higher for the Zn-amended poultry manure, as compared to $ZnSO_4$. In another study, Singh *et al.* (122) investigated the effect of applications of 12.5, 25, and 50 kg/ha of $ZnSO_4$ alone and in conjuction with 10T of poultry compost/ha on rice yield and Zn uptake. Application of both Zn and compost significantly increased Zn content and uptake by the rice over the control. Maximum Zn concentration and uptake were recorded at 50 kg $ZnSO_4$ mixed with 10T of compost/ha. These results confirm earlier reports that organic manures contain complexing agents that form metallo-organic complexes with Zn, thereby facilitating its utilization by crops.

While most work with soil applications of Cu has been performed with inorganic Cu compounds, organic sources are available that can and have been soil applied to correct plant deficiencies. CuEDTA, Cu-lignosulfonates and Cu-polyflavenoids at application rates of 1 to 5.5 kg/ha have been recommended as band or broadcast applications.

Recent unpublished studies (12) have shown that Mn-amended poultry and cattle manure, as well as Mn-enriched peat in granulated or powdered forms, can significantly increase the Mn content of cotton and barley when grown on deficient soils.

5.5. Stability constants of metal complexes with humic and fulvic acids

An important characteristic of a metal-organic complex is its formation (stability) constant, the value of which provides a quantitative measure of the affinity of the metal for the ligand. Accurate values for stability constants of metal-humate and metalfulvate complexes are required for a full understanding of the factors affecting the availability and partition of micronutrient cations in soil. They are also needed for use in computational models for predicting the speciation of trace elements in the soil solution from thermodynamic data of organic and inorganic complexes (90).

Numerous problems are encountered in determining stability constants of trace element complexes with humic and fulvic acids. Humic substances, from whatever source, are heterogeneous with respect to shape and size, and a pH effect dictates the degree of ionization of acidic groups and thereby the number of sites available for binding (51, 102, 130, 133, 134). Another complication is that several classes of binding sites may be present, in which case the class forming the most stable complex will be the first to react. Humic and fulvic acids may also contain combining sites that are identical, but which react in such a way that binding at one site affects binding at subsequent sites.

5.5.1. General considerations

The overall reaction between a metal ion, M, and an organic ligand, L, is given by:

$$aM + bL \rightleftharpoons M_aL_b \qquad [1]$$

where a is the number of moles of metal ion combined per complex molecule and b is the number of moles of organic ligand molecules. The terms M, L, and M_aL_b represent molar concentrations of free M, free L and the complex, respectively. The over-all formation (equilibrium) constant is:

$$K = \frac{(M_aL_b)}{(M)^a(L)^b} \qquad [2]$$

Calculations of K from equation [2] require that accurate values be obtained for L and M, the concentration of free ligand and metal ion, respectively. In many studies, allowance has not been made for side reactions involving the metal ion or ligand. Side reactions involving the ligand include protonation of the dissociated form of the reactive site $(L^- + H^+ \rightarrow HL)$. To avoid problems with the formation of chloro complexes (i.e., MCl^+ and MCl_2), nitrate has often been used as the supporting electrolyte. Sposito and Holtzclaw (127) concluded that perchlorate is the preferred media.

Irrespective of the approach for calculating stability constants, some determination must be made for free and bound forms of the metal ion or ligand. Only rarely has the ligand concentration (L) been given in molar units (26, 109-111, 128). Most often, this parameter has been expressed in terms of reactive site concentration (nL_t). Methods for estimating nL_t of humic substances include potentiometric titration (129, 130, 136), spectrophotometric titration and bioassay (16, 17), spectrofluorimetry (109), equilibrium dialysis (160), and cation exchange with synthetic resins (42).

A résumé of methods for measuring metal-humic matter reaction parameters is given in Table 5.3. The molar concentration of the free metal ion, M, has usually been the measured value, thereby leading to estimates for the amount of metal tied up in the complex. Early work was done by competition of the metal ion with a cation exchange resin (see review of Stevenson and Ardakani (133)). Other techniques (see Table 5.3.) have included equilibrium dialysis, gel filtration, bioassay, voltammetry, and fluorescence spectrometry. Increasing use is being made of ion-selective electrodes (ISE) and anodic stripping voltammetry (ASV). Advantages and limitations of ISE and ASV have been discussed by Bresnahan *et al.* (26) and Tuschall and Brezonik (140).

The vast majority of measurements has been done using ion selective electrodes, for which four (Cu, Pb, Cd, and Ca) have been widely used.

Table 5.3. Methods used to measure metal-humic acid reaction parameters.

	Method	Value measured	Reference
Two-step process: Separation and analysis	1. Ion-exchange/AA, colorimetry	M or M_b	Schnitzer and Hansen (117) van den Berg and Kramer (144)
	2. Gel filtration/UV visible	M	Mantoura and Riley (88) Mantoura *et al.* (87)
One-step process: Analysis in situ	1. Spectrofluorimetry	L_b or L	Saar and Weber (109); Ryan and Weber (108)
	2. Spectrophotometry	L	Blaser *et al.* (16); Langford and Khan (74)
	3. Ion-selective electrode (ISE)	M	Bresnahan *et al.* (26); Buffle *et al.* (29); Cheam (32); Cheam and Gamble (33); Saar and Weber (109-111); Sposito and Holtzclaw (127); Sposito *et al.* (128)
	4. Anodic stripping voltammetry (ASV)	M	Brezonik *et al.* (27); Greter *et al.* (58); O'Shea and Mancy (100); Shuman and Woodward (121)

sponse curves tend to flatten out below about $10^{-5.5}$ M free metal, which restricts estimates for binding to rather high levels of metal ion additions. Some of the stronger (and important) binding sites are thereby missed and the stability constants thus obtained reflect binding at the weaker sites. Two procedures have been used in attempts to extend the lower limits of detection, namely, computer-fitting of a polynominal to the calibration curve (109-111) and linear extension of the standard curve (29, 50, 51).

5.5.2. Modeling approaches

Several types of reactions of metal ions with naturally occurring macromolecules may be identified. The simplest case is 1:1 binding (a=1, b=1). More complex, but mathematically solvable, are cases of mononuclear complexes with two or more binding substrates. The central molecule may be the macromolecule (M_aL) or the metal ion (ML_b). Both approaches have been used in metal binding

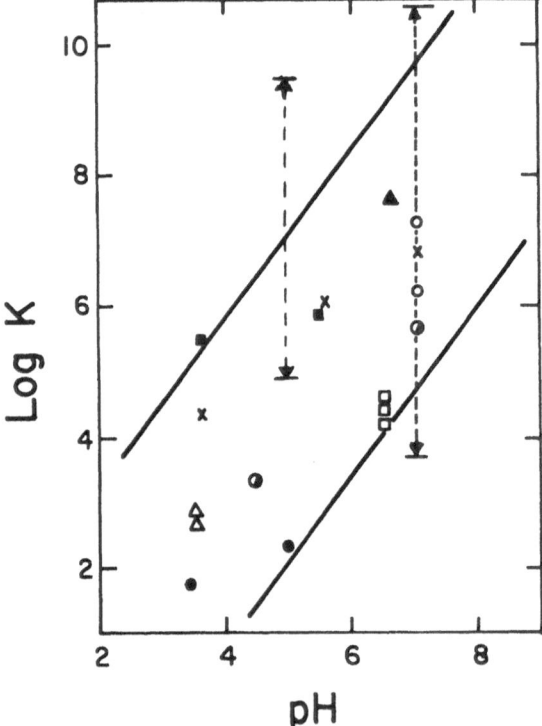

Fig. 5.4. Effect of pH on stability of Zn-organic matter complexes as summarized by Stevenson and Ardakani (133). Reproduced from Micronutrients in Agriculture, 1972, pages 79-114 by permission of the Soil Science Society of America. F.J. Stevenson and M.S. Ardkani, authors, Figure 10 page 109.

studies of humic substances. A more difficult system to handle mathematically is one in which polynuclear binding occurs (M_aL_b).

Early work on stability constants of metal complexes with humic substances was done using the ion-exchange equilibrium method of Schubert. This work has been reviewed by Stevenson and Ardakani (133). The continuous variation method of Job has also been applied. The two approaches have been critically examined by Clark and Turner (39) and MacCarthy and Mark (85), respectively.

An indication of the range of conditional and over-all stability constants (log K) that have been obtained for humic substances (assumed 1:1 complexes) are illustrated for some select micronutrient cations (pH range 5 to 8) in Table 5.4. A pronounced pH dependency is shown, an effect further illustrated for Zn-organic matter complexes in Fig. 5.4.

The results of Table 5.4. and Fig. 5.4. show that widely divergent log K values have been recorded for the same pH. The extent to which these differences reflect variations in the chemical properties of humic substances from various sources is unknown. Matsuda and Ito (89) concluded that the absorption strength of Zn for humic substances increased with increasing degree of humification, an observation that requires confirmation.

5.5.2.1. *Macromolecule as the central group*

Recent studies on metal ion binding by humic substances have been based on the assumption that the macromolecule is the central group (formation of LM, LM_2---LM_n complexes). The reaction can be described by n stability constants.

$$K_1 = \frac{(LM)}{(L)(M)} \quad , K_2 = \frac{(LM_2)}{(LM)(M)} \quad , K_n = \frac{(LM_n)}{(LM_{n-1})(M)} \quad [3]$$

where the parentheses designate concentration.

The extent of binding is expressed in terms of a formation function, ν, defined as:

$$\nu = \frac{\text{sites bound}}{\text{polymer concentration}} = \frac{M_b}{L_t} = \frac{(LM) + 2(LM_2) + \ldots n\,(LM_n)}{L + (LM) + \ldots (LM_n)} \quad [4]$$

The quantity ν represents the average number of metal ions associated with each macromolecule, L. When all combining sites are identical, ν is related to the intrinsic or microscopic binding constant, K_o, by the equation.

$$\nu = \frac{nK_o(M)}{1 + K_o(M)} \quad [5]$$

Table 5.4. Conditional and over-all stability constants for the complexes of Co, Cu, Mn, Ni, and Zn with humic substances from various sources

Metal ion	Source**	Supporting electrolyte	pH			Reference
			5	6-7	8	
Co	SFA	0.1M KCl	4.10			Schnitzer and Hansen (177)
	PFA	0.02M Tris			4.51	Mantoura et al. (87)
	LHC(4)	,,			4.67-4.90	ibid
	MHC(4)	,,			4.29-4.83	ibid
	SDHC	,,			4.91	ibid
Cu	SFA	0.1M KCl	4.00			Schnitzer and Hansen (117)
	SFA	0.1M NaClO$_4$	4.35			Cheam and Gamble (33)
	SFA	0.1M NaNO$_3$	4.00			Buffle et al. (29)
	SFA	0.1M KNO$_3$	4.68	5.03-5.45		Ryan and Weber (108)
	WFA	0.01M KNO$_3$		7.82		van den Berg and Kramer (144)
	LHC(4)	,,		4.51-6.72		Shuman and Woodward (121)
	PFA	0.02M Tris			7.85	Mantoura et al. (87)
	RHC(2)	,,			8.42-9.83	ibid
	LHC(4)	,,			9.48-9.58	ibid
	MHC(4)	,,			8.89-10.21	ibid
	SDHC(4)	,,			9.91-11.37	ibid

Table 5.4. Continued

Metal ion	Source**	Supporting electrolyte	pH 5	6-7	8	Reference
Mn	SFA	0.1M KCl	3.70			Schnitzer and Hansen (117)
	PFA	0.02M Tris			4.17	Mantoura et al. (87)
	LHC(2)	,,			4.30-4.85	ibid
	MHC(2)	,,			4.45, 4.51	ibid
Ni	SFA	0.1M KCl	4.20			Schnitzer and Hansen (117)
	PFA	0.02M Tris			4.98	Mantoura et al. (87)
	LHC(2)	,,			5.14, 5.27	ibid
	MHC(4)	,,			5.19-5.51	ibid
Zn	SHA(4)	0.1M KCl		2.82-4.93		Ardakani and Stevenson (7)
	SHA(29)			4.20-10.83		Matsuda and Ito (89)
	SFA(32)			3.88-9.30		ibid
	SFA	0.1M KCl	3.70			Schnitzer and Hansen (117)
	PFA	0.02M Tris			4.83	Mantoura et al. (87)
	RHC(2)	,,			5.36, 5.41	ibid
	LHC(4)	,,			5.03-5.31	ibid
	MHC(2)	,,			5.27, 5.31	ibid
	SDHC(2)	,,			4.99, 5.87	ibid

* Values at other pH's can be found in references cited.

** SHA = soil humic acid; SFA = soil fulvic acid; PFA = peat fulvic acid; LHC =lake humic colloid; RHC = river humic colloid MHC = marine humic colloid; SDHC = sediment humic colloid.

When more than one class of sites occurs on the macromolecule, ν is the sum of the ν's for each site:

$$\nu = \frac{n_1 K_1(M)}{1+K_1(M)} \frac{n_2 K_2(M)}{1+K_2(M)} \ldots + \frac{n_j K_j(M)}{1+K_j(M)} \qquad [6]$$

where n_j is the number of sites of class j and K_j is the stability constant for class j.

When the molecular weight is unknown, the formation function can be expressed in terms of binding site concentration, nL_t.

Define:

$$\theta = \frac{\text{molar conc. of bound metal ion}}{\text{total number of reactive sites}} = \frac{M_b}{nL_t} = \frac{\nu}{n} \qquad [7]$$

By combining equations [5] and [7], the following is obtained:

$$\theta = \frac{K_0(M)}{1+K_0(M)} \qquad [8]$$

Equations [5] and [8] can be arranged in a number of ways to yield information about the number of classes of sites and the stability constant for each class (K_1, K_2, etc.). Plotting can be done as a Scatchard plot (ν/M vs ν; θ/M vs θ); a reciprocal plot (as M/ν vs M), a double reciprocal plot ($1/\nu$ vs $1/M$), and, in some cases, as a Hill plot ($\log[\theta/(1-\theta)]$ vs log M). The various equations and plotting variables are listed in Table 5.5.

An approach proposed by Buffle et al. (29) to study the binding of Cu(II) to humic substances is similar to a double reciprocal plot (50, 51). This plotting method is generally considered to be less desirable for modeling binding data than the Scatchard plot method because undue weight is given to high free M values (46).

Zunino and Martin (160) derived the following equation to determine apparent stability constants for the reaction of Cu(II) with humic acid.

$$\text{Log} \frac{(M_b)}{MBA - (M_b)} = \log K + n \log [M] \qquad [9]$$

where K is the apparent stability constant for the reaction: $nM + L \rightleftharpoons M_n L$. The MBA (maximum binding ability) was the total binding site concentration as determined by a Langmuir plot of the binding data. To obtain Log K, a plot was made of the left side of the equation vs log [M]. This equation can be shown to have the same form as the Hill plot.

Fitch and Stevenson (50, 51) pointed out that stability constants obtained by the Hill plot method are invalid when binding at one site decreases binding affini-

Table 5.5. Plotting approaches for analyzing experimental binding data.

Plot title	Form of the equation	Plot	
		Y	X
Scatchard	$\nu/M = nK_0 - \nu K_0$	ν/M	ν
	or		
	$\theta/M = K_0 - \theta K_0$	θ/M	θ
Reciprocal (Langmuir)	$M/\nu = M/n + 1/nK_0$	M/ν	M
	or		
	$M/M_b = M/nL_t + 1/nL_tK_0$	M/M_b	M
Double reciprocal	$1/\nu = 1/n + 1/nK_0M$	$1/\nu$	1/M
	or		
	$1/M_b = 1/nL_t + 1/nL_tK_0M$	$1/M_b$	1/M
Hill	$\log[\theta/(1-\theta)] = \log K^* + n\log M$	$\log[\theta/(1-\theta)]$	Log M

ty at subsequent sites, which is the case for metal complexes with humic substances, as noted below.

The Scatchard plot approach has been the method of choice in most studies on metal ion binding by humic substances (26, 109-111, 128, 140). Rearrangement of equation [8] gives:

$$\frac{\theta}{M} = K_0 + K_0 \theta \qquad [10]$$

A plot of θ/M vs θ yields K_0 as the slope or intercept.

A typical Scatchard plot is given in Fig. 5.5. Nonlinear curves were obtained, from which stability constants for binding at two "classes" of sites (log K_1 and log K_2) have been calculated.

The dissection of Scatchard plots into two straight line segments is somewhat arbitrary and additional "sites" can be found by assigning linear segments to the Scatchard plot (50, 51, 140). In practice, a continium of binding sites may be present (102). Bresnahan *et al.* (26) postulated that differences for the two classes of sites were due to the geometry of the site rather than to the type of donor atom. In other work, Buffle *et al.* (29) attributed the curvilinearity of their plots to the formation of ML_2 complexes (*i.e*, metal ion is the central group).

Stability constants for the binding of Cu^{2+} by fulvic acids from several sources are given in Table 5.6. The values are recorded in terms of binding at the strongest (log K_1) and weakest (log K_n) sites, although most constants are recorded for

104

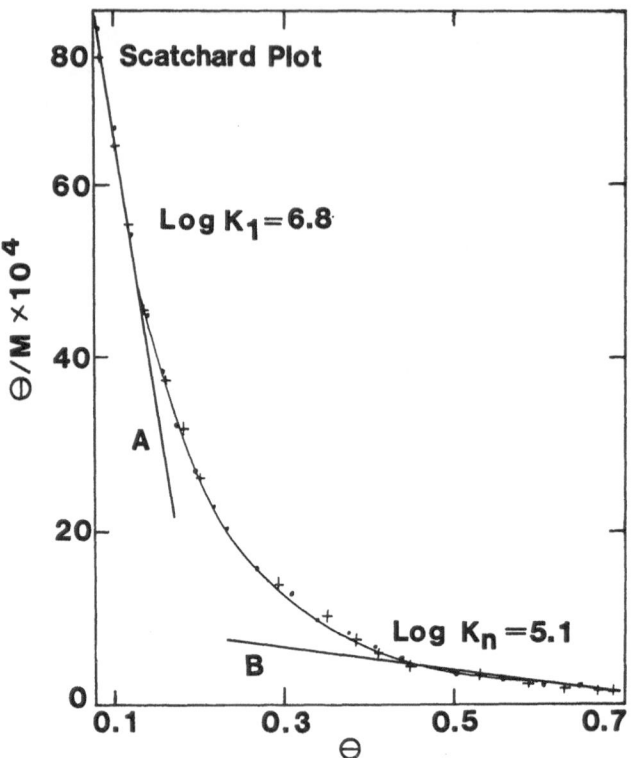

Fig. 5.5. Scatchard plot for the binding of Cu^{2+} by a soil humic acid. From Fitch and Stevenson(50). Reproduced from Soil Science Society of America Journal, Volume 48, 1984, pages 1044-1150 by permission of the Soil Science Society of America A. Fitch and F.J. Stevenson, author Figure 5, page 1048.

data
-
a 2-site system. As expected, log K values increase with increasing pH. Insufficient data are available for valid comparisons to be made between humic substances from various sources.

5.5.2.2. Metal ion as the central group

When the metal ion is the central group, a series of species of the type ML_b are formed. In this case, complex formation can be regarded as a competitive reaction between the metal ion and H^+ for the reactive site.

$$HL + M^{2+} \rightleftharpoons ML^+ + H^+ ; \quad HL + ML^+ \rightleftharpoons ML_2 + H^+ \qquad [11]$$

with b_1 over the first reaction and b_2 over the second.

Table 5.6. Comparison of stability constants for binding of Cu^{2+} at the strongest (log K_1) and weakest (log K_n) sites for the fulvic acids from several sources, including peat. The values were calculated using the Scatchard plot approach.

Source	pH	Supporting Electrolyte	Log K_1	Log K_n	Reference
Water	8.0	0.01M NaCl	8.80	8.05	Mantoura and Riley (88)
Peat	8.0	ibid	8.50	7.16	ibid
Water*	6.25	---	8.11	5.34	Tuschall and Brezonik (141)
Water*	6.25	---	7.82	5.26	ibid
Water	6.0	0.01M KNO$_3$	6.11	3.85	Bresnahan *et al.* (26)
Soil	6.0	ibid	6.30	3.07	ibid
Water	5.0	ibid	5.95	3.70	ibid
Soil	5.0	ibid	6.00	4.08	ibid
Sewage	5.0	0.1M KClO$_4$	3.85	2.09	Sposito *et al.* (129)
Soil	4.0	0.1M KNO$_3$	5.60	3.95	Bresnahan *et al.* (26)
Water	4.0	ibid	5.48	4.00	ibid

* Constants obtained for binding at three sites.

The two successive constants are given by:

$$b_1 = \frac{(ML^+)(H^+)}{(HL)(M^{2+})} \; ; \quad b_2 = \frac{(ML_2)(H^+)}{(HL)(ML^+)} \qquad [12]$$

The over-all constant ν_2 is given by:

$$\beta_2 = b_1 b_2 = \frac{(ML_2)(H^+)^2}{(HL)^2(M^{2+})} \qquad [13]$$

A functional relationschip exists between constants obtained by this approach and those obtained for the reaction of the metal with the dissociated form of the reactive site: $k'_1 = (MA^+)/(L)(M^{2+})$ and $k'_2 = (ML_2)/(L)(ML^+)$. The relationship is given by:

$$b_i = K_a k'_i \qquad [14]$$

where K_a is the ionization constant of the acidic functional group. The prime is used to indicate that the metal ion is the central group.

The technique, as applied to humic substances, involves the determination of bound and unbound forms of the ligand from base titration measurements. A formation function, n, is determined, which can be defined as the average number of ligand sites bound to the metal ion:

$$\bar{n} = \frac{ML^+ + 2(ML_2) + \ldots n(ML_n)}{(M^{2+}) + (ML^+) + \ldots(ML_n)} = \frac{L_t \cdot HL \cdot L^-}{M_t} = \frac{L_b}{M_t} \qquad [15]$$

or

$$\bar{n} = \frac{\beta_1\,[HL/H^+] + 2\beta_2\,[HL/H^+]^2 + \ldots n\beta_n\,[HL/H^+]^n}{1 + \beta_1\,[HL/H^+] + \beta_2\,[HL/H^+]^2 \ldots \beta_n\,[HL/H^+]^n} \qquad [16]$$

For 2:1 complexes, the desired constants can be obtained from the following equation by regression analysis:

$$\frac{(2 \text{-} \bar{n})\,(HL/H^+)^2}{n} = \frac{b_1\,(\bar{n}\text{-}1)}{\beta_2\,\bar{n}}\,(HL/H^+) + \frac{1}{\beta_2}_2 \qquad [17]$$

In the potentiometric titration method, values for HL are obtained from the amount of base consumed during titration (HL = L_t · KOH · H^+ + OH), and L^- is determined from the ionization constant for the acidic functional group [L^- = HA(K_a/H^+)]. Several values for $\bar{n}$ are then calculated using equation [15], thereby providing the necessary values for calculating b_1 and β_2 by equation [17].

A complication in using titration data for calculating stability constants is that titrations carried out in the presence of metal ions result in horizontal displacement of the titration curves, apparently due to release of an otherwise nontitratable H^+ from the humic acid and/or protons from hydration water of the metal bound in 1:1 complexes (129, 130). Modifications have been made in the titration procedure in attempts to solve this problem (130).

Over-all stability constants, log K_2, for the Cu^{2+} and Zn^{2+} complexes of a soil humic acid as a function of ionic strength (I) are shown in Fig. 5.6. Regression equations relating log K_2 to I at pH 5 was as follows:

Cu^{2+} Log K_2 = 8.9 - 4.9 $\sqrt{I}$

Zn^{2+} Log K_2 = 6.7 - 5.4 $\sqrt{I}$

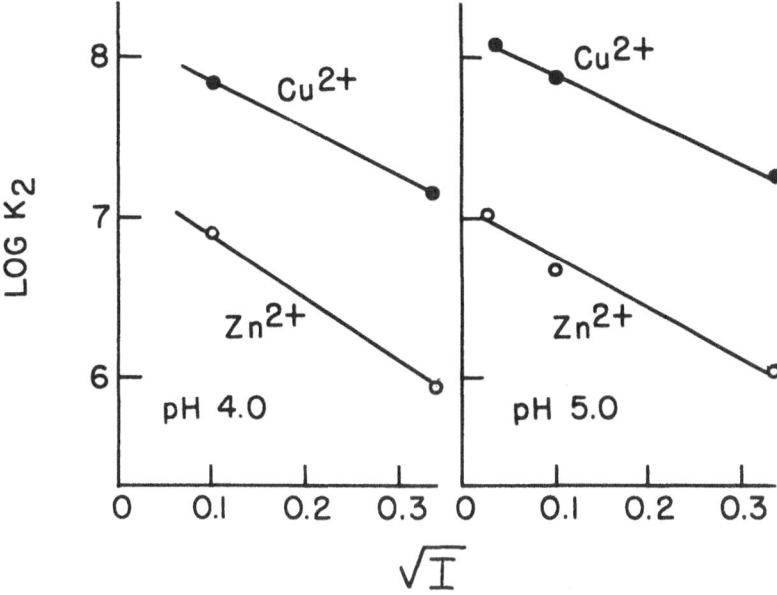

Fig. 5.6. Over-all stability constants (Log K_2) for the Cu^{2+} and Zn^{2+} complexes of a soil humic acid at two pH values at affected by ionic strength (I). Adapted from Stevenson (130).

Log K_2 values obtained at low I are more likely to represent conditions prevailing in the soil solution than those at high I. On this basis, the values obtained are within the range of log K values for complexes of many naturally occurring biochemical compounds. Nevertheless, it would appear that amino acids and hydroxy carboxylic acids secreted by plant roots may compete favorably with humic acids for metal ions and promote their uptake by plant roots. Log K values ranging from 5 to 15 have been recorded for Cu^{2+} complexes of amino acids and simple organic acids. It should be noted that constants for the known compounds were not all determined at the same I or under the same experimental conditions; accordingly, they could vary over a log unit and are not strictly comparable. The competitive nature of soil ligands for micronutrient cations deserves further study.

Young *et al.* (158) used the potentiometric method to study the binding of Cu(II) to some soluble "polycarboxylates" from a peat and a mineral topsoil. Constants were calculated for the equilibria:

$$2L^- + Cu^{2+} \rightleftharpoons CuL_2; \quad \beta_{Cu} = (CuL_2)/(Cu^{2+})(L^-)^2$$
and
$$2HL + Cu^{2+} \rightleftharpoons CuL_2 + H^+; \quad \beta_{Cu}^H = (CuL_2)(H^+)^2/(Cu^{2+})(HL)^2$$

Values for log β_{Cu}^{H} varied from -1.26 to -1.80 and were unaffected by the degree of dissociation of acidic functional groups, or pH. In contrast, log β_{Cu} values dramatically increased with increasing dissociation of acidic functional groups (RCOOH $\rightleftharpoons$ RCOO$^-$ + H$^+$), being of the order of 0.5 to 2.8 for the completely protonated sample (low pH) and 7.6 to 9.3 for the fully dissociated sample (high pH). The value for log β_{Cu} at a high degree of dissociation were of the same order of magnitude as those recorded by Stevenson (129) for the Cu^{2+} complexes of humic acid.

5.5.2.3 Polynuclear complexes

Complexes which have more than one central group are termed polynuclear. They can be further classified as homopolynuclear, where only one type of metal ion is involved, and heteropolynuclear where more than one type of metal is present. The possibility that humic acids and fulvic acids form polynuclear complexes with metal ions deserves greater attention in the future.

Considerable data implicate polynuclear complex formation in metal-fulvic acid reactions. With humic acids, di- and trivalent metal ions have been found to result in an increase in the average molecular weight (54, 70). Underdown *et al.* (143) and Bhat and Weber (14) observed that an increase in metal ion binding by fulvic acid led to a continuous increase in aggregation well before the onset of observed flocculation.

5.6. Summary and conclusions

Organic matter plays a key role in the behavior of micronutrients in soil. Both soluble and insoluble complexes are formed. Substantial evidence has accumulated to indicate that complexing agents (*e.g.*, defined biochemical compounds and humic substances) play a prominent role in the dissolution of micronutrients and their transport to plant roots.

Whereas natural organic substances show promise as carriers of trace elements in micronutrient deficient soils, their use has not been widely adopted. Further research and technological advancements in fertilizer manufacture may lead to more extensive use in the future.

Stability constants for metal complexes with humic and fulvic acids are highly variable and differ by as much as three orders of magnitude for the same experimental conditions (pH, ionic strength, *etc.*). In many instances, comparisons cannot be made because of differences in approaches for modeling the experimental data. A variety of modeling methods have been applied, but, thus far, there is no sound basis for selecting one approach at the exclusion of all others. Consideration needs to be given to standardization of methods so that results from various laboratories can be compared. In the future, attention should be given to competitive binding (1) between metal ions for reactive sites on humic and fulvic acids, (2) between naturally occurring ligands (*e.g.*, humic acids, fulvic acids, biochemical compounds) for the metal ion, and (3) between soluble and insoluble organic matter for the metal ion.

Acknowledgement   This research was supported by a grant from BARD - United States - Israel Agricultural Research and Development Fund.

5.7. References

1 Abdou F.M. and El-Nennah M. 1980 Effect of irrigating loamy sand by liquid sewage sludge on its contents of some micronutrients. Plant Soil 561, 53-57.

2 Abdul-Halim A.L., Evans J.C., Rowlands C.C. and Thomas J.H. 1981 An EPR spectroscopic examination of heavy metals in humic and fulvic acid soil fractions. Geochim. Cosmochim. Acta 45, 481- 487.

3 Aiken G. R., Thurman E.M. and Malcolm R.L. 1979 Comparison of XAD macroporous resins for the concentration of fulvic acid from aqueous solution. Anal. Chem. 51, 1799-1803.

4 Akers H.A. 1983 Multiple hydroxamic acid microbial iron chelators (siderophores) in soils. Soil Sci. 135, 156-159.

5 Alexsandrova L.N. 1960 The use of sodium pyrophosphate for isolating free humic substances and their organic-mineral compounds from the soil. Soviet Soil Sci. 1960, 190-197.

6 Anderson W.B. 1964 Effect of synthetic chelating agents as sources of zinc for calcareous soils. Ph. D. Thesis, Colorado State University Fort Collins.

7 Ardakani M.S. and Stevenson F.J. 1972 A modified ion-exchange technique for the determination of stability constants of metal-soil organic matter complexes. Soil Sci. Soc. Am. Proc. 36, 884-890.

8 Aso S. and Sakai I. 1963 Studies on the physiological effects of humic acid. Part 1. Uptake of humic acid by crop plants and its physiological effects. Soil Sci. Plant Nutrition 99, 85-91.

9 Baham J., Ball N.B. and Sposito G. 1978 Gel filtration studies of trace metal-fulvic acid solutions extracted from sewage sludges. J. Environ. Qual. 7, 124-127.

10 Barber S.A., Halstead E.H. and Follett B.F. 1966 Significant mechanisms controlling the movement of manganese and molybdenum to plant roots growing in soil. p. 299-304. *In* Int. Soil Sci. Soc. (Aberdeen, Scotland) Trans. Comm. II and IV.

11 Barber S.A., Walker J.M. and Vasey E.H. 1963 Mechanisms for the movement of plant nutrients from the soil and fertilizer to the plant root. J. Agric. Food Chem. 11, 204-207.

12 Bar-Ness E. 1985 The effect of Fe, Mn and Zn amended naturally occurring substances on yield and micronutrient uptake by plants. M.S. Thesis. The Hebrew Univ. of Jerusalem.

13 Bar-Tal A., Bar-Yosef B. and Chen Y. 1986 The effects of various organic chelating agents and pH on zinc transport in soils. Soil Sci. (submitted).

110

14 Bhat G.A. and Weber J.H. 1982 Cadmium (II) binding by soil-derived fulvic acid measured by anodic stripping voltammetry. Anal. Chim. Acta 141, 95-103.
15 Blaschke H. 1979 Leaching of water-soluble organic substances from coniferous needle litter. Soil Biol. Biochem. 11, 581-584.
16 Blaser P., Flühler H. and Polomski, J. 1980 Metal binding properties of leaf litter extracts: I. Soil Sci. Soc. Am. J. 44, 709-716.
17 Blaser P., Landolt W. and Flühler, W. 1980 Metal binding properties of leaf litter extracts: II. Soil Sci. Soc. Am. J. 44, 717-720.
18 Blondel D. 1970 Induction of iron chlorosis in sandy soil (Dior) by irrigation waters rich in calcium and magnesium. Agron. Trop., Paris 25, 555-560. (Fr.).
19 Bloom P.R. and Mcbride M.B. 1979 Metal ion binding and exchange with hydrogen ions in acid-washed peat. Soil Sci. Soc. Am. J. 43, 687-692.
20 Bloom P.R., McBride M.B. and Weaver R.M. 1979 Aluminum organic matter in acid soils: Buffering and solution aluminum activity. Soil Sci. Soc. Am. J. 43, 488-493.
21 Boehl J., Jr and Lindsay W.L. 1969 Micronutrients - The fertilizer shoenails–Zinc. Fert. Solns. 13, 6-8, 10, 12.
22 Boyd S.A., Sommers L.E. and Nelson D.W. 1979 Infrared spectra of sewage sludge fractions: Evidence for an amide metal binding site. Soil Sci. Soc. Am. J. 43, 893-899.
23 Boyd S.A., Sommers L.E. and Nelson D.W. 1981 Copper (II) and iron (III) complexation by the carboxylate group of humic acid. Soil Sci. Soc. Am. J. 45, 1241-1242.
24 Boyd S.A., Sommers L.E., Nelson D.W. and West D. X. 1981 The mechanism of copper (II) binding by humic acid: An electron spin resonance study of Cu(II)-humic acid complex and some adducts with nitrogen donors. Soil Sci. Soc. Am. J. 45, 745-749.
25 Boyd S.A., Sommers L.E., Nelson D.W. and West D.X. 1983 Copper(II) binding by humic acid extracted from sewage sludge: An electron spin resonance study. Soil Sci. Soc. Am. J. 47, 43-46.
26 Bresnahan W.T., Grant C.L. and Weber J.H. 1978 Stability constants for the complexation of copper(II) ions with water and soil fulvic acids measured by an ion selective electrode. Anal. Chem. 50, 1675-1679.
27 Brezonik P.L., Brauner P.A. and Stumm W. 1976 Trace metal analysis by anodic stripping voltammetry: Effect of sorption by natural and model organic compounds. Water Res. 10, 605-612.
28 Bruckert S. and Jacquin F. 1969 Interaction entre la mobilite de plusieurs acides organiques et de divers cations dans un sol a mull et dons un soil a mor. Soil Biol. Biochem. 1, 275-294.
29 Buffle J., Greter F-L. and Haerdi W. 1977 Measurement of complexation properties of humic and fulvic acids in natural waters with lead and copper ion-selective electrodes. Anal. Chem. 49, 216-222.
30 Bureau R.G., White R.G. and MacGregor J.M. 1960 Uptake of applied iron by soybeans from calcareous soil treated with peat-based humates and synthetic chelates. 7th Intern. Congress Soil Sci. Madison, Wis. 3, 544-553.
31 Campbell P.G.C., Bissom M., Bouqie R., Tessier A. and Villeneuve J-P. 1983 Speciation of aluminum in acidic freshwaters. Anal. Chem. 55, 2246-2252.
32 Cheam V. 1973 Chelation study of copper(II)-fulvic acid system. Can. J. Soil Sci. 53, 377-382.
33 Cheam V. and Gamble D.S. 1974 Metal-fulvic acid chelation equilibrium in aqueous NaNO$_3$ solution. Hg(II), Cd(II), and Cu(II) fulvate complexes. Can. J. Soil Sci. 54, 413-417.
34 Chen Y. and Barak P. 1983 Iron-enriched peat and lignite as iron fertilizer. Proc. 2nd Intl. Symp. Peat in Agric. pp. 195-202. Bet-Dagan, Israel.

35 Chen Y., Navrot J. and Barak P. 1982 Remedy of lime-induced chlorosis with iron-enriched muck. J. Plant Nutr. 5, 927-960.

36 Chen Y., Steinitz B., Cohen A. and Elber Y. 1982 The effect of various iron-containing fertilizers on growth and propagation of *Gladiolius grandiflorus*. Scientia Hort. 18, 169-175.

37 Chesnin L. 1968 Flavenoid and lignin compounds derived from forest products as micronutrient carriers for plant growth. pp. 411-419. *In*: Isotopes and radiation in soil organic matter studies. Intern. Atomic Energy Agency, Vienna.

38 Clark A.L. and Graham E.B. 1968 Zinc diffusion and distribution coefficients in soil as affected by soil texture, zinc concentration and pH. Soil Sci. 105, 409-418.

39 Clark J.S. and Turner R.C. 1969 An examination of the resin exchange method for the determination of metal-soil organic matter complexes. Soil Sci. 107, 8-11.

40 Cline G.R., Powell P.E., Szaniszlo P.J. and Reed C.P.P. 1983 Comparison of the abilities of hydroxamic and other natural organic acids to chelate iron and other ions in soil. Soil Sci. 136, 145-157.

41 Cronan C.S., Reiners W.A., Reynolds R.C., Jr and Lang G.E. 1978 Forest floor leaching: Contribution from mineral, organic, and carbonic acids in New Hampshire subalpine forests. Science 200, 309-311.

42 Crosser M.L. and Allen H.E. 1977 Determination of complexation capacity of soluble ligands by ion exchange equilibrium. Soil Sci. 123, 176-181.

43 Davies M.I., Cheshire M.V. and Graham-Bryce I. J. 1969 Retention of low levels of copper by humic acid. J. Soil Sci. 20, 65-71.

44 De Kock P.C. and Strmecki E.L. 1954 An investigation into the growth promoting effects of a lignite. Physiol. Plant. 7, 503-512.

45 De Coninck F. 1980 Major mechanisms in formation of spodic horizons. Geoderma 24, 101-128.

46 Dowd J.E. and Riggs D.S. 1965 A comparison of estimates of Michaelis-Menten kinetic constants from various linear transformations. J. Biol. Chem. 240, 863-869.

47 Dubach P. and Mehta N.C. 1963 The chemistry of soil humic substances. Soils Fertilizers 26, 293-300.

48 Ellis B.G., Knezek B.D. and Jacobs L.W. 1983 The movement of micronutrients in soils. pp. 109-122. *In* Nelson D.W., Elrick D. E. and Tanji K.K. Chemical Mobility and Reactivity in Soil Systems. SSSA Special Publication Number 11, Soil Science Soc. of America, Madison, Wisc.

49 Enis M.T. and Brogan J.C. 1961 The availability of copper from copper humic acid complexes. Irish J. Agr. Res. 1, 35-42.

50 Fitch A. and Stevenson F.J. 1984 Comparisons of models for determining stability constants of metal complexes with humic substances. Soil Sci. Soc. Am. J. 48, 1044-1050.

51 Fitch A. and Stevenson F.J. 1983 Stability constants of metal-organic matter complexes: Theoretical aspects and mathematical models. pp. 645-649. *In* S.S. Augustithis (ed.) The significance of trace elements in solving petrogenetic problems and controversies. Theoprastus Publications S.A., Athens, Greece.

52 Francis H.J., Rajagopal C.K. and Krishnamoorthy K.K. 1979 Effect of organically complexed iron on the available iron content in soil and uptake by sorghum CSH5 in two different soils at successive growth stages. Mysore J. Agric. Sci. 13, 159-164.

53 Gamble D.S., Schnitzer M. and Hoffman I. 1970 Cu^{2+} -fulvic acid chelation equilibrium in 0.1 M KCl at 25°C. Can. J. Chem. 48, 3197-3204.

112

54 Ghosh K. and Schnitzer M. 1981 Fluorescence excitation spectra and viscosity behavior of a fulvic acid and its copper and iron complexes. Soil Sci. Soc. Am. J. 45, 25-29.

55 Goodman B.A. *and Cheshire M.V.* 1973 Electron paramagnetic resonance evidence that copper is complexed in humic acid by the nitrogen of porphyrin groups. Nature (London) 244, 158-159.

56 Goodman B.A. and Cheshire M.V. 1976 The occurrence of copperporphyrin complexes in soil humic acids. J. Soil Sci. 27, 337-347.

57 Graustein W.C., Cromack K., Jr and Sollins P. 1977 Calcium oxalate: Occurrence in soils and effect on nutrient and geochemical cycles. Science 198, 1252-1254.

58 Greter F.L., Buffle J. and Haerdi W. 1979 Voltammetry study of humic and fulvic substances. J. Electroanal. Chem. 101, 211-279.

59 Griffith S.M., Silver J. and Schnitzer M. 1980 Hydrazine derivatures at Fe^{3+} sites in humic materials. Geoderma 23, 299-302.

60 Guy R.D. and Chakrabarti C.L. 1976 Analytical techniques for speciation of heavy metal ions in the aquatic environment. Chemistry in Canada 28, 26-29.

61 Halstead E.H., Barber S.A., Warncke D.I. and Bole J.B. 1968 Supply of Ca, Sr, Mn and Zn to plant roots growing in soil. Soil Sci. Soc. Am. Proc. 32, 69-72.

62 Hargrove W.L. and Thomas G.W. 1979 Effect of organic matter on exchangeable aluminum and plant growth in acid soils. ASA Special Publication 40, 151-166. Soil Sci Soc. Am. Inc., Madison, Wis.

63 Hayes M.H.B. and Swift R.S. 1978 The chemistry of soil organic colloids. pp. 179-230. *In* D.J. Greenland and R.S. Swift (eds.). The Chemistry of Soil Constituents. Wiley Interscience, Chichester, England.

64 Hodgson J.F. 1963 Chemistry of the micronutrient elements in soils. Adv. Agron. 15, 119-159.

65 Hodgson J.F., Geering H.R. and Norvell W.A. 1965 Micronutrient cation complexing in soil solution. I. Soil Sci. Soc. Am. Proc. 29, 665-669.

66 Hodgson J.F., Lindsay W.L. and Trierweiler J.F. 1966 Nutrient cation complexing in soil solution: II. Soil Sci. Soc. Am. Proc. 30, 723-726.

67 Jalali V.K. and Takkar P.N. 1979 Evaluation of parameter for simultaneous determination of micronutrient cations available to plants from soil. Indian J. Agric. Sci. 49, 622-626.

68 Katase T. 1981 Distribution of different forms of p-hydroxybenzoic, p-coumaric and ferulic acids in forest soils. Soil Sci. Plant Nutr. 27, 365-371.

69 Kononova M.M. 1966 Soil Organic Matter. Pergamon, New York, 544 p.

70 Kribeck B., Kaigl J. and Oruzinsky V. 1977 Characterization of di and trivalent metalhumic acid complexes on the basis of their molecular weight distributions. Chem. Geol. 19, 73-81.

71 Kuchenbuck R. and Jungk A. 1982 A method for determining concentration profiles at the soil-root interface by thin slicing rhizospheric soil. Plant Soil 68, 391-396.

72 Lakatos B., Korecz L. and Meisel J. 1977 Comparative study of Mössbauer parameters of iron humates and polyuronates. Geoderma 19, 149-157.

73 Lakatos B., Tibai T. and Meisel J. 1977 ESR spectra of humic acids and their metal complexes. Geoderma 19, 319-338.

74 Langford C.H. and Khan T.R. 1975 Kinetics and equilibrium of binding of Fe^{3+} by a fulvic acid. Can. J. Chem. 53, 2979-2984.

75 Lavy T.L. and Barber S.A. 1964 Movement of molybdenum in the soil and its effect on availability to the plant. Soil Sci. Soc. Am. Proc. 28, 93-97.

76 Lee Y. S. and Bartlett R.J. 1976 Stimulation of plant growth by humic substances. Soil Sci. Soc. Am. J. 40, 876-879.

77 Lewis D.G. and Quirk J.P. 1967 Phosphate diffusion in soil and uptake by plants. III. P^{31} movement and uptake by plants as indicated by P^{32} autoradiography. Plant Soil 26, 445-453.

78 Lindsay W.L. 1974 Role of chelation in micronutrient availability. *In* E.W. Carson (ed.): The Plant Root and its Environment, pp. 507-524. University Press of Virginia, Charlottesville.

79 Lindsay W.L. and Schwab A.P. 1982 The chemistry of iron in soils and its availability to plants. J. Plant Nutr. 5, 821-840.

80 Linehan D.J. 1978 The uptake by plants of polymeleic acid: a polycarboxylic acid structurally related to those of soils. Plant and Soil 50, 625-632.

81 Linehan D.J. 1978 Humic acid and iron uptake by plants. Plant Soil 50, 663-673.

82 Linehan D.J. and Shepherd H. 1979 A comparative study of the effects of natural and synthetic ligands on iron uptake of plants. Plant Soil 52, 281-289.

83 Little R.C. 1971 Treatment of iron deficiency. pp. 45-61. *In* Trace elements in soils and crops. Tech. Bull. 21. Min. Agr., Fisheries and Food. Her Majesty's Stationery Office, London.

84 Low P.F. 1962 Effect of quasi-crystalline water on rate processes involved in plant nutrition. Soil Sci. 93, 6-15.

85 MacCarthy P. and Mark H.B., Jr 1976 An evaluation of Job's method of continuous variation as applied to soil organic matter-metal ion interaction. Soil Sci. Soc. Amer. J. 40, 267-276.

86 Mann M.S., Takkar T.S., Bansal R.L. and Randhawa N.S. 1978 Micronutrient status of soil and yield of maize and wheat as influenced by micronutrient and farmyard manure application. J. Indian Soc. Soil Sci. 26, 208-214.

87 Mantoura R.F.C., Dickson A. and Riley J.P. 1978 The complexation of metals with humic materials in natural waters. Est. Coast. Mar. Sci. 6, 387-408.

88 Mantoura R.F.C. and Riley J.P. 1975 The use of gel filtration in the study of metal binding by humic acids and related compounds. Anal. Chim. Acta. 78, 193-200.

89 Matsuda K. and Ito S. 1970 Adsorption strength of zinc for soil humus: III. Soil Sci. Plant Nutr. (Tokyo) 16, 1-10.

90 Mattigod S.V., Sposito G. and Page A.L. 1981 Factors affecting the solubilities of trace metals in soils. ASA Special Publ. 40, 203-221. Soil Sci. Soc. Am. Inc., Madison, Wis.

91 McBride M.B. 1978 Transition metal bonding in humic acid: An ESR study. Soil Sci. 126, 200-209.

92 Melton J.R., Mahtab S.K. and Swoboda A.R. 1973 Diffusion of zinc in soils as a function of applied zinc, phosphorus, and soil pH. Soil Sci. Soc. Am. Proc. 37, 279-381.

93 Mengel K. and Kirby E.A. 1982 Principals of Plant Nutrition. International Potash Institute, Bern, Switzerland.

94 Moghimi A., Tate M.E. and Oades J.M. 1978 Characterization of rhizosphere products including 2-ketogluconic acid. Soil Biol. Biochem. 10, 283-287.

95 Mylonas V.A. and McCants C.B. 1980 Effects of humic and fulvic acids on growth of tobacco. I. Root initiation and elongation. Plant Soil 54, 485-490.

96 Murphy L.S. and Walsh L.M. 1972 Correction of micronutrient deficiencies with fertilizers. pp. 347-387. *In* Mortvedt, J.J., Giordano P.M. and Lindsay W.L. (eds.). Micronutrients in Agriculture. Soil Sci. Soc. Am. Inc. Madison, Wis.

97 Nye P.H. 1966 The measurement and mechanism of ion diffusion in soil. I. The relation between self-diffusion and bulk diffusion. J. Soil Sci. 17, 16-23.

98 O'Connor G.A., Lindsay W.L. and Olsen S.R. 1971 Diffusion of iron and iron chelates in soil. Soil Sci. Soc. Am. Proc. 35, 407-410.

114

99 Oliver S. and Barber S.A. 1966 Mechanisms for the movement of Mn, Fe, B, Cu, Zn, Al, and Sr from one soil to the surface of soybean roots. Soil Sci. Soc. Am. Proc. 30, 468-470.

100 O'Shea T.A. and Mancy K.H. 1976 Characterization of trace metal-organic interactions by anodic-stripping voltammetry. Anal. Chem. 48, 1603-1607.

101 Parsa A.A. and Wallace A. 1979 Organic solid wastes from urban environment as iron sources for sorghum. Plant Soil 53, 455-461.

102 Perdue E.M. and Lytle C.R. 1983 Distribution model for binding of protons and metal ions by humic substances. Environ. Sci. Tech. 17, 654-661.

103. Piccolo A. and Stevenson F.J. 1981 Infrared spectra of Cu^{2+}, Pb^{2+}, and Ca^{2+} complexes of soil humic substances. Geoderma 27, 195-208.

104 Powell P.E., Szaniszlo P.J., Cline G.R. and Reid C.P.P. 1982 Hydroxamate siderophores in the iron nutrition of plants. J. Plant Nutr. 5, 653-673.

105 Rashid M.A. 1971 Role of humic acids of marine origin and their different molecular weights in complexing di- and tri-valent metals. Soil Sci. 111, 298-306.

106 Richardson G. 1967 Iron deficiency in sorghum. The micronutrient manual. Farm Tech. 23, No. 6

107 Rovira A.D. 1969 Plant root exudates. Bot. Rev. 35, 35-57.

108 Ryan D.K. and Weber J.H. 1982 Fluorescence quenching titration for determination of complexing capacities and stability constants of fulvic acid. Anal. Chem. 54, 986-990.

109 Saar R.A. and Weber J.H. 1980 Comparison of spectrofluorometry and ion-selective electrode potentiometry for determination of complexes between fulvic acid and heavy-metal ions. Anal. Chem. 52, 2095-2100.

110 Saar R.A. and Weber J.H. 1980 Lead(II)-fulvic acid complexes: Conditional stability constants, solubility, and implications for Pb(II) mobility. Environ. Sci. Tech. 14, 877-880.

111 Saar R.A. and Weber J.H. 1980 Lead(II) complexation by fulvic acid: How it differs from fulvic acid complexation of copper(II) and cadmium(II). Geochim. Cosmochim Acta 44, 1381-1384.

112 Salardini A.A. and Murphy L.S. 1978 Grain sorghum (*Sorghum bicolor* Pers.) responses to organic iron on calcareous soils. Plant Soil 49, 57-70.

113 Sanders J.R. 1982 The effect of pH upon the copper and cupric ion concentrations in soil solution. J. Soil Sci. 33, 679-689.

114 Sanders J.R. 1983 The effect of pH on the total and free ionic concentrations of manganese, zinc, and cobalt in soil solutions. J. Soil Sci. 34, 315-323.

115 Schnitzer M. 1977 Recent findings on the characterization of humic substances extracted from soils from widely differing climatic zones. pp. 117-131. *In* Proc. Symposium on Soil Organic Matter. FAO/IAEA, Vienna.

116 Schnitzer M. 1969 Reaction between fulvic acid, a soil humic compound, and inorganic soil constitutents. Soil Sci. Soc. Am. Proc. 33, 75-81.

117 Schnitzer M. and Hansen E.H. 1970 Organo-metallic interactions in soils: 8. Soil Sci. 109, 333-340.

118 Schnitzer M. and Khan S.U. 1972 *Humic Substances in the Environment*. Marcel Dekker, New York, 327 pp.

119 Schnitzer M. and Rauthan B.S. 1981 Effects of a soil fulvic acid on the growth and nutrient content of cucumber (*Cucumus sativus*) plants. Plant Soil 63, 491-495.

120 Senesi M., Griffith S.M. and Schnitzer M. 1977 Binding of Fe^{3+} by humic materials. Geochim. Cosmochim. Acta 41, 969-976.

121 Shuman M.S. and Woodward G.P., Jr 1977 Stability constants of copper-organic chelates in aquatic samples. Environ. Sci. Technol. 11, 809-813.

122 Singh A.P., Sakal R. and Singh B.P. 1982 Effect of zinc enriched compost and other methods of zinc application on zinc nutrition of rice in calcareous soil. J. Indian Soc. Soil Sci. 30, 572-573.

123 Singh S.P., Sinha M.K. and Randhawa N.S. 1979 Effect of zinc-amended poultry manure and zinc sulphate on the growth and uptake of zinc by corn (*Zea mays* L.). Plant Soil 52, 501-505.

124 Sladky Z. 1959 The effect of extracted humus substances on growth of plant roots. Biol. Plant. Praha 1, 142-150.

125 Smidova M. 1960 The influence of humic acid on the respiration of plant roots. Biol. Plant. Praha 2, 152-164.

126 Smith W.H. 1976 Character and significance of forest tree root exudates. Ecol. 57, 324-331.

127 Sposito G. and Holtzclaw K.M. 1979 Copper(II) complexation by fulvic acid extracted from sewage sludge as influenced by nitrate versus perchlorate background ionic media. Soil Sci. Soc. Am. J. 43, 47-51.

128 Sposito G., Holtzclaw K.M. and LeVesque-Madore C.S. 1979 Cupric ion complexation by fulvic acid extracted from sewage sludge-soil mixtures. Soil Sci. Soc. Am. J. 43, 1148-1155.

129 Stevenson F.J. 1976 Stability constants of Cu^{2+}, Pb^{2+}, and Cd^{2+} complexes with humic acids. Soil Sci. Soc. Amer. J. 40, 665-672.

130 Stevenson F.J. 1977 Nature of divalent transition metal complexes of humic acids as revealed by a modified potentiometric titration procedure. Soil Sci. 123, 10-17.

131 Stevenson F.J. 1979 Humates: facts and fantasies on their value as commercial soil amendments. Crops and Soils 31, 14-16.

132 Stevenson F.J. 1982 Humus chemistry: Genesis, composition, reactions. Wiley Interscience. 465 pp.

133 Stevenson F.J. and Ardakani M.S. 1972 Organic matter reactions involving micronutrients. p. 79-114. *In* J.J. Mortvedt, P.M. Giordano, and W.L. Lindsay (eds.) Micronutrients in agriculture. Am. Soc. Agron., Madison, Wisc.

134 Stevenson F.J. and Fitch A. 1981 Reactions with organic matter. pp. 69-95. *In* J.F. Loneragan, A.D. Robson, and R.D. Graham (eds.) Copper in Soils and Plants. Academic Press, New York.

135 Strickland R.C., Chaney W.R. and Lamoreaux R.S. 1979 Organic matter influences phytotoxicity of cadmium to soybeans. Plant Soil 52, 393-402.

136 Takamatsu T. and Yoshida T. 1978 Determination of stability constants of metal-humic acid complexes by potentiometric titration and ion-selective electrodes. Soil Sci. 125, 377-386.

137 Takkar P.N. 1969 Effect of organic matter on soil iron and manganese. Soil Sci. 108, 108-112.

138 Tan K.H. and Nopamornbodi V. 1979 Effect of different levels of humic acids on nutrient content and growth of corn (*Zea mays* L.). Plant Soil 51, 283-287.

139 Thomas J.D. and Mathers A.C. 1979 Manure and iron effects on sorghum growth on iron-deficient soil. Agron. J. 71, 792-794.

140 Tuschall J.R. Jr and Brezonik P.L. 1981 Evaluation of the copper anodic stripping voltammetry complexometric titration from complexing capacities and conditional stability constants. Anal. Chem. 53, 1986-1989.

141 Tuschall J.R. jr and Brezonik P.L. 1983 Application of continuous-flow ultrafiltration

and competing ligand/differential spectrophotometry for measurement of heavy metal complexation by dissolved organic matter.Anal. Chim. Acta. 149, 47-58.

142 Tyler L.D. and McBride M.B. 1982 Influence of Ca, pH and humic acid on Cd uptake. Plant Soil 64, 259-262.

143 Underdown A.W., Langford C.H. and Gamble D.S. 1981 Light scattering of a polydisperse fulvic acid. Anal. Chem. 53, 2139-2140.

144 van den Berg C.M.G. and Kramer J.R. 1979 Determination of complexing capacities of ligands in natural waters and conditional stability constants of the copper complexes by means of manganese dioxide. Anal. Chim. Acta. 106, 113-120.

145 van Dijk H. 1971 Cation binding of humic acids. Geoderma 5, 53-67.

146 Vaughan D. and MacDonald I.R. 1976 Some effects of humic acid on cation uptake by parenchyma tissue. Soil Biol. Biochem. 8, 415-441.

147 Vinkler P., Lakatos B. and Meisel J. 1976 Infrared spectroscopic investigations of humic substances and their metal complexes. Geoderma 15, 231-242.

148 Walker D.R. and Smith R.L. 1967 Iron deficiency in deciduous fruit. The micronutrient manual. Farm Tech. 23(6).

149 Wallace A. and Ashcroft R.T. 1956 Ammonium lignin sulfonate as a chelating agent for supplying soluble iron to plants. Soil Sci. 82, 233-236.

150 Wang T.S.C., Cheng S.Y. and Tung H. 1967 Dynamics of soil organic acids. Soil Sci. 104, 138-144.

151 Warncke D.D. and Barber S.A. 1972 Diffusion of zinc in soil: I. The influence of soil moisture. Soil Sci. Soc. Am. Proc. 36, 39-42.

152 Warncke D.D. and Barber S.A. 1973 Diffusion of zinc in soils: III. Relation to zinc adsorption isotherms. Soil Sci. Soc. Am. Proc. 37, 355-358.

153 White M.C. and Chaney R.L. 1980 Zinc, cadmium and manganese uptake by soybean from two zinc- and cadmium- amended coastal plain soils. Soil Sci. Soc. Am. J. 44, 308-313.

154 Whitehead D.C., Dibb H. and Hartley R.D. 1981 Extractant pH and release of phenolic compounds from soils, plant roots and leaf litter. Soil Biol. Biochem. 13, 343-348.

155 Whitehead D.C., Dibb H. and Hartley R.D. 1982 Phenolic compounds in soil as influenced by the growth of different plant species. J. Appl. Ecol. 19, 579-588.

156 Wilkinson H.F., Loneragan J.F. and Quirk J.P. 1968 The movement of zinc to plant roots. Soil Sci. Soc. Am. Proc. 32, 831-833.

157 Wilkinson H.F., Loneragan J.F. and Quirk J.P. 1968 Calcium supply to plant roots. Science 161, 1245-1246.

158 Young S.D., Bache B.W. and Linehan D.J. 1982 The potentiometric measurement of stability constants of soil polycarboxylate-Cu^{2+} chelates. J. Soil Sci. 33, 467-475.

159 Zunino H., Peirano P., Aquilera M. and Escobar I. 1972 Determination of maximum complexing ability of water-soluble complexants. Soil Sci. 114, 414-416.

160 Zunino H. and Martin J.P. 1977 Metal-binding organic macromolecules in soil: 2. Characterization of the maximum binding ability of the macromolecule. Soil Sci. 123, 188-202.

6. Effect of soil redox conditions on microbial oxidation of organic matter.*

K.R. REDDY, T.C. FEIJTEL and W.H. PATRICK Jr.

6.1. Introduction

Soils undergo varying redox changes as a result of restricted gaseous exchange, increased soil-water content due to poor drainage, and flooding or incorporation of highly O_2 demanding carbonaceous wastes. Depending on the intensity of these conditions, soil O_2 can decrease to negligible concentrations, thus decreasing the aerobic soil volume and increasing the anaerobic soil volume. The microbial populations which thrive on O_2 decrease and facultative anaerobes and obligate anaerobes which rely on other sources of electron acceptors predominate. The rate at which these bacteria obtain energy for their growth and cell maintenance depends on the oxidation-reduction reactions utilizing specific inorganic or organic molecules as electron acceptors and substrate availability. Depending on the redox status of the soil, two general types of microbial metabolisms are found: (1) processes utilizing inorganic substances (O_2, nitrogen oxides such as NO_3^-, NO_2^-, NO, N_2O, manganic compounds, ferric oxyhydroxide compounds, SO_4^{2-}, CO_2, and H_2), and (2) fermentative processes in which organic molecules (succinate) are utilized as electron acceptors. Under substrate nonlimiting conditions and in the absence of competition among electrons, these types of microbial metabolisms can occur simultaneously in different soil zones of the same soil. For example, in a typically well-drained soil, O_2 can be used as an electron acceptor during respiration of aerobic bacteria, while in anaerobic microzones, NO_3^- and manganese compounds are used as electron acceptors during respiration of facultative anaerobic bacteria. In a flooded paddy soil, lake sediment, or estuarine sediment aerobic, facultative anaerobic, and obligate anaerobic metabolism will occur simultaneously in different soil zones (Fig. 6.1.). Rate of organic matter decomposition under these conditions will depend on the bacterial efficiency and the capacity of the soil system to supply electron acceptors.

The purpose of this review is to quantitatively describe (1) the rate of organic matter degradation as influenced by the type of microbial metabolism, (2) identify the soil and environmental factors influencing the process, and (3) the influence of aerobic/anaerobic metabolism on soil biochemical processes.

* Florida Agricultural Experiment Stations, Journal Series No 5316.

118

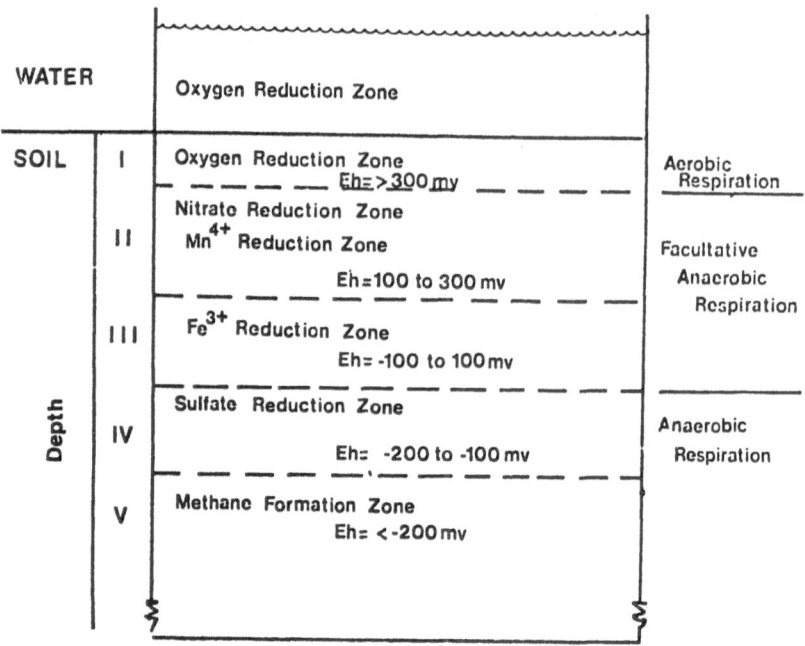

Fig. 6.1. Schematic presentation of a flooded soil showing the zones with different microbial metabolism.

6.2. Sources and types of organic matter

The organic matter is an important constituent of soil and the average content has been estimated to equal 4.5% with carbon as the main constituent (approximately 45% by weight) (68). Organic matter is thermodynamically unstable in soils and undergoes a dynamic decomposition. The original material and their decomposition products undergo a continuous microbial attack. This is associated with a loss of the carboneous materials, the generation of new microbial tissue, and the release of mineral nutrients (*e.g.* N and P). The dissolved inorganic nutrients are now available for the primary producers, although microbial immobilization is likely to occur as their content of essential elements is relatively high compared to plant tissue.

Each year enormous quantities of organic residues are deposited into terrestrial and aquatic environments. Williams and Gray (161) estimated that plant litter alone accounts for between 1.0 and 15.3 tons of organic matter ha[-1] yr[-1]. The root biomass in the top 30 cm of soil is estimated on the order of 440 to 1575 g m[-2] (5). In addition, animals and their excreta and dead microbial cells provide a significant residue. In order to maintain a steady state condition between assimilation and mineralization a continuous degradation is needed. The enormous flexibility as to the nature and the number of substrates microorganisms can utilize is responsible for the continuity of nutrient cycles.

6.2.1. Soil organic matter

The product of the assimilative and decomposing activities of the microflora is often called humus or organic fraction of the soil. The native organic fraction originates from the original plant debris entering the soil and the microorganisms within the soil body. As organic matter is mineralized, the portion remaining becomes increasingly resistant to microbial breakdown. A dynamic chemical alteration due to microbial and abiotic processes causes an accumulation of the more resistant humic substances. Soil organic matter can be divided into three major components: (1) humic acid, which is soluble in dilute alkali but insoluble upon acidification of the alkali extract; (2) fulvic acid, which is soluble in dilute alkali and remains in solution after acidification of the soil extract; and (3) humin, which is not soluble in dilute acid or alkali (152). Humic substances have an increasing turnover time as a function of age, which could explain the large pool of dead organic matter characterizing many ecosystems (37). Because of the heterogeneity, different components of humic materials display different susceptibilities to microbial attack or to a biological release and therefore have different residence times in soil. A frequent estimate is that between 2 and 5% of the humus carbon is mineralized each year (5). This process can apparently be accelerated through the addition of easily degradable substrates (52).

6.2.2. Root exudates

The root system of higher plants is associated with the organic and inorganic soil environment and with a vast community of metabolically active microorganisms. In the rhizosphere and rhizoplane, excretory soluble organic materials and slough-off root tissue are found which induce a highly favorable habitat for the proliferation and metabolism of numerous microbial types. Radio-labeled $^{14}CO_2$ photosynthate studies showed that up to 11% of the dry matter production by cereals can be exuded through the roots under aspetic conditions, but this could increase to 18% in the presence of microorganisms (9). The root exudate material consisted usually of about 90% carbohydrates and 10% amino acids (8). Most of the root exudates are easily biodegradable under aerobic or anaerobic conditions.

6.2.3. Added substrates

The carbon substrate for saprophytes in forest ecosystems is leaf litter and the counterpart in agriculture is crop residues. In most fresh water and estuarine environments the greatest input of particulate matter originates from macrophyte tissue (37). Farther offshore, dead phytoplankton cells play an increasing role. The organic matter persists in the environment because it is continually renewed through photosynthesis. Other sources of substrate C to the terrestrial or aquatic systems include sewage effluents, industrial organics, and animal manures.

Very often decomposition of the substrate C is initiated before it enters the soil by a predominantly fungal flora associated with living or senescent plant surfaces. In the soil, leaching of soluble low molecular weight compounds takes place which are used by the primary saprophytic "sugar fungi" i.e. mainly yeasts (70). The further succession of the microflora depends on the environmental conditions. The secondary colonizers use the more complex material, such as polysaccharides where tertiary colonizers are capable of metabolizing some of the complex polymers such as lignin, but as these are not readily assimilable energy sources their growth is quite slow (4).

Simple substrates that are added to soil are readily metabolized but almost always with an apparent lag period prior to the maximal oxidation rate. It frequently results in a flush of microbial growth and biosynthesis of new compounds (130). The chemical and physical characteristics of the added substrates dictates the rate of decomposition in soil. The effect of added substrates on the kinetics of decomposition is discussed in the latter part of the paper.

6.3. Role of inorganic redox couples on microbial respiration

The metabolism of all living cells is an open system which is characterized by a continuous input and output of matter and energy. Each cell is endowed with a system that transforms the chemical and physical energy taken up into biologically useful energy (ATP) and utilizes the latter to perform work (146). It also should be noticed that energy utilization does not occur with 100% efficiency. Yoshida (167) reviewed the microbial metabolic activities functioning in flooded rice soils.

A great variety of organic compounds can serve as hydrogen or electron donors for microbial respiration. Depending on the source, approximately 10%(22) to 20%(130) of the organic matter in soil occurs as carbohydrate and most of that is as polysaccharide. In the following discussion, glucose will serve as the electron donor in order to assess the role of the different redox systems. The microbial degradation of organic matter involves the consumption of a certain amount of electron acceptors either in the case of respiratory metabolism or indirectly in the case of fermentative metabolism. Oxidants used in microbial metabolism are O_2, NO_3^-, manganic oxides, ferric oxyhydroxides, SO_4^{2-}, and CO_2. The reduction of the inorganic redox systems can be described in both intensity and capacity terms. The intensity factor determines the relative ease of the reduction, whereas the capacity factor denotes the amount of the redox system undergoing reduction. The capacity factor of a redox system probably can be best described in terms of its O_2 equivalent (Fig. 6.2., Table 6.3.). The intensity factor can be represented by the free energy of the reduction, or more commonly by the equivalent electromotive force (EMF) of the reactions. For a cell in which n electrons are lost from a reductant in one-half cell per mole of reductant oxidized, and are received by one mole of an oxidant in the second half cell:

$$-\Delta Gr = nF \, \Delta \, Eh \qquad [1]$$

ΔGr = change in Gibbs free energy for the reaction in J mol^{-1}
  n = number of electrons transferred between the two couples per mole of oxidant and reductant involved
  F = Faraday constant = 23.061 kcal $volt^{-1}$ equivalent $^{-1}$
ΔEh= EMF or voltage difference of both half cells

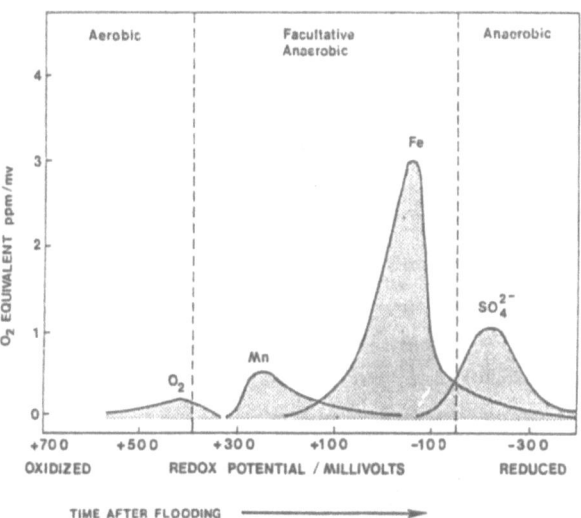

Fig. 6.2. The capacity of different redox systems to accept electrons decomposing organic matter on the basis of oxygen equivalents.

In natural systems, the oxidation-reduction or redox potential is ordinarily used to denote the intensity of reduction. The concept and measurement of redox potential in natural environments is well established. This is mainly due to the fact that an internal thermodynamic equilibrium is reasonably approached within the subsystem formed by the main inorganic redox couple involved in the energy yielding metabolism of the microorganisms (7, 140, 157). Microorganisms oxidize inorganic and organic substrate in order to obtain energy for growth and maintenance. The organic matter which is metabolized by heterotrophic organisms is consumed for energy generation, which in turn is used to convert another portion of the organic matter into cells. The relative proportions can be assessed by a mass balance for the system as described by McCarty (74) and Sawyer and McCarty (121).

$$f_s R_c + f_e R_e - R_d = R \tag{2}$$

where R is the overall reaction, R_c represents the half reaction for synthesis of bacterial cells, R_e represents the half reaction for the electron acceptor and R_d for the electron donor. The values f_s and f_e represent the portion of the electron donor used for synthesis and energy, respectively, and the sum of the two must equal 1.0 (Table 6.2.).

6.3.1. Aerobic respiration

Oxygen is the terminal electron acceptor in aerobic systems and is reduced while organic electron donors are being oxidized (Fig. 6.3a,b,c.). This reduction of O_2 to H_2O is carried out by true aerobic microorganisms and CO_2 is evolved as a waste product. Therefore, a supply of oxidizable organic compounds, as well as a supply of O_2 and some means of removing CO_2 produced are indispensable for aerobic respiration to occur. Oxygen availability can be expressed either by its concentration or its rate of supply (flux). Soil redox potentials can also help to identify changes of O_2 availability because redox potential is closely linked to the presence or absence of O_2.

Aerobic soils have values of Eh between 0.3 and 0.8 V and usually between 0.4 and 0.6 V, and the CO_2 evolved during aerobic respiration diffuses relatively quickly when the air filled porosity is large. The glycolysis converts glucose to pyruvate with the production of two ATP molecules. The oxidative decarboxylation of pyruvate to form acetyl CoA is the link between the glycolysis and citric acid cycle. The citric acid cycle operates only under aerobic conditions because it requires a supply of NAD_3 and FAD. These electron acceptors are regenerated

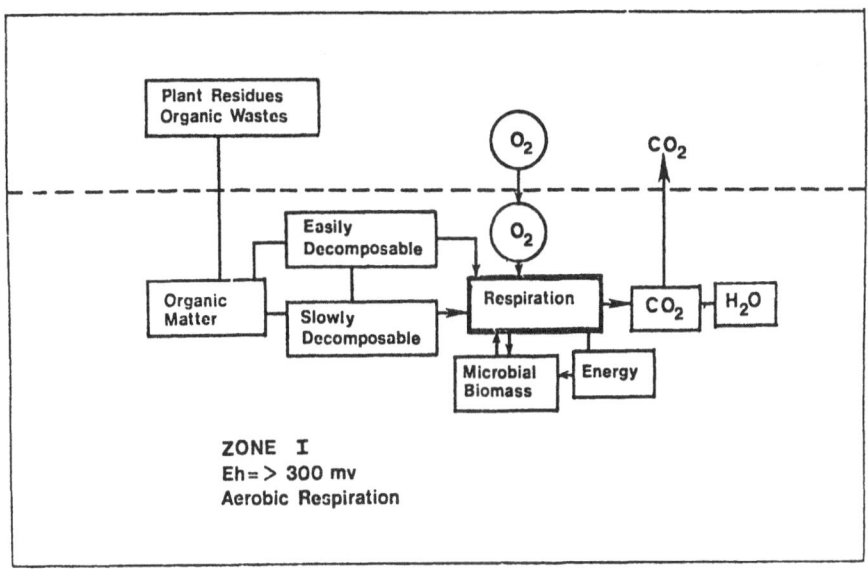

Fig. 6.3a. Pathways of organic matter decomposition during aerobic respiration.

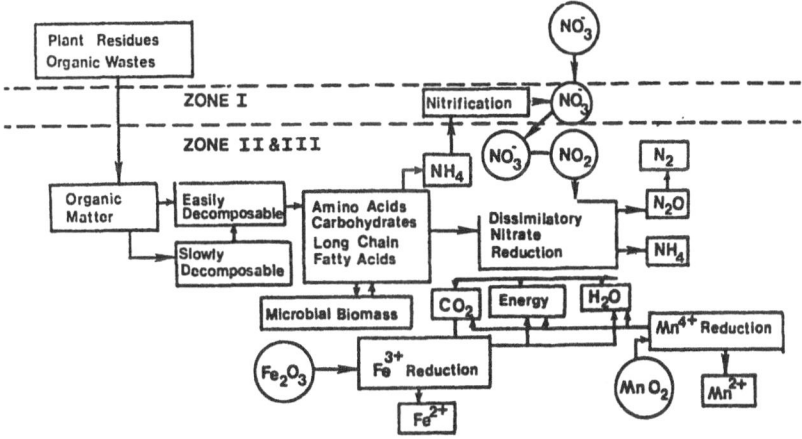

Fig. 6.3b. Pathways of organic matter decomposition during facultative anaerobic respiration.

124

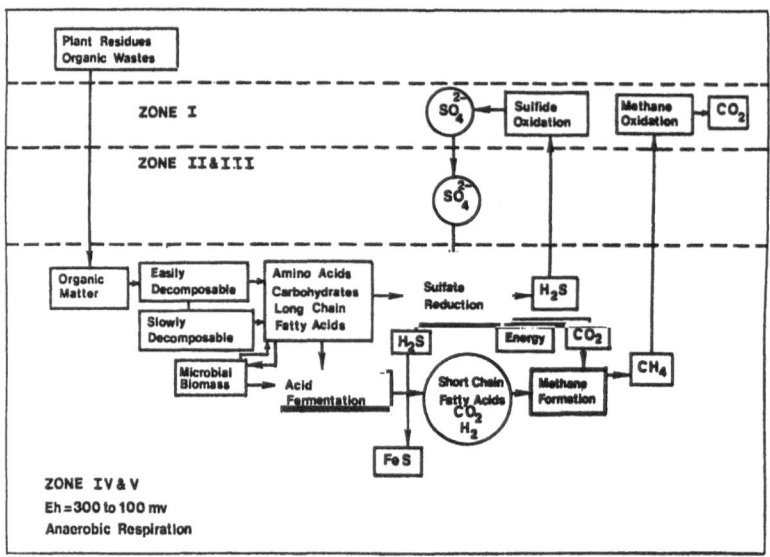

Fig. 6.3c. Pathways of organic matter decomposition during anaerobic respiration.

when NADH and $FADH_2$ transfer their electrons to O_2 through the electron-transport chain, with the concomitant production of ATP. The overall breakdown of glucose with different electron acceptors is shown in Table 6.1.

Table 6.1. Glucose respiration with different electron-acceptors.

1. $C_6H_{12}O_6 + 6\,O_2 \rightarrow 6\,CO_2 + 6\,H_2O$	$\Delta G_r^\circ 1 = -686.4$
2. $5\,C_6H_{12}O_6 + 24\,NO_3^- + 24\,H^+ \rightarrow 30\,CO_2 + 12\,N_2 + 42\,H_2O$	$\Delta G_r^\circ 1 = -649.0$
3. $C_6H_{12}O_6 + 12\,MnO_2 + 24\,H^+ \rightarrow 6\,CO_2 + 12\,Mn^{2+} + 18\,H_2O$	$\Delta G_r^\circ 1 = -457.8$
4. $C_6H_{12}O_6 + 24\,Fe(OH)_3 + 48\,H^+ \rightarrow 6\,CO_2 + 24\,Fe^{2+} + 66\,H_2O$	$\Delta G_r^\circ 1 = -100.4$
5. $C_6H_{12}O_6 + 3\,SO_4^{2-} \rightarrow 6\,CO_2 + 3\,S^{2-} + 6\,H_2O$	$\Delta G_r^\circ 1 = -91.0$

$\Delta G_r^\circ 1 = $ change is standard Gibbs free energy units: kcal/mol. Based on ΔG_f published by U.S. National Bureau of Standards.

Table 6.2. Synthesis (fs) and Energy (fe) factors for biological reactions.

Electron donor	Electron acceptor	fs (max)	fe (min)
Carbohydrate	O_2	0.72	0.28
Carbohydrate	NO_3^-	0.60	0.40
Carbohydrate	MnO_2	[0.60 - 0.30]	[0.40 - 0.70]
Carbohydrate	$Fe(OH)_3$		
Carbohydrate	SO_4^{2-}	0.30	0.70
Carbohydrate	CO_2	0.28	0.72

Oxygen is preferentially used as electron acceptor because of its high affinity for electrons in comparison with other electron acceptors or its high intensity of reduction (Table 6.1.). This is coupled with a greater energy yield, which in general results in higher values for f_s (max) (Table 6.2.).

6.3.2. Facultative anaerobic respiration

When the supply of O_2 to the soil is cut off, the obligate aerobes can no longer function and so either die or go into a resting stage. The microbial community shifts to facultative anaerobic bacteria and when anoxic conditions persist obligate anaerobes also proliferate, while the fungal population decreases (Fig. 6.3b).

As O_2 is depleted, NO_3^- will be used as electron acceptor followed by oxidized manganese compounds and then followed by ferric iron compounds. The order of these reductions is the same as that indicated by thermodynamic considerations (Table 6.1.).

6.3.3. Nitrate respiration

Many microorganisms can utilize NO_3^- as terminal hydrogen acceptor instead of O_2, a process called NO_- respiration or dissimilatory NO_3^- reduction. This process primarily occurs in the absence of O_2. The general pathway of this reaction is shown below.

$$2NO_3 \rightarrow 2NO_2 \rightarrow 2[HNO] \rightarrow H_2N_2O_2 \rightarrow NO \rightarrow N_2O \rightarrow N_2$$
$$\downarrow \leftarrow \qquad\qquad \text{denitrification} \qquad\qquad \rightarrow$$

$2NH_2OH$

$\rightarrow$ Dissimilatory reduction

$\downarrow$ of NO_3 to NH_4

$2NH_4$

Dissimilatory NO_3^- reduction to gaseous end products is commonly known as denitrification. Detailed reviews on this process have been reported (32, 38, 62, 99, 100, 138). The intermediate N-oxides of this process can also potentially be used as electron acceptors. Nitrate reduction through denitrification involves 5 e^- compared to the acceptance of 8 e^- during the reduction of NO_3^- to NH_4^+. Several researchers (60, 64, 129) have shown that dissimilatory reduction of NO_3^- to NH_4^+ can potentially occur in systems which are anaerobic for long periods. Some examples include sediments, anaerobic digestors, and rice paddy soils. In soils where temporary anaerobic conditions exist and soils that are less intensively reduced, denitrification can be the major pathway of removing NO_3^-. In both the pathways, NO_3^- is primarily used as the terminal electron acceptor during the oxidation of organic matter.

Nitrate respiration to N_2 occurs at Eh values of 200 to 300 mv (90) while NO_3^- respiration to NH_4^+ occurs at Eh values of less than -100 mv (20). All denitrifiers can grow with O_2 and no obligate denitrifier has been discovered so far. However, obligate anaerobes such as *Clostridium* can be responsible for dissimilatory NO_3^- reduction to NH_4^+ (24). Significant correlations were observed between soil organic matter decomposition (expressed as CO_2 production) and NO_3^- respiration to N_2 indicating the use of NO_3^- as electron acceptor (114). No data are available on the significance of dissimilatory NO_3^- reduction to NH_4^+ on organic matter decomposition. The energy yield during NO_3^- reduction is slightly less than for O_2 as terminal electron acceptor (Table 6.3.) and in order to release

Table 6.3. O_2 equivalents and relative energy yields

Electron acceptor	Moles oxidant per mole glucose	Energy yield kcal/mole	Energy equivalent on mole basis	O_2 equivalent on mole basis
O_2	6	-684.4	1.000	1.0
NO_3^-	24/5	-649.0	0.948	0.8
MnO_2	12	-457.8	0.669	2.0
$Fe(OH)_3$	24	-100.4	0.147	4.0
SO_4^{2-}	3	- 91.0	0.133	0.5

Table 6.4. Energy equivalents of different electron acceptors

Electron acceptor	Energy equivalent on mole basis	O_2 equivalent ppm/ppm O_2	Energy equivalent ppm/ppm O_2
O_2	1.000	1.00	1.000
NO_3^-	0.948	1.55	1.635
MnO_2	0.669	5.43	8.116
$Fe(OH)_3$	0.147	13.36	90.884
SO_4^{2-}	0.133	1.50	11.278

the same amount of energy 1.635 (1.55/0.948) times the amount of O_2 would be needed using NO_3^- (Table 6.4.). Since this energy availability is quite large, it is not surprising to discover that there are a number of denitrifying bacteria capable of exploiting the use of NO_3^- as terminal electron acceptor.

Microbial growth during NO_3^- respiration is less efficient than O_2 respiration, primarily because of low energy yield. Oxygen respiration was found to be about 1.7 times more efficient than NO_3 respiration in studies with *Pseudomonas denitrificans* and *P. stutzeri* (34, 63). However, much lower energy yield (2.6 to 2.9 times less than O_2) was reported by Bryan (18) for *P. aeroginosa* and *P. stutzeri*. However, the relative efficiencies of NO_3^- and O_2 respiration in a soil system with mixed microbial population have not been documented.

The role of NO_3^- in organic matter oxidation in O_2 deficient soils depends on the concentration of NO_3^- in the soil, in addition to the soil and environmental factors required for NO_3^- respiration to function at an optimum rate. In mineral soils, NO_3^- concentration usually does not exceed 100 mg N ℓ^{-1}, and during short-term flooding, NO_3^- not subjected to leaching will be involved as an electron acceptor in organic matter oxidation. However, in permanently waterlogged soils, NO3- can also be derived from 1) external sources, such as wastewaters, drainage effluents rich in NO_3^-, and 2) internal sources, *i.e.* NH_4^+ diffusion from anaerobic soil layer to overlying aerobic soil layer and floodwater, and subsequent oxidation of NH_4^+ to NO_3^- and NO_3^- reduction in the O_2 free zone (95). Under these conditions, NO_3^- respiration can play a significant role in microbial oxidation of soil organic matter.

6.3.4. Manganese respiration

As the demand for electron acceptors increases, facultative anaerobes can utilize manganic compounds as electron acceptors during their metabolic activity while using either endogenous or externally supplied substrates as electron donors (141, 147, 150).

$$Mn\, O_2 + 2e + 2H^+ \rightleftharpoons Mn\, (OH)_2$$
$$\underline{Mn(OH)_2 + 2H^+ \rightleftharpoons Mn^{2+}\, 2H_2O}$$
$$MnO_2 + 2e + 4H^+ \rightleftharpoons Mn^{2+} + 2H_2O$$

A wide range of heterotrophic anaerobes were found utilizing MnO_2 as terminal electron acceptor (147, 148). The oxidized Mn compound buffers Eh values in the range of +200 to +300 mv, until all of the bioreducible Mn has been converted to the Mn^{2+} form. Manganese reduction occurs at relatively high redox potentials (Eh = 200 mv), and their energy yield is about 67% relative to the O_2 (Table 6.4). Studies have shown that Mn can be reduced in the presence of low levels of NO_3^- (141). The extent of organic matter decomposition by the facultative anaerobic bacteria utilizing Mn^{2+} as an electron acceptor depends on 1) reducible concentration Mn^{4+} compounds, 2) amount of easily decomposable substrate, and 3) the relative ratios of both and the presence of other factors favorable for microbial activity such as temperature and absence of substances that inhibit microbial activity.

Although the bioreducible Mn content is usually lower than Fe content of many soils, the capacity of Mn system to accept electrons per mole of oxidant reduced is higher than Fe. For example, during Mn^{4+} reduction 2 electrons are accepted as compared to one electron during the reduction of Fe.

6.3.5. Iron respiration

Two pathways of Fe reduction have been proposed: 1) direct biological reduction of Fe^{3+} by iron-reducing bacteria containing two electron donating system: A NO_3^- reductase system and a ferrireductase system in cytoplasmic membrane (45) and 2) indirect microbial reduction of Fe^{3+} by certain substances released from heterotrophic bacteria (117, 133). However, direct biological reduction is probably the most significant process in Fe^{3+} reduction.

Enzymatic Fe^{3+} reduction by facultative anaerobic bacteria occurs during the metabolic activity, where Fe^{3+} is used as an electron acceptor and organic matter is used as an electron donor. The net result of this process is synthesis of cell biomass, ATP formation and end products such as CO_2 and Fe^{2+}. Kamura and Takai (59), using mixtures of antibiotics added to preincubated soil, demonstrated that Fe^{2+} reduction is a microbially mediated enzymatic process. In the presence of

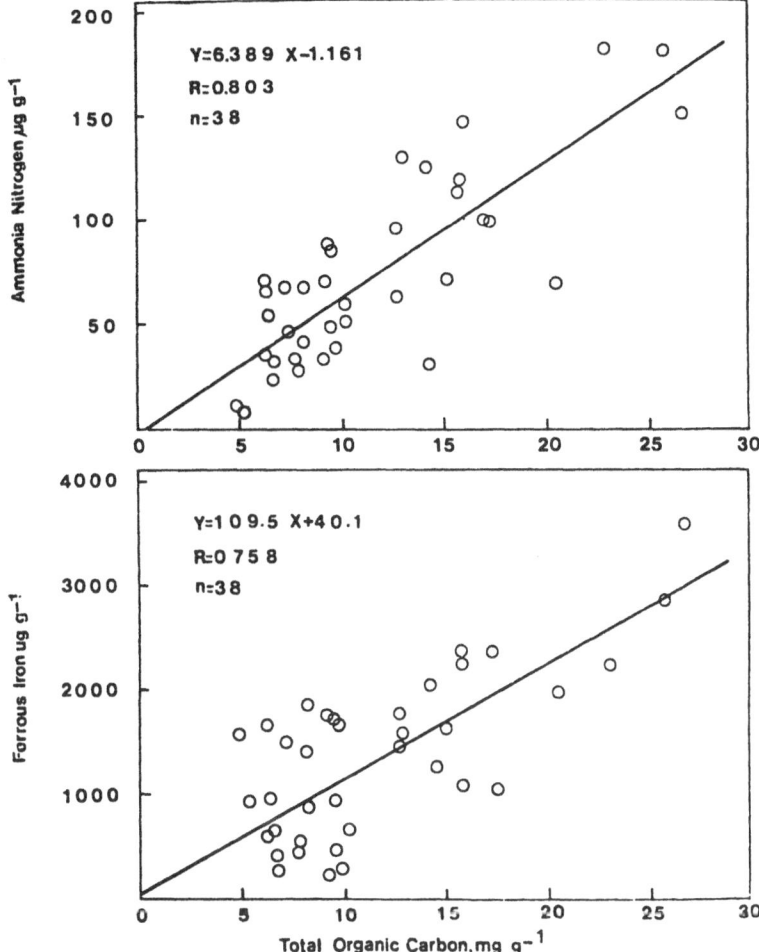

Fig. 6.4. Relationship between total organic carbon and ferrous iron and ammonium N accumulation under anaerobic conditions (115).

H^+ donors, organisms such as *Bacillus circulans, Bacillus megaterium,* and *Aerobacter aerogenes* can reduce ferric compounds to Fe^{2+}. Bromfield (17) showed that $Fe(OH)_3$ was reduced when the chelator, 2.2'-dipyridyl, was present with bacteria and a suitable energy source. Asami and Takai (6) showed a direct relationship between Fe^{2+} formation and CO_2 production in an anaerobic soil amended with Fe_2O_3, thus indicating the enzymatic reduction of Fe^{3+} during microbial respiration. Iron reduction was found to be directly proportional to the organic matter content of 38 mineral soils of Louisiana (Fig. 6.4.). The higher the concentrations of easily decomposable organic carbon, and lower the amount of NO_3^- and Mn^{4+}-compounds, the more intensive is the reduction of Fe^{3+}. However, the capacity of the system to utilize Fe as an electron acceptor during microbial metabolism also depends on the concentration of easily reducible Fe compounds. Most soils contain a wide variety of Fe^{3+}-oxides ranging from completely amorphous to highly crystalline (goethite, hematite). Crystalline Fe^{3+} oxides are more stable and exothermous than their noncrystalline counterparts (85). Ottow (85) has experimentally demonstrated the preferential reduction of amorphous Fe oxides over crystalline Fe-compounds. This mechanism was postulated by Ottow (85) as the enzymes (reductases) in electron transfer are specific enough to select between the amorphous and crystalline Fe compounds, probably by the difference in the standard free energy formation.

The Fe^{3+} compounds buffer the Eh of the system in the range of -100 to +100 mv. The buffering effect of Fe is essentially depleted at about -150 mv, and it is in this range that facultative anaerobes cease to function (91). For many mineral soils, the capacity of the Fe system is usually much greater than the Mn system. This is experimentally demonstrated for Crowley silt loam soil by Patrick (92). These results show the relatively high redox buffering capacity of the Fe systems and explain why soils with large amounts of bioreducible Fe do not undergo rapid decreases in Eh. For example, the ferric oxyhydroxide reduction occurs at the Eh values shown above and its energy yield is only about 15% as compared to the ideal electron acceptor (O_2). This indicates the low efficiency of Fe respiration as compared to the aerobic (O_2) respiration.

6.3.6. Anaerobic respiration

Anaerobic respiration is defined as the metabolic activity of true anaerobes which utilize organic matter as energy source, and SO_4^{2-} and CO_2 as electron acceptors (Fig. 6.3c). Anaerobic respiration is predominant in soils which are under continuous flooding and coastal areas that receive sea water containing large amounts of SO_4^{2-}.

6.3.7. Sulfate respiration

In anaerobic soils and sediments, two genera of strict anaerobes, i.e., Desulfovibrio and Desulfotomaculum are capable of utilizing SO_4^{2-} as terminal electron acceptor for catabolic oxidation processes (105). This process involves the oxidation of substrate C (electron donor) and transfer of electrons to SO_4^{2-} (electron acceptor), a process termed as dissimilatory SO_4^{2-} respiration. This mechanism results in a net release of free energy which is utilized for microbial cell synthesis. The metabolism of sulfur in anaerobic systems has been the subject of recent reviews by Krouse and McCready (67), Thauer and Badziong (145), and Trudinger (149). Only a small amount of reduced sulfur is assimilated by the organism and virtually all is released into the external environment as sulfide. The oxidation of organic matter during SO_4^{2-} reduction does not proceed beyond the level of acetate (153).

$$2 \text{ lactate} + SO_4^{2-} \rightarrow 2 \text{ acetate} + 2 H_2O + 2 CO_2 + S_2.$$
$$\Delta G°_f = \text{-88 kJ/mole}$$

However, recently Widdel and Pfennig (158) have isolated a *Desulfotomaculum* species which can grow on acetate as C source and SO_4^{2-} as an electron acceptor.

$$\text{Acetate} + 3H^+ + SO_4^{2-} \rightarrow H_2S + 2CO_2 + 2H_2O$$
$$\Delta G°_f = \text{-63 kJ/mole}$$

The catabolic pathway of SO_4^{2-} respiration is given as follows (145):

$$\downarrow 2e^- \qquad\qquad \downarrow 6e\text{-}$$

$$SO_4^{2-} \longrightarrow APS \longrightarrow SO_3^{2-} \longrightarrow H2S$$
$$\quad\; 2ATP \qquad\qquad\qquad\qquad 3ATP$$

Sulfate is a unique terminal electron acceptor in the way that it has to be activated with 2 moles of ATP and is converted to adenosine phosphosulfate (APS) and a total of 3 moles of ATP are generated during SO_3^{2-} reduction to H_2S. So in order to have a net energy yield, the organism invests some energy before the mechanism becomes functionable. The ATP economy balances exactly as one molecule of ATP is generated from lactate at the substrate level and one is consumed at the sulfate reduction level. During the process of SO_4^{2-} reduction to H_2S, there is a net gain in energy, since the process is coupled with the oxidation of organic matter. It follows that the organism could not possibly grow by these reactions unless they had another mechanism of phosphorylation. Yet they do grow and the respiratory chain is the only known alternative system for ATP generation.

The SO_4^{2-} reduction occurs when the redox potential drops below -100 mv and the equilibrium Eh at 15°C and pH 7 is around -320 mv. In coastal areas that receive sea water containing large amounts of SO_4^{2-} the reduction of SO_4^{2-} provides a few species of true anaerobes with a respiratory system that can support considerable oxidative activity. Sulfate has about the same O_2 equivalent as NO_3^- (Table 6.4.) and if the energy yields are compared on a ppm basis it can be noted that about 11 times more SO_4^{2-} is needed to release the same amount of energy in the system than when the organic matter is oxidized with O_2. The difference between anaerobic and aerobic yield is stored in their reaction products - reduced inorganic sulfur compounds and partially metabolized C-substrates. Howarth (47) and Howarth and Teal (48) reported that in a salt marsh, SO_4^{2-} reduction degraded about 12 times more organic matter than O_2 respiration and denitrification combined. Under very reducing conditions, the anaerobic microflora will tend to be replaced by fermentative anaerobes where organic molecules are reduced in order to allow the glucose to be oxidized.

In nature SO_4^{2-} reduction occurs in soils and sediments which have adequate amounts of easily decomposable organic matter and a supply of SO_4^{2-} ions and anoxic conditions free of O_2, NO_3^-, Mn^{4+}, and Fe^{3+}.

6.3.7.1 Fermentation

The exact biocenosis responsible for the liquefaction and fermentation in nature, or in technical CH_4 fermentations, is still unknown. The three groups of bacteria (i.e. hydrolytic, acetogenic, and methanogenic) form a biocenosis within the same ecological niche to assert an environment equilibrated to a neutral pH (167). Reviews on the ecology, microbiology, and biochemistry of CH_4 formation has been published recently by Mah et al. (71), Wolfe (165), and Zeikus (168). The aspects of energy conservation were covered by Thauer et al. (146), and methane cycling in aquatic environments was discussed by Rudd and Taylor (118).

The strict anaerobes which obtain their energy from coupling the oxidation of a limited range of substrates to the reduction of CO_2 to CH_4 are called methanogenic bacteria. The major in situ substrates are acetate and H_2 - CO_2 (168), although formate and methanol have also been shown as possible substrates for methanogenesis. The relative importance of acetate and H_2 - CO_2 as CH_4 precursors in organotrophic ecosystems depends largely on environmental conditions. Lake sediment studies by Cappenberg and Prins (23) and by Strayer and Tiedje (139) showed that approximately 70% of the CH_4 originates from the methyl position of acetate. Belyaev et al. (12) reported that 32 to 98% of the CH_4 produced was formed from CO_2 by microbial reduction with H^+, depending on the specific lake studied. Winfrey et al. (164) demonstrated that CO_2 reduction by H^+ was impor-

tant in sediment methanogenesis; however, CH_4 production was limited by the a-
vailability of H^+. In marine sediments of the Santa Barbara Basin, the major CH_4
precursor would be CO_2 (29). *In vitro* experiments showed that only 2-11% was
derived from acetate. This supports the hypothesis of Claypool and Kaplan (29)
whose isotopic data ($\delta^{13}C$) indicated that CH_4 was derived chiefly from CO_3
in marine environments.

In the anaerobic fermentation of glucose to CH_4 metastable intermediates are
formed. The subsequent four reactions in which CH_4 is formed are mediated by
individual different bacteria species. The overall reaction yields more energy than
any combination of reactions in the series; however, no organism has been found
which can mediate it. This fact has in no way precluded the occurrence of the less
favorable reactions nor has it prohibited the production of metastable intermediates (146, 168). ATP is generated by ETP for CH_4 bacteria using H_2 gas as e^-
donor and the chemiosmotic hypothesis is thought to be the mechanism of the
ETP (77, 78). Methanogenesis in marine sediments does not occur until SO_4^{2-} is
depleted and it has been speculated that the inhibition of methanogenesis by
SO_4^{2-} was due to a competition for available H^+ or related to the relative free energy yields of CO_3^{2-} and SO_4^{2-} reduction (29, 73). Winfrey and Zeikus (163) indicated that the carbon and electron flow are altered when SO_4^{2-} is added to sediments and that the competition for both acetate and hydrogen appears to be responsible for inhibition of methane production by SO_4^{2-}.

6.4. Kinetics of microbial organic matter oxidation

6.4.1. Rate of reaction

An evaluation of the kinetics of the substrate organic C decomposition in soils
with varying redox conditions is important in order to understand the biogeochemical cycles and to formulate mathematical models to predict the organic matter
decomposition. It is generally accepted that decomposition of substrate organic C
compounds can be described using first order kinetics with respect to either total
organic matter present in the system or biodegradable part of the organic matter
(15, 40, 50, 54, 116, 119, 124, 131).

$$R = -\frac{dt}{dC} = kC \qquad [3]$$

where k is the decomposition rate constant, time^{-1}, and C are the amount of organic matter present in the system. Integration of the above equation gives:

$$C = C_0 \cdot \exp(-kt) \qquad [4]$$

where C_0 = the concentration of C at t = o

As the organic matter consists of different chemical compounds such as soluble sugars, cellulose, hemicellulose, and lignin with varying degrees of biodegradability, Hagin and Amberger (44), Jorgensen (57), and Berner (13) proposed first order kinetics for each of the carbon compounds.

$$R = \in_i k_i C_i \qquad [5]$$

where C_i refers to a particular class of organic compound, k_i is the kinetic rate constant of decomposition of organic compounds during bacterial respiration.

The total organic matter present in the soil or sediment is:

$$C = \in_i C_i \qquad [6]$$

A comparison of the relative rate of aerobic decomposition of soil organic matter fractions is shown in Table 6.5. Similarly, first-order kinetics is also widely used to describe the biological oxygen demand (BOD) of the wastewater effluents.

Using the first order kinetic approach, Gilmour et al. (40), Hunt (50), and Reddy et al. (113) described substrate organic C decomposition in two to three phases. The soluble and easily degradable C compounds were assumed to be decomposed during initial phases of decomposition, while more complex C compounds decomposed at slower rates. However, no mention was made to any specific organic compounds. First order rate constants for rice straw decomposition under aerobic conditions were found to be 0.0054 day^{-1} for Phase I and 0.0013 day^{-1} for Phase II, while under anaerobic conditions, these constants were 0.0024 day^{-1} for Phase I and 0.0003 day^{-1} for Phase II (113). Easily decomposable frac-

Table 6.5. First order rate constants for aerobic decomposition of soil organic matter fractions (Hagin and Amberger (44).

Organic matter fraction	First order rate constant	Half-life
	---k day^{-1}---	---days---
Sugars	1.1500	0.6
Hemicellulose	0.1035	6.7
Cellulose	0.0495	14.0
Lignin	0.0019	364.7

tion was found to be significantly correlated with C/N ratio of the substrate organic C (50, 113).

Rate of carbon decomposition was also expressed by an exponential relationship (16, 28, 30, 49, 86) in the form of:

$$C = Kt^m \qquad [7]$$

where C is the carbon loss at time t, and k and m are constants. Pal and Broadbent (86) calculated turn over times (time required for complete loss of added substrate C) using the above equation. These values range from 0.8 to 3.4 years. Jenkinson and Rayner (54) calculated the half lives of several forms of C compounds using the data obtained from 1) the long term Rothamsted experiments of 10-100 years; 2) ^{14}C-labeled plant residue decomposition experiments conducted for a period of 1-10 years; and 3) ratio carbon dating. The calculated half lives were 0.165 years for easily decomposable plant material, 2.31 years for resistant plant material, 1.69 years for soil biomass, 49.5 years for physically stabilized organic matter, and 1980 years for chemically stabilized organic matter. The decomposition rate of each fraction was assumed to be proportional to its actual content.

During microbial degradation of C, C-compounds enter CO_2, cell biosynthesis, and to some extent low molecular weight metabolites (97). These researchers suggested that rate constants measured using CO_2 evolution or C remaining in the soil should be corrected for the microbial biosynthesis. The efficiency for micro organisms under aerobic conditions was considered to be between 40 and 60% (98). A critical mathematical review of organic matter decomposition and nutrient cycling is presented by Smith (126).

Rate of decomposition is influenced by a number of soil and environmental factors. Detailed discussion is presented in the later part of this paper. Microbial metabolic efficiency decreases significantly when soil redox conditions change from aerobic to anaerobic. In the absence of O_2, the rate of organic matter decomposition depends on the capacity of the system to supply electron acceptors. We propose the use of a first order kinetic rate equation with respect to concentration of substrate C and electron acceptor (the oxidant).

$$\frac{dc}{dt} = -k \, [C]_{e_d^-} \cdot [O]_{e_a^-} \qquad [8]$$

where $[C]_{e^-}$ is the amount of organic matter in the system, $[O]_{e_a^-}$ = electron acceptor as O_2 equivalent, e_d^- = electron donor, e_a^- = electron acceptor, and k is the rate constant measured with respect to C and oxidant concentration. Very limited data are available to test the validity of equation [6].

6.4.2. *Soil and environmental factors*

Rate of microbial oxidation of soil organic matter and consumption of oxidants is influenced by 1) soil moisture, 2) oxidant supply and type of oxidants present, 3) pH, 4) temperature, 5) available plant nutrients, 6) microflora, 7) soil texture and structure, and 8) An evaluation of other factors that might inhibit microbial activity. An evaluation of these factors is important in order to understand the biogeochemical cycles functioning in soils and sediments, and will aid in developing mathematical models to predict the fate of organic matter. A brief discussion of some of the most important factors influencing the metabolic activity of soil microorganisms is presented.

6.4.2.1. *Soil moisture*

Increase in soil-water content decreases the fraction of soil pores filled with air. The tolerance of soil microorganisms to varying soil-water tension widely varies with the species. Bacteria are more sensitive to low water content than fungi and actinomycetes (33, 65). Decreasing soil water potential results in a decrease in the amount of CO_2 evolved during organic matter decomposition. Optimum soil-water potential to achieve maximum rate of decomposition was found to be in the range of 0.1 to 0.5 atm (76, 82, 159). As the soil-water content is decreased below optimum range, water rather than O_2 becomes a limiting factor for the heterotrophic microorganisms. In soils with water content higher than optimum range, O_2 diffusion to the reaction site limits the rate of decomposition. Under these conditions, aerobic organisms will function in soil zones where O_2 supply is readily available, while facultative anaerobes utilize NO_3^- and Mn^{4+} -oxides as electron acceptors in the zones where O_2 supply is curtailed. The rate of microbial respiration in soils will be controlled by the ratio of aerobic/anaerobic soil volume. Rate of organic matter decomposition and oxidant supply can also be influenced by the depth of the water table (Fig. 6.5.). Decomposition rates in organic soils were significantly lowered when the water table depth was raised from 25 to 5 cm (151). Pal and Broadbent (87) have shown a maximum decomposition rate for plant residue decomposing at 60% of the water holding capacity, and rates were shown to decrease at either 30 or 150% of the water holding capacity. Utilizing the data presented by Miller and Johnson (76), Reddy *et al.* (113) obtained the following relationship for soils with water content of 0.02 to 0.33 bars:

$$Fm = 1.223 + 0.201 \ln \psi; 0.02 \leqslant \psi \leqslant 0.33 \qquad [9]$$

where Fm = relative rate of decomposition (ranges from 0 to 1) ψ = soil-water potential, bars.

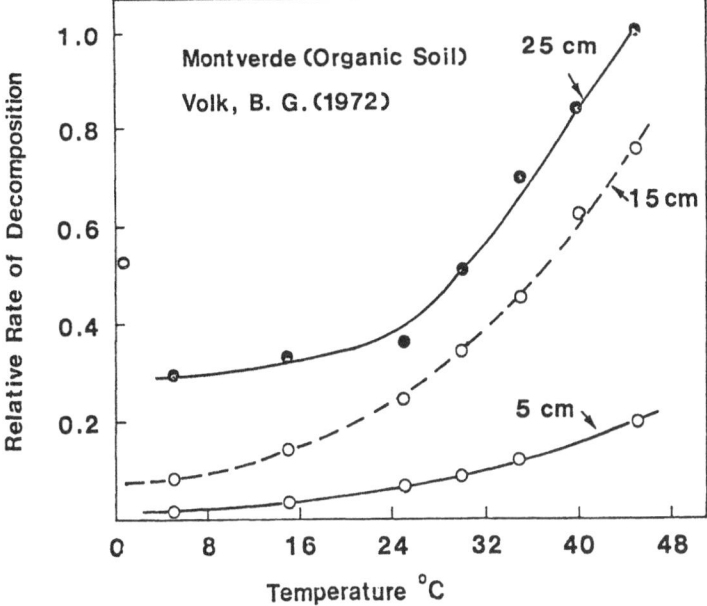

Fig. 6.5. Decomposition of soil organic matter as influenced by temperature and soil moisture (151).

6.4.2.2. Oxidant supply

The concentration and type oxidant available in the soil system determines the types of microorganisms involved in the decomposition process. Percent of the added C decomposed was shown to increase exponentially by increasing the O_2 concentration of the system (89) (Fig. 6.6.). However, no information is available on the effect of added NO_3^-, Fe^{3+}, Mn^{4+} and SO_4^{2-} on the decomposition of added organic matter. Organic matter decomposition was also found to be influenced by the alternate wetting and drying of a soil or creating alternate aerobic and anaerobic environment. Reddy and Patrick (110) found that the extent of decomposition decreased as the length of the anaerobic period increased. After 128-day incubation, decomposition rates were found to be the same when soil was incubated either under completely aerobic conditions (O_2 in the only source of e^- acceptor) or with 2-2 days or 4-4 days aerobic anaerobic conditions (Fig. 6.7.). The most probable e^- acceptors during short-term anaerobic conditions were probably NO_3^- and Mn^{4+}. In a waterlogged soil, oxidant supplied from external sources is utilized by microorganisms in the surface soil layers. The thickness of aerobic zone (O_2 reduction zone) will vary from a few millimeters to 1 to 2 cm. Reduc-

138

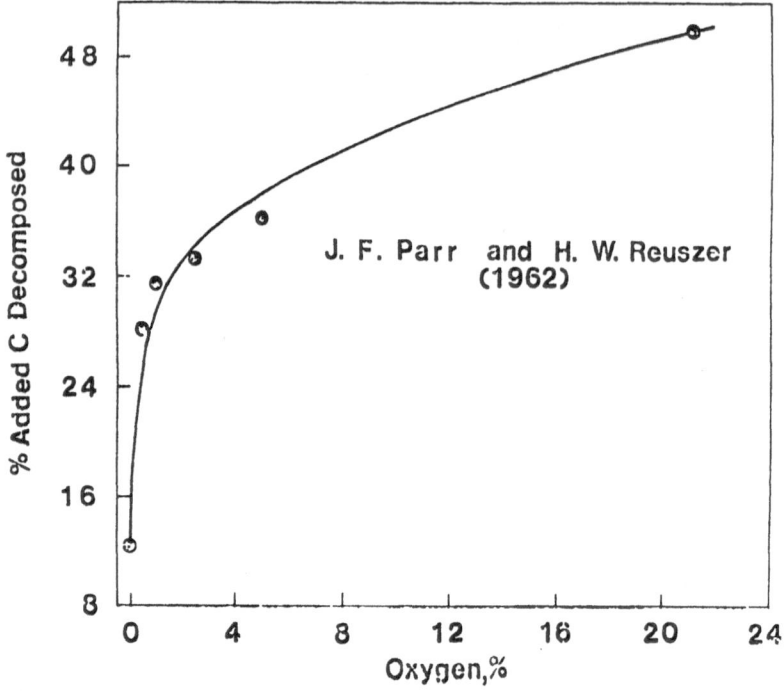

Fig. 6.6. Decomposition of wheat straw carbon in Clermont silt loam, as influenced by the oxygen concentration in the aerative gas (89).

tion of NO_3^-, Mn^{4+}, and Fe^{3+} occurs below this zone, while the reduction of SO_4^{2-} occurs in deeper soil zones (Fig. 6.1.).

6.4.2.3. *Temperature*

Rate of microbial respiration is usually accelerated by increased temperature (Fig. 6.5.). It should be noted that each microbial species and each strain has its own minimum, optimum, and maximum temperature. Optimal temperature for maximum microbial activity may vary depending on the interactive factors occurring in the field (33). Maximum rates were observed at temperatures higher (45-65°C) than those that occur in soils and sediments. A linear relationship was observed between Q_{10} (decomposition rate ratio at an interval of a 10°C rise in temperature) and the reciprocal of absolute temperature below 45°C (82). Several research workers have shown a strong interaction between temperature and soil

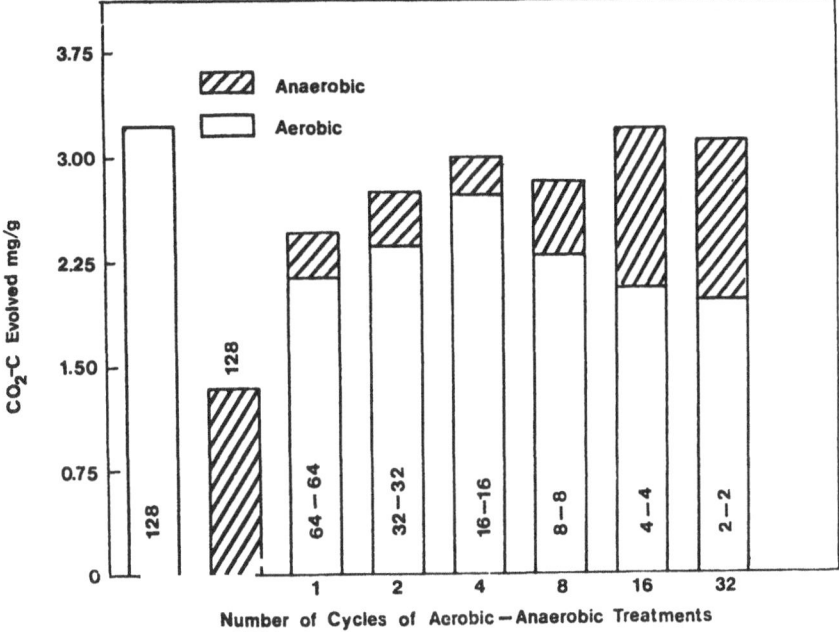

Fig. 6.7. Decomposition of native and added organic matter, as influenced by the frequency of alternate aerobic and anaerobic conditions (110).

moisture on rate of decomposition (11, 27, 82, 151, 159). At low temperature (6 to 12°C), the effect of soil moisture was less significant than at more optimum temperatures (18 to 24°C). Pal *et al.* (88) have shown that prolonged lag phase at low temperature (7.2°C) resulted in negligible amounts of CO_2 evolution during straw decomposition. When incubation temperature was increased to 22°C, decomposition rate was increased by a Q_{10} value of 12 during the first 2 days. Further increase in temperature to 37°C resulted in a Q_{10} value of about 1.5. These data suggest that lag phase was very brief at optimal temperatures. Pal *et al.* (88) also showed that Q_{10} values were highest after 2 day decomposition and decreased sharply after decomposition and remained at a constant value for the remaining 120-day decomposition period.

6.4.3. Substrate factors

Organic matter is added to the soil in various forms, *i.e.* 1) crop residues, 2) organic wastes such as sewage sludge and animal manure, 3) root exudates, 4) dead microbial cell biomass. Large additions of these substrates increases the microbial activity in the soil, thus accelerating the rate of organic matter decomposition, a mechanism commonly known as "priming effect" (52). Pal and Broadbent (86) showed significant priming effects in soils treated with rice straw. These workers quantified this effect by calculating the priming ratios (soil C loss with straw added/soil C loss without added straw). Priming ratios in this study were found to be in the range of 1.25 to 2.06. Priming effect was more striking during the first week of decomposition of added straw, indicating the easily decomposable fraction of the added straw was the main contributing factor for the enhanced decomposition of soil organic matter. Addition of organic substrate can have pronounced effects on the rate of soil organic matter decomposition at any redox condition of the soil. Substrate additions to the soil increases the biological electron acceptor demand on the system. Heavy loadings of organic wastes to the soil can deplete soil O_2, creating favorable conditions for facultative anaerobic or obligate anaerobic respiration.

In a waterlogged soil, addition of easily decomposable C substrates can result in an increased accumulation of organic acids, such as acetic, formic, butyric, lactic, and succinic acids (25, 106, 125). These organic acids function as energy source to certain bacteria and eventually are broken down into either CO_2 or CH_4.

The rate of added substrate C decomposition is also dependent on the chemical composition of the substrate itself. The organic components of substrate organic matter comprises of 1) 5-30% dry wt of water soluble simple sugars, amino acids, aliphatic acids, 2) cellulose 10-50% dry wt, 3) hemicellulose 10-30% dry wt, 4) lignin 5-30% dry wt, 5) fats, waxes, oils, and resins, 1-8% dry wt, 6) proteins, 1-20% dry wt (5). Rate of organic matter breakdown depends on the relative proportion of each of the fraction and the ease of decomposition. Tenny and Waksman (143) established the order of rapidity of decomposition as: the water soluble organics 〉 hemicelluloses 〉 celluloses 〉 lignin. Although relative rates of decomposition of these organic components in aerobic soils have been well documented, decomposition rates in O_2 free environments are limited. However, Tenny and Waksman (144) observed that the three chemical constituents, cellulose, hemicellulose, and lignin, were found to have decomposed more slowly under anaerobic conditions than aerobic conditions. The aerobic degradation of substrates in soil may be reasonably efficient. It has been estimated that during the primary decomposing cycles, the microflora converts between 40 to 60% of the original carbon to microbial tissues (53, 123). Anaerobes, by comparison, may assimilate less than 5% of their carbon supply. The different types of microorganisms considered in this discussion can be classified as heterotrophic organotrophes, which utilize organic molecules as both electron donor and carbon substrate (Table 6.1.).

Decomposition rates were found to be significantly correlated with the initial

C/N ratio or the initial lignin content of the substrate (75). Several researchers have demonstrated that the lower the N content or a wide C/N ratio of the substrate, the slower the rate of decomposition. This is expected because of the dependency of microflora on N. More recently, Herman *et al.* (46) found that under aerobic conditions CO_2 production and percent loss of carbohydrate were inversely related to $[(C:N) (\% \text{ lignin})]$ (carbohydrate $^{-1/2}$). Percent loss of lignin; however, was directly proportional to this factor. Using the data reported by Pinck *et al.* (102). Hunt (50) obtained the following equation to calculate the easily decomposable fraction of plant residues:

$$S_o = 0.070 + 1.11 \sqrt[3]{(N/C);} \quad R^2 = 0.98 \qquad [10]$$

where S_o = initial proportion of easily decomposable C fraction, and N/C = nitrogen to carbon ratio.

Decomposition of detritus plant tissue in aquatic systems and subsequent N release were also found to be dependent on the C/N ratio of the plants (36, 84).

6.5. Influence of aerobic/anaerobic respiration of organic matter on soil biochemical processes

Utilization of a chemical energy source, whether organic or inorganic, leads to the release from the cell of both the reduced electron acceptor and the products of metabolism of the energy source. The release of plant-nutrients will be dictated by the concentration of plant nutrients in the substrate carbonaceous material. This is most important as it influences both the rate of decomposition of the organic material and the amount of nutrients liberated during decomposition.

The role of the electron acceptor can be assessed in such a way that aerobic microbial decomposition is associated with high energy release and the decomposition of substrate and synthesis of cell substances proceed at a rapid rate. The anaerobic decomposition, on the other hand, operates at a much lower energy level and decomposition and cell synthesis occurs at a much slower rate (144).

It is important to stress that organotrophic metabolisms generate a flux of electrons to the subsystem formed by mineral redox couples, while chemolithotrophic metabolism tends to restore this internal equilibrium. Oxidation of the reduced mineral species at the expense of oxidized ones will occur when this becomes thermodynamically feasible.

Not all the energy released during oxidation is captured by decomposing microbes, a considerable amount being dissipated as heat. A large amount of carbon also escapes from the system as CO_2, but the mineralized nutrients are not usually lost and are available again and again for microbial use. Unless the energy supply is renewed therefore, by the evolution of fresh substrates in the form of plant debris, a stage will be reached when inorganic ions accumulate and we can speak of net mineralization. In these circumstances, nutrients will be readily available for plant uptake. Conversely, if readily decomposable substrates are added to the system, there tends to be a surplus of energy, the demands of the microflora for inorganic ions become greater than the mineralization outflow and any source of inorganic ions in the soil will be drawn upon by microorganisms, resulting in a net immobilization (5, 83).

6.5.1. Nitrogen

The major biochemical processes involving N transformations in soils and sediments are mineralization, immobilization, nitrification, denitrification, and N_2 fixation. These processes were described in detail elsewhere (19, 112, 120, 135).

The chemical nature of about half the total organic bound soil N remains obscure but 20-50% is known to occur as bound amino acids and 5-10% as combined amino-sugar N (5, 101, 136). The nitrogenous compounds present in the soil organic fraction persist for long periods in nature which is reflected in the small proportion of the N reservoir which is mineralized each growing season. In differentiating aerobic and anaerobic decomposition, the difference in mineralization rate was even more pronounced. For corn stalks high in water-soluble substances, including reducing sugars and N compounds, 20% of the total material was lost under anaerobic conditions after 27 days, whereas 37% disappeared during the same period of time under aerobic conditions (144). Williams et al. (162) reported that the N-requirement for the decomposition of rice straw in submerged soils was one-third (0.54 vs. 1.5%) the average requirement of N required for aerobic decomposition of plant residues. This was in agreement with earlier studies (1, 2, 3, 125) and with the typical low N-requirement feature of anaerobic bacterial degradation. The greater net release of NH_4^+ during anaerobic decomposition is extremely important for lowland rice culture. Waring and Bremner (155) observed a more rapid release of inganic N under waterlogged than under aerobic conditions in a number of soils. Stanford and Smith (131) showed a significant relationship between total organic carbon and N mineralized under aerobic conditions. Similarly, data presented by Reddy et al. (115) also showed significant relationships between total organic C and NH_4 accumulation under anaerobic conditions (Fig. 6.4.).

During the mineralization process, organic bound N is liberated in the form of NH_4. The transformation of inorganic N-compounds into the organic state (soil biomass) is defined N-immobilization. The nature of the organic substrate or the ratio of readily available energy to readily available N which can be assessed by the C/N ratio will indicate if net mineralization or net immobilization will occur (51). The concept of a critical C/N ratio of about 20/1 was developed for agricultural soils and net mineralization will take place when the C/N ratio is lower than about 20. It is clear that the C/N ratio for anaerobic decomposition will not be as critical and in order for net mineralization to occur a C/N ratio lower than 70 is required. During aerobic decomposition of organic matter, mineralization end product is NO_3^-. The NO_3^- thus formed can undergo several transformations or leached into ground water. Under anaerobic conditions, mineralization end product is NH_4^+, and NH_4^+ thus formed can diffuse upwards into overlying waters and undergo various transformations (112).

Dissimilatory NO_3^- reduction either to gaseous end products such as N_2O and N_2 or to NH_4^+ is influenced by the decomposition of organic matter. Significant correlations were reported between dissimilatory NO_3 reduction and available C as evaluated either by glucose equivalent C (132), by water-soluble C (21, 116), or by mineralizable C (21, 114).

Nitrogen fixation by heterotrophic bacteria is significantly influenced by the addition of substrates with wide C/N ratios (107, 111). These bacteria can utilize a number of carbohydrates, alcohols, organic acids and aromatic compounds as energy sources. Since 8 electrons and 17 moles of ATP are required to fix one mole of N_2, the efficiency will depend on the metabolic pathway of bacteria to utilize the available substrates (43). The largest amounts of N_2 can be fixed by a-erobic-N_2 fixers at reduced O_2 levels (81).

6.5.2. Phosphorus

The main part of the total P in soils is organic and varies from 15 up to 85%. Inositol phosphates, nucleic acids, phospholipids, and related compounds are derived from the surface vegetation, microbial protoplasm and metabolic products making up the humus fraction. Of the inorganic forms, large amounts are locked up in insoluble P minerals such as Ca, Fe, and Al phosphates. The dissolved P is in the form of orthophosphate and in the form of organic phosphate esters in part colloidal. Sommers et al. (128) in an attempt to characterize lake sediments observed positive correlations between major organic P and organic matter. These relationships suggest that organic P transformations are closely tied to the organic matter dynamics in soils and sediments. The solubility and the release of organic P increases with reduction and chelation and is directly proportional to the amount of utilizable organic matter (14).

Microbial mineralization and immobilization reactions are important features in the P cycle. The organic P which cannot be utilized by plants is mineralized by the action of bacteria, fungi, and actinomycetes with the conversion to inorganic P forms. The mineralization and immobilization of P are closely related to the analogous transformations of N and it can be said that P release is most rapid under conditions favoring ammonification (5). Bacteria, algae, and higher plants take up dissolved orthophosphate. Much of the P taken up by plants is provided by the weathering of primary minerals while the microbial oxidation of organic substrates is an important supplementary source of inorganic P.

The reduction of soil by waterlogging and associated changes in pH have been shown to cause marked changes in the amount of plant available P (103) and sometimes in the P sorption properties of soils (61, 94). Depending on the pH, increased available P may be caused by increases in labile P or decreases in buffer capacity, but both of these effects are mainly mediated through changes in the Fe chemistry of the soil. As such the P concentration is subject to the oxidation reduction state of the soil. The changes in the Fe and P chemistry of anaerobic soils will be time dependent and the increase of Fe in the soil solution can come from the dissolution of ferric hydrous oxides and possibly ferric P. The release of occluded and/or absorbed forms will also increase the labile P pool. Strengite undergoes partial dissolution upon flooding and the greatest release of P and Fe occurs under conditions of low oxidation reduction potential in combination with low pH (96).

6.5.3. Potassium, calcium, and magnesium

Change in redox conditions from aerobic to anaerobic increases the concentration of K, Ca, and Mg in the soil solution. The increased soil solution concentration of K, Ca, and Mg in the absence of O_2 is due to the release of Fe^{2+} and Mn^{2+} from the insoluble Fe^{3+} and Mn^{4+} oxide hydrates during facultative anaerobic respiration (103). This process is enhanced by the addition of organic substrates. Despite the potential agricultural importance of K, Ca, and Mg, very little is known on the microbial transformations of these elements in soils and sediments. Active biodegradation coupled with an extensive CO_2 production can lead to the dissolution of Ca and Mg carbonates. In order to correct soil acidity, Ca and Mg are supplied as liming compounds which increases the decay of organic matter and increases the availability of essential plant nutrients.

6.5.4. Sulfur

Since sulfur, in the form of cysteine, methionine, and other organic molecules is an essential component of living matter, all organisms play a role in the biological S cycle. The major reserve of the elements in soils is the organic fraction and in decomposition there is a good resemblance to the transformations affecting N availability. Sulfur can be utilized in its most highly oxidized naturally occurring form, SO_4^{2-}, though reduction is needed for the formation of S containing amino acids and proteins.

Under aerobic conditions, the reductive part of the cycle is the assimilatory SO_4^{2-} reduction of microorganisms and green plants. The reduced organic S compounds which are ultimately liberated during aerobic decomposition of organic substances from plants and animals become reoxidized to SO_4^{2-} by the aerobic chemoautotrophic and micotrophic S bacteria.

The mineralization of the organic S-compounds 20 to 65% in ester SO_4^{2-} and 5 to 35% in amino acids in humus is often quite slow. The mineralization rate increases with increased aeration and temperature and is highly affected by the moisture content of the soil (160). In the absence of O_2, H_2S and odoriferous meraptans accumulate. In aerobic environments a portion of the SO_4^{2-} is used for microbial assimilation while the remainder is released into the environment. The S content of most microorganisms lies between 0.1 and 1.0% of the dry weight which implies that the critical C:S ratio of a carbonaceous material lies around 50:1 up to 400:1 (10, 137).

Under anaerobic conditions, the S cycle is entirely microbial. Sulfate is used as the electron acceptor in the energy-conserving, dissimilatory SO_4^{2-} reducing bacteria. This reduction will occur at redox potentials lower than -200 mv in the absence of O_2 and NO_3^-. Ferric and Mn_4^+ compounds also retared SO_4^{2-} reduction. The S^{2-} arising from this process is either reoxidized to SO_4^{2-} under anaerobic conditions in the light by the phototrophic green or purple S bacteria or the H_2S reaches aerobic regions and is then chemically or biologically oxidized *via* S and thiosulfate to SO_4^{2-}.

All dissimilatory SO_4^{2-} reducing bacteria species (*Desulfovibrio* and *Desulfotomaculum*) oxidize organic substrates incompletely. When either lactate pyruvate or ethanol served as electron donor and carbon source acetate will accumulate as regular end product (104).

6.5.5. *Micronutrients and heavy metals*

The effects of microbial respiration on micronutrients and heavy metals present in the soil is numerous and difficult to assess. There is evidence for direct or indirect biological alterations in the availability, solubility, or oxidation-reduction state of Fe, Mn, Zn, Cu, Al, Mo, Co, Se, As, Hg, Cd, Si, Ni, U, Pb, and various others. It is beyond the scope of this paper to discuss all possible transformations of these elements.

The most evident transformation is the release of inorganic ions upon the decomposition of organic materials. Humic substances, the major organic constituents of soils and waters, have the ability to interact with metal ions, metal oxides, and hydroxides and more complex soil minerals. The formation of metal-organic associates of widely differing chemical and biological stabilities and characteristics is a well-known fact, though relatively little is known about the chemistry of metal-humic interaction products. The ability of inorganic surfaces to catalyze organic reactions may be important in the synthesis, alteration, and degradation of humic materials (122). Conversely, the continuous changes during decomposition will influence the formation of those metal-organic associations.

The accumulation of humic material is higher under anaerobic soil conditions than aerobic conditions. Delaune *et al.* (31) and Reddy (108) observed increased concentration of soluble organic C and soluble organic N in soil solutions of an anaerobic organic soil as compared to an aerobic organic soil. However, no mention was made on the types of soluble organic compounds present in soil solution. Humic material in anaerobic systems are usually characterized by large molecular weights and a greater structural complexity, thus resulting in increased metal retention capacity. Gambrell *et al.* (39) observed a decrease in stability of Cu, Pb, and Cd complexes with insoluble organics, as anaerobic sediments were subjected to aerobic conditions.

Addition of organic substrates increases the reducing conditions, thus affecting the solubility of Fe and Mn compounds as a result of facultative anaerobic respiration. In the absence of adequate electron acceptors (O_2 and NO_3^-) microbial activity may lead to the reduction of first MnO_2 and then $Fe(OH)_3$ or $FePO_4$ with the solubilization of Mn^{2+} and Fe^{2+}, respectively. Different authors found evidence that $Fe(OH)_3$ and MnO_2 could act as terminal electron acceptors (147,148, 166). The reduction of Mn^{4+} and Fe^{3+} compounds can also be the result of the organic reduced metabolites produced during the anaerobic decomposition of organic matter. Manganese oxides are reduced between +200 and +300 mV at pH levels between 6.0 and 8, while the critical redox potentials for iron reduction and consequent dissolution lay between +300 and 100 mV at pH 6 and 7 and -100 mV at pH 8 (41, 42).

Aerobic decomposition of organic matter favors the stability of the sparingly soluble oxides and hydroxides of Fe and Mn, thus decreasing their availability to the plants. Under anaerobic conditions these compounds are solubilized and result in greater accumulation in the plant tissue. Typical examples are Mn and Fe, plant availability of these metals were severely affected by the change in redox status of the soil (72). Several other researchers have demonstrated the increased plant availability of Fe and Mn under flooded conditions, as compared to nonflooded conditions (26, 55, 156).

Although the solubility of several other metallic cations are not subject to direct oxidation-reduction reactions, their availability is influenced by the redox intensity and addition of easily degradable organic substrates. Solubility of most of the metals is affected by the formation of stable, insoluble metal sulfide precipitates under anaerobic conditions (38, 66, 79). Sulfide accumulation in anaerobic soils is enhanced by the addition of organic substrates. Metal precipitation with sulfide in soil can decrease the toxic levels of both metals and sulfides (142).

6.6. Agronomic and environmental significance

Agronomic significance of organic matter decomposition in soils with varying redox conditions is well documented for its role in plant nutrition. In well drained mineral soils where aerobic respiration predominates, soil organic matter decomposition can potentially release about 10 to 20% of that N as inorganic N which can be used (131) by crops. Organic soils (Histosols) can release about 410 to 1250 kg N ha^{-1} yr^{-1} and 38 to 185 kg P ha^{-1} yr^{-1} as a result of aerobic decomposition of soil organic matter. Shifting soil redox conditions from aerobic to facultative anaerobic or obligate anaerobic metabolism decreases the rate of decomposition. Low nutrient requirements of anaerobic bacteria will have beneficial effects on the plant available nutrient accumulations in O_2 deficient soils. In O_2 deficient soils, the inorganic redox systems such as NO_3^- Mn^{4+} oxides, Fe^{3+} oxyhydroxides support the organic matter decomposition, similar to the decomposition supported by the O_2, but at lower efficiency. However, reduction of Mn^{4+} and Fe^{3+} to soluble forms increases the plant available Mn^{2+} and Fe^{2+}. Reduction of NO_3^- to gaseous end products during facultative anaerobic respiration is not a beneficial process with respect to N economy of agricultural soils. As the metabolic activity shifts to obligate anaerobes, decomposition of organic matter can result in the accumulation of organic acids, aldehydes, sulfides and organic sulfur compounds that under certain conditions are toxic to plants.

In soils treated with organic wastes, aerobic decomposition can result in NO_3^- accumulation in the soil and subsequent accumulation in the plants and leaching into ground waters. In organic soils, decomposition of organic matter contributes a discharge of 12 to 56 kg N ha^{-1} yr^{-1} into drainage effluent (108). During aerobic respiration of organic soils, NO_3^- in the drainage effluent is a serious problem since these soils are often drained to keep the water table down and excess drainage water is being discharged into adjacent water bodies. As a result of continuous organic matter decomposition, Florida's organic soils are subsiding at a rate of 3 cm per year (134). To reduce soil loss it is recommended to return a portion of the soil profile to its original flooded state (127, 134). Although increasing the water table or complete flooding shifts the metabolic activities to facultative anaerobes or obligate anaerobes, these anaerobic processes solubilize the organic matter, thus increasing the soluble organic C, soluble organic N, and soluble P in the drainage effluents as compared to aerobic conditions (108, 109). Discharge of these effluents into adjacent water bodies can enhance the eutrophication process.

In permanently waterlogged soils, noncultivated coastal marshes where most of the active Mn and Fe are in the reduced form, it has been estimated that much of the respiratory activity is due to SO_4^{2-}. Jorgenson (56) showed that about 53% of the total mineralization of organic matter in the coastal marine sediment was due to SO_4^{2-} respiration.

In aquatic systems, aerobic respiration primarily occurs in the water column

148

and at the sediment-water interface. Aerobic bacteria are involved in the breakdown of detritus plant material and subsequent nutrient release supports the growth of aquatic biota. Organic acids and mineral nutrients accumulated in anaerobic zones of the sediments diffuse upward into the aerobic zone and subsequently are oxidized. Oxygen reaching sediment surface probably supports only a small fraction of organic matter decomposition. Nitrates and SO_4^{2-}, which are continuously formed are likely to support major parts of the organic matter decomposition in aquatic sediments.

6.7. References

1 Acharya C.N. 1935 Studies on the anaerobic decomposition of plant materials. I. Anaerobic decomposition of rice straw. Biochem J. 29, 528-541.

2 Acharya C.N. 1935 Studies on the anaerobic decomposition of plant materials. II. Some factors influencing the anaerobic decomposition of rice straw. Biochem J. 29, 953-960.

3 Acharya C.N. 1935 Studies on the anaerobic decomposition of plant materials. III. Comparison of the course of decomposition under anaerobic, aerobic, and partially aerobic conditions. Biochem J. 29, 1116-1120.

4 Alexander M. 1973 Nonbiodegradable and other recalcitrant molecules. Biotechnol. Bio. Engr. 15, 611-647.

5 Alexander M. 1977 Introduction to soil microbiology. John Wiley and Sons, Inc., NY.

6 Asami T. and Takai Y. 1970 Behavior of free iron oxide in paddy soils. IV. Relation between reduction of free iron oxide and formation of gases in paddy soils. Nippon Dojohiryo Gaku Zasshi 41, 48-55 (In Japanese).

7 Baas Becking L. G.M., Kaplan J.R. and Moore 1960 Limits of the natural environment in terms of pH and oxidation-reduction potential. J. Geology 68, 243-284.

8 Barber D.A. and Gunn K.B. 1974 The effect of mechanical forces on the exudation of organic substances by the roots of cereal plants grown under sterile conditions. New Phytol. 73, 39-45.

9 Barber D.A. and Martin J.K. 1976 The release of organic substances by cereal roots in soil. New Phytol. 76, 69-80.

10 Barrow M.J. 1960 A comparison of the mineralization of nitrogen and sulfur from decomposing materials. Aust. J. Agri. Res. 11, 960-969.

11 Bartholomew W.V. and Norman A.G. 1946 The threshold moisture content for active decomposition of some mature plant materials. Soil Sci. Am. J. 11, 270-279.

12 Belyaev S.S., Finkel'shtein Z I and Ivanov M.V. 1975 Intensity of bacterial methane formation in ooze deposits of certain lakes. Microbiol. 44, 272-275.

13 Berner R.A. 1980 A rate model for organic matter decomposition during bacterial sulfate reduction in marine sediments. In Biogeo Chemic de la Matiere Organique a' l'interface Eau-sediment Marine. pp. 35-45. Collogne Inter du Centre National de la Recherche Scientifigue CMRS Paris.

14 Bigander L.E. and Schippel F. 1973 Chemical dynamics of Baltic sediments-phosphate and sulphate. In The Chemical Microbiological Dynamics of the Sediment-water Interface. R.O. Hallberg (ed.) ASKO Laboratory, Univ. of Stockholm, Sweden 2, 25-48

15 Billen G. 1982 Modeling the processes of organic matter degration and nutrients recy-

ing in sedimentary systems. *In* Sediment Microbiology. Eds. D.B. Nedwell and C.M. Brown, pp. 15-52 Acad. Press.

16 Broadbent F.E., Jackman R.H. and McNicoll J. 1964 Mineralization of carbon and nitrogen in some New Zealand allophanic soils. Soil Sci. 98, 118-128.

17 Bromfield S.M. 1954 Reduction of ferric compounds by soil bacteria. J. Gen. Microbiol. 11, 1-6.

18 Bryan B.A. 1980 Cell yield and energy characteristics of denitrification with *Pseudomonas stutzeri* and *Pseudomonas aeruginosa.* Ph. D. Thesis, Univ. California, Davis, Univ Microfilms, Ann Arbor, MI (Diss. Abstr. 80, 27-39).

19 Buresh, R.J., Casselman M.E. and Patrick Jr. W.H. 1980 Nitrogen fixation in flooded soil systems. A review. Advan. Agron. 33, 149-192.

20 Buresh R.J. and Patrick Jr. W.H. 1981 Nitrate reduction to ammonium and organic nitrogen in an estuarine sediment. Soil Biol. Biochem 13, 279-283.

21 Burford J.R. and Bremner J.M. 1975 Relationships between the denitrification capacities of soils and total water-soluble and readily decomposable soil organic matter. Soil Biol. Biochem 7, 389-394.

22 Burns R.G. 1982 Carbon mineralization by mixed cultures. *In* Microbial Interactions and Communities. Vol 1. Eds. A.T. Bull and J.H. Slater. Acad. Press.

23 Cappenberg T. and Prins H. 1974 Interrelations between sulfate reducing and methane producing bacteria in bottom deposits of a freshwater lake. III. Experiments with ^{14}C-labeled substrates. J. Microbiol. Serol. 40, 457-469.

24 Caskey W.H. and Tiedje J.M. 1980 The reduction of nitrate to ammonium in soils. J. Gen. Microbiol. 119, 217-223.

25 Chandrasekaram S. and Yoshida Y. 1973 Effect of organic acid transformations in submerged soils on growth of rice plant. Soil Sci. Plant Nutr. (Tokyo) 19, 39-45.

26 Clark F., Nearpass D.C. and Specht A.W. 1957 Influence of organic additions and flooding on iron and manganese uptake by rice. Agron. J. 49, 586-

27 Clark M.D. and Gilmour J.T. 1983 The effect of temperature on decomposition at optimum and saturated soil-water contents. Soil Sci. Soc. Am. J. 47, 927-929.

28 Chase F.E. and Gray P.H.H. 1957 Application of the Warburg respirometer in studying respiratory activity in soil. Can. J. Microbiol. 3, 335-349.

29 Claypool G. and Kaplan I. 1974 The origin and distribution of methane in marine sediments. pp. 99-140 *In* Natural Gases in Marine Sediments. Ed. I.R. Kaplan Plenum Press, NY, NY.

30 Corbet A.S. 1934 Studies on tropical soil microbiology. I. The evolution of carbon dioxide from soil and the bacterial growth curve. Soil. Sci. 37, 109-115.

31 Delaune R.D. Reddy C.N. Patrick W.H. Jr. 1981 Effect of pH and redox potential on concentration of dissolved nutrients in an estuarine sediment. J. Environ. Qual. 10, 276-279.

32 Delwiche C.C. and Bryan B.A. 1976 Denitrification. Ann Rev. Microbiol. 30, 241-262.

33 Dommergues Y.R., Belser L.W. and Schmidt E.L. 1978 Limiting factors for microbial growth and activity in soil. Adv. Microbiol Ecol. 2, 49-104.

34 Elliott R.G. and Gilmour C.M. 1971 Growth of *Pseudomonas stutzeri* with nitrate and oxygen as terminal electron acceptors. Soil Biol. Biochem 3, 331-335.

35 Engler R.M. and Patrick Jr. W.H. 1975 Stability of sulfides of manganese, iron, zinc, copper, and mercury in flooded and non-flooded soil. Soil Sci. 119, 217-

36 Fenchel T.M. and Jorgensen B.B. 1977 Detritus food chains of aquatic ecosystems: The role of bacteria. Adv. Microbial Ecol. 1, 1-58.

37 Fenchel T.M. and Blackburn T.H. 1979 Bacteria and mineral cycling. Acad. Press, NY, NY.

150

38 Firestone M.K. 1982 Biological denitrification. *In* Nitrogen in Agricultural Soils. Agron. 22, 289-326. Amer. Soc. Agron., Madison, WI.

39 Gamrell R.P., Khalid R.A., Verloo M.G. and Patrick Jr. W.H. 1977 Transformation of heavy metals and plant nutrients in dredged sediments as affected by oxidation-reduction potential and pH. Part II. Materials and methods, results and discussion. Report No. DACW-39-74-C-0076. Office of Dredged Material Research, U.S. Engineer Waterways Experiment Station, Vicksburg, MS.

40 Gilmour, C.M., Broadbent F.E. and Beck S.M. 1977 Recycling of carbon and nitrogen through land disposal of various wastes. *In* Soils for Management of Organic Wastes and Wastewaters. Eds. L.F. Elliott and F.J. Stevenson. Am. Soc. of Agron., Madison, WI, pp. 173-194.

41 Gotoh S. and Patrick Jr. W.H. 1972 Transformation by manganese in a waterlogged soil as affected by redox potential and pH. Soil Sci. Soc. Am. Proc. 36, 738-742.

42 Gotoh S. and Patrick Jr. W.H. 1974 Transformation of iron in a waterlogged soil as influenced by redox potential and pH. Soil Sci. Soc. Am. Proc. 38: 66-71.

43 Granhall U. 1981 Biological nitrogen fixation in relation to environmental factors and fuctioning of natural ecosystems. *In* Terrestrial Nitrogen Cycles. Eds. F.E. Clark and T. Rosswall. Ecol. Bull (Stockholm) 33, 131-144.

44 Hagin J. and Amberger A. 1974 Contribution of fertilizers and manures to the N- and P-load of waters. A computer simulation. Final Rept. to the Deutsche Forschungs Gemeinschaft from Technion, Israel, 123 pp.

45 Hammann R. and Ottow J.C.G. 1974 Reductive dissolution of Fe_2O_3 by *Saccharolytic clostridia* and *Bacillus polymyxa* under anaerobic conditions. z. Pfl. Bodenkd, 137, 108-115.

46 Herman W.A., McGill W.B. and Dormaar J.F. 1977 Effects of initial chemical composition on decomposition of roots of three grass species. Can. J. Soil Sci. 57, 205-215.

47 Howarth R.W. 1979 Pyrite: its rapid formation in salt marsh and its importance to ecosystem metabolism. Science 204, 49-51.

48 Howarth R.W. and Teal J.M. 1979 Sulfate reduction in a New England salt marsh. Limnol. Oceanogr. 24, 999-1013.

49 Hsieh Y.P., Douglas L.A. and Motto H.L. 1981 Modeling sewage sludge decomposition in soil: I. organic carbon transformation. J. Environ. Qual. 10, 54-59.

50 Hunt H.W. 1977 A simulation model for decomposition in grasslands. Ecol. 58, 469-484.

51 Jansson S.L. 1966 Use of isotopes in soil organic matter studies. p. 415-422. Pergamon Press, NY, NY.

52 Jenkinson D.S. 1966 The priming action. *In* The Use of Isotopes in Soil Organic Matter Studies. pp. 199-208. Pergamon Press, Oxford.

53 Jenkinson D.S. 1968 Chemical tests for potentially available nitrogen in soil. J. Sci. Food Agric. 19, 160-168.

54 Jenkinson D.S. and Rayner J.H. 1977 The turnover of soil organic matter in some of the Rothamsted classical experiments. Soil Sci. 123, 298-305.

55 Jones H.E. and Etherington J.R. 1970 Comparative studies of plant growth and distribution in relation to waterlogging. I. The survival of *Erica cinerea* L and *E. tetralix* L and its apparent uptake in waterlogged soil. J. Ecol. 58, 487-

56 Jorgenson B.B. 1977 The sulfur cycle of a coastal marine sediment. Limnol. Oceanogr. 22, 814-832.

57 Jorgenson B.B. 1978 A comparison of methods for the quantification of bacterial sulfate reduction in coastal marine sediments. II. Calculation from mathematical models. Geo. Microbiol. J. 1, 29-47.

58 Jugsujinda A. 1975 Growth and nutrient uptake by rice under controlled oxidation-reduction and pH conditions in a flooded soil. Ph. D. Dissertation., Louisiana State Univ, Baton Rouge, LA.

59 Kamura T. and Takai Y. 1961 The microbial reduction mechanisms of ferric iron in paddy soils (Part I) Nippon Dojohiryo Gaku Zasshi 32, 135-138.

60 Keeney D.R., Chen R.L. and Graetz D.A. 1971 Importance of denitrification and nitrate reduction in sediments to the nitrogen budgets in lakes. Nature 233,6.

61 Khalid R.A., Patrick Jr. W.H. and R.D. DeLaune 1977 Phosphorus sorption characteristics of flooded soils. Soil Sci. Soc. Am. J. 41, 305-

62 Knowles R. 1981 Denitrification. *In* Terrestrial Nitrogen Cycles. F.E. Clark and T. Rosswall (eds.) Ecol. Bull. (Stockholm) 33, 315-329.

63 Koike I and Hattori A. 1975 Growth yield of denitrifying bacterium. *Pseudomonas denitrificans* under aerobic and denitrifying conditions. J. Gen. Microbiol. 88, 1-10.

64 Koike I and Hattori A. 1978 Denitrification and ammonia formation in anaerobic coastal sediments. Appl. Environ. Microbiol. 35, 278-282.

65 Kouyeas V. 1964 An approach to the study of moisture relations of soil fungi. Plant and Soil 20, 351.

66 Krauskopf K.P. 1956 Factors controlling the concentration of thirteen rare metals in sea water. Geochim. Cosmochim Acta 9, 1-

67 Krouse H.R. and McCready R.G.L. 1979 Biogeochemical cycling of sulfur. *In* Biogeochemical Cycling of Mineral Forming Elements. Eds. P.A. Trudinger and D.J. Swaine. pp. 401-403, Elsevier Publ., NY, NY.

68 Lindsay W.L. 1979 Chemical equilibria in soils. John Wiley & Sons, Inc., NY, NY.

69 Loveley D.R. and Klung M.J. 1983 Methanogenesis from methanol and methylanoimes and acetogenesis from hydrogen and carbon dioxide in the sediments to an eutrophic lake. Appl. Environ, Microbiol. 45, 1310-1315.

70 Lynch J.M. and Poole M.J. 1979 Microbial Ecology: A conceptual approach. John Wiley & Sons, NY, NY.

71 Mah R.A., Ward D.M., Baresi L. and Glass T.L. 1977 Biogenesis of methane. Ann. Rev. Microbiol. 31, 309-341.

72 Mandal L.N. 1962 Levels of iron and manganese in soil solution and the growth of rice in waterlogged soils in relation to the oxygen status of soil solution. Soil Sci. 94, 387-

73 Martens C.S. and Berner R.A. 1974 Methane production in the interstitial waters of sulfate depleted marine sediments. Science 185, 1167-1169.

74 McCarty P.L. 1975 Stoichiometry of biological reactions. Progr. Water Technol. 7, 157-172.

75 Melillo J.M., Aber J.D. and Muratore J.F. 1982 Nitrogen and lignin control of hardwood leaf litter decomposition dynamics. Ecol. 63, 621-626.

76 Miller R.H. and Johnson D.D. 1964 The effect of soil moisture tension on carbon dioxide evolution, nitrification, and nitrogen mineralization. Soil Sci. Soc. Am. Proc. 28, 644-646.

77 Mitchell P. 1972 Chemicosmotic coupling in energy transduction: a logical development of biochemical knowledge. Bioenergetics 3, 54-

78 Mitchell P. 1973 Performance and conservation of osmotic work by proton coupled solute porter systems. Bioenergetics 4, 63-91.

79 Morcel F.M.M. Westall J.C.O'Melia C.R. and Morgan J.J. 1975 Fate of trace metals in Los Angeles Country wastewater discharge. Environ. Sci. Technol. 9, 756-

80 Mortimer C.H. 1941 The exchange of dissolved substances between water and mud in lakes. J. Ecol. 29, 280-329.

81 Mulder E.G. and Brotonegoro S. 1974 Free-living heterotrophic nitrogen fixing bacteria. *In* The Biology of Nitrogen Fixation. Ed. A. Quispel. pp. 37-85. Amsterdam: North-Holland Publ. Co. and Oxford: Plenum Press.

82 Nyhan J.W. 1976 Influence of soil temperature and water tension on the decomposition rate of carbon-14 labeled herbage. Soil Sci. 121, 288-293.

83 Odum E.P. 1971 Fundamentals of Ecology. 3rd Ed. W.B. Saunders Co., pp. 574.

84 Ogwada R.A. Reddy K.R. and Graetz D.A. 1984 The effects of aeration and temperature on nutrient regeneration from selected aquatic macrophytes. J. Environ. Qual (*In* press).

85 Ottow J.C.G. 1981 Mechanisms of bacterial iron reduction in flooded soils. *In* Proc. Symp. on Paddy Soils (Ed. Inst. of Soil Sci. Acad. Sinica, China) Springer Verlag, NY p. 330-343.

86 Pal D. and Broadbent F.E. 1975a Kinetics of rice straw decomposition in soils. J. Environ. Qual. 4(2), 256-260.

87 Pal D. and Broadbent F.E. 1975b Influence of moisture on rice straw decomposition in soils. Soil Sci. Soc. Am. Proc. 39, 59-63.

88 Pal D., Broadbent F.E. and Mikkelsen D.S. 1975 Influence of temperature on rice straw decomposition in soils. Soil Sci. 442:449.

89 Parr J.F. and Reuszer H.W. 1962 Organic matter decomposition as influenced by oxygen level and flow rate of gases in the constant aeration method. Soil Sci. Soc. Am. Proc. 26, 552-556.

90 Patrick Jr. W.H. 1960 Nitrate reduction rates in a submerged soil as affected by redox potential. 7th Int'l Congress of Soil Sci., Madison, WI. 2, 494-500.

91 Patrick Jr. W.H. 1964 Extractable iron and phosphorus in a submerged soil at controlled redox potential. Trans. 8th Int'l Cong. Soil Sci. 3, 650.

92 Patrick Jr. W.H. 1981 The role of inorganic redox systems in controlling reduction in paddy soils. *In* Proc. Symp. on Paddy Soils (Ed. Inst. of Soil Sci., Acad., Sinica, China) Springer-Verlag, NY. pp. 107-117.

93 Patrick Jr. W.H. and Delaune R.D. 1972 Characterization of oxidized and reduced zones in flooded soil. Soil Sci. Soc. Am. Proc. 36, 573-575.

94 Patrick Jr. W.H. and Khalid R.A. 1974 Phosphate release and sorption by soils and sediments: Effect of aerobic and anaerobic conditions. Science 186, 53-55.

95 Patrick Jr. W.H. and Reddy K.R. 1976 Nitrification-denitrification reaction in flooded soils and sediments: Dependence on oxygen supply and ammonium diffusion. J. Environ. Qual. 5, 469-472.

96 Patrick Jr. W.H., Gotoh S. and Williams B.G. 1973 Strengite dissolution in flooded soils and sediments. Science 179, 564-565.

97 Paul E.A. and VanVeen J.A. 1978 The use of tracers to determine the dynamic nature of organic matter. *In* Vol. 3. Symposia papers. 11th Int'l Cong. of Soil Sci. p. 61-102. Edmonton, Canada.

98 Payne W.J. 1970 Energy yields and growth of heterotrophs. Ann. Rev. Microbiol. 24, 17-52.

99 Payne W.J. 1973 Reduction of nitrogenous oxides by microorganisms. Bacteriol Rev. 37, 409-452.

100 Payne W.J. 1981 Denitrification. John Wiley & Sons, NY, NY. P. 214.

101 Pfennig M. and Wioldel F. 1981 Ecology and physiology of some anaerobic bacteria from the microbial sulfer cycle. pp. 169-178. *In* Biology of Inorganic Nitrogen and Sulfur. Eds. H. Bothe and A. Trebst. Springer-Verlog.

102 Pinck L.A., Allison F.E. and Sherman M.S. 1950 Maintenance of soil organic matter: II.

Losses of carbon and nitrogen from young and mature plant materials during decomposition in soil. Soil Sci. 69, 391-401.

103 Ponnamperuma F.N. 1972 The chemistry of submerged soils. Adv. Agron 24, 29-96.

104 Postgate J.R. 1979 The sulfate reducing bacteria. Cambridge Univ. Press. London. p.

105 Postgate, J.R. and Cambell L.L. 1966 Classification of *Desulfovibrio* species, the non-sporulating sulfate reducing bacteria. Bacteriol. Rev. 30, 732.

106 Rao D.N. and Mikkelsen D.S. 1977 Effect of acetic, propionic, and butyric acids on ground rice seedling's growth. Agron. J. 69, 923-928.

107 Rao R.V. 1976 Nitrogen fixation as influenced by moisture content, ammonium sulfate and organic sources in a paddy soil. Soil Biol. Biochem 8, 445-448.

108 Reddy K.R. 1982 Mineralization of nitrogen in organic soils. Soil Sci. Soc. Am. J. 46, 561-566.

109 Reddy K.R. 1983 Soluble phosphorus release from organic soils, agriculture, ecosystems, and environment 9, 373-382.

110 Reddy K.R. and Patrick Jr. W.H. 1975 Effect of alternate aerobic and anaerobic conditions on redox potential, organic matter decomposition and nitrogen loss in a flooded soil. Soil Biol. Biochem. 7, 87-94.

111 Reddy K.R. and Patrick Jr. W.H. 1979 Nitrogen fixation in flooded soil. Soil Sci. 128(2), 80-85.

112 Reddy K.R. and Patrick Jr. W.H. 1984 Nitrogen transformations and loss in flooded soils and sediments. CRC Critical Reviews in Environ. Control 13, 273-309.

113 Reddy K.R., Khaleel R. and M.R. Overcash 1980 Carbon transformations in the land areas receiving organic wastes in relation to nonpoint source pollution: A conceptual model. J. Environ. Qual. 9(3), 434-442.

114 Reddy K.R., Rao P.S.C. and Jessup R.E. 1982 The effect of carbon mineralization on denitrification kinetics in mineral and organic soils. Soil Sci. Soc. Am. J. 46, 62-68.

115 Reddy K.R., Rao P.S.C. and Patrick Jr. W.H. 1980 Factors influencing the oxygen consumption rates in flooded soils. Soil Sci. Soc. Am. J. 44, 741-744.

116 Reddy K.R., Khaleel R. and Overcash M.R. 1980 Nitrogen, phosphorus, and carbon transformations in a coastal plain soil treated with animal wastes. Agric. Wastes Int'l. J. 2, 225-238.

117 Roberts J.L. 1947 Reduction of ferric hydroxide by strain of *Bacillus polymyxa*. Soil Sci. 63, 135-140.

118 Rudd J.W.M. and Taylor C.D. 1980 Methane cycling in aquatic environments. Adv. Aquatic Microbiol. 2, 77-150.

119 Russell J.S. 1975 A mathematical treatment of the effect of cropping systems on soil organic nitrogen in two long-term sequential experiments. Soil Sci. 120, 37-44.

120 Savant N.K. and DeDatta S.K. 1982 Nitrogen transformations in wetland rice soils. Adv. Agron. 35, 241-302.

121 Sawyer C.N. and McCarty P.L. 1978 Chemistry for sanitary engineers McGraw Hill Co., NY, NY.

122 Schnitzer M. 1978 Reactions of humid substances with minerals in the soil environment. *In* Environmental Biogeochemistry and Geomicrobiology. Vol. 2. Ed. W.E. Krumbein Ann Arbor Sci. pp 397-717.

123 Shields J.A., Paul E.A., Lowe W.E. and Parkinson D. 1973 Turnover of microbial tissue in soil under field conditions. Soil Biol. Biochem 5, 753-764.

124 Sinha M.K., Sinha D.P. and Sinha H. 1977 Organic matter transformations in soils. V. Kinetics of carbon and nitrogen mineralization in soils amended with different organic materials. Plant and Soil 46, 579-590.

154

125 Sircar S.S.G., De S.C. and Bhownick H.D. 1940 Microbiological decomposition of plant materials. Ind. J. Agric. Sci. 10, 119-151.

126 Smith O.L. 1982 Soil microbiology: A model of decomposition and nutrient cycling. *In* CRC series in mathematical models in microbiology. Ed. M.J. Bazin. CRC Press, Inc., Boca Raton. Fl p. 273.

127 Snyder G.H., Burdine H.W., Crockett J.R., Gascho G.J., Harrison D.S., Kidder G., Mishoe J.W., Myhre D.L., Pate F.M. and Shih S.F. 1978 Water table management for organic soil conservation and crop production in the Florida Everglades. Bull. 801. Agric. Expt. Sta., IFAS, Univ. of Fla, Gainesville, Fl.

128 Sommers L.E., Harris R.F., Williams J.D.H., Armstrong and Syers J.K. 1972 Fractionation of organic phosphorus in lake sediments. Soil Sci. Soc. Am. Proc. 36, 51-54.

129 Sorensen J. 1978 Capacity for denitrification and reduction of nitrate to ammonia in coastal sediment. Appl. Environ. Microbiol. 35, 301-305.

130 Sparling G.P., Cheshire M.V., Mundie C.M. and Murayama S. 1981 The transformation of ^{14}C labeled glucose in sterilized soil inoculated with selected microorganisms. Rev. Ecol. Biol. Sol. 18, 447-457.

131 Stanford G. and Smith S.J. 1972 Nitrogen mineralization potentials of soils. Soil Sci. Soc. Am. Proc. 36, 465-472.

132 Stanford G., Vander Pol R.A. and Dzienia S. 1975 Denitrification rates in relation to total and extractable soil carbon. Soil Sci. Soc. Am. Proc. 39, 284-289.

133 Starkey R.L. and Halvorson H.O. 1927 Studies on the transformation of iron in nature. II. Concerning the importance of microorganisms in the solution and precipitation of iron. Soil Sci. 24, 381-402.

134 Stephens J.C. 1969 Peat and muck drainage problems. J. Irr. Drain. Div. Am. Soc. Civil Eng. 95, 285-305.

135 Stevenson F.J. 1982 Nitrogen in agricultural soils. Agron. 22 Amer. Soc. Agron, Madison WI.

136 Stevenson F.J. 1982 Organic forms of soil nitrogen. *In* Nitrogen in Agricultural Soils. Agron. 22, 67-122. Amer. Soc. Agron., Madison, WI.

137 Stewart B.A., Porter L.K. and Viets F.G. 1966 Effect of sulfur content of straw on rate of decomposition and plant growth. Soil Sci. Soc. Am. Proc. 30, 355-358.

138 Stouthamer A.H., Van't Riet J. and Oltmann L.F. 1980 Respiration with nitrate as acceptor. *In* Diversity of Bacterial Respiratory Systems. Vol. 2. Ed. C.J. Knowles. CRC Press, Inc., Boca Raton, FL. pp 19-48.

139 Strayer R.F. and Tiedje J.M. 1978 Kinetic parameters of the conversion of the methane precursors to methane in a hypereutrophic lake sediment. Appl. Environ. Microbiol. 36, 330-346.

140 Stumm W. 1966 Redox potential as an environmental parameter: Conceptual significance and operational limitation. Proc. Int'l. Water Poll. Res. Conf. p. 283-308.

141 Takai Y. and Kamura T. 1966 The mechanism of reduction in waterlogged paddy soil. Folia Microbiologia 11, 304-313.

142 Tanaka A. and Navasero S.A. 1966 Interaction between iron and manganese in the rice plant. Soil Sci. Plant Nutri. 12, 197-201.

143 Tenny F.G. and Waksman S.A. 1929 Composition of natural organic materials and their decomposition in the soil: IV. The nature and rapidity of decomposition on the various organic complexes in different plant materials under aerobic conditions. Soil Sci. 28, 55-84.

144 Tenny F.G. and Waksman 1930 Composition of organic materials and decomposition in the soil. V. Decomposition of various chemical constituents in plant materials under

anaerobic conditions. Soil Sci. 30, 143-160.

145 Thauer R.K. and Badziong W. 1980 Respiration with sulfate as electron acceptor. *In* Diversity of Bacterial Respiratory Systems. Vol. 2. Ed. C.J. Knowles. CRC Press, Inc., Boca Raton, FL p. 66-85.

146 Thauer R.K., Jungermann K. and Decker K. 1977 Energy conservation in chemotrophic anaerobic bacteria. Bacteriol. Rev. 41, 100-180.

147 Trimble R.B. and Ehrlich H.L. 1968 Bacteriology of manganese nodules. III. Introduction of MnO_2 by two strains of nodule bacteria. Appl. Microbiol. 16, 695-702.

148 Trimble R.B. and Ehrlich H.L. 1970 Bacteriology of manganese nodules IV. Reduction of an MnO_2-reductase system in a marine Bacillus. Appl. Microbiol. 19, 966-972.

149 Trudinger P.A. 1979 The biological sulfur cycle. *In* Biogeochemical Cycling of Mineral Forming Elements. Eds. P. A. Trudinger and D.J. Swaine. Elsevier Sci. Publ. Co., NY, NY. pp 293-368.

150 Turner F.T. and Patrick Jr. W.H. 1968 Chemical changes in waterlogged soils as a result of oxygen depletion. Trans. 9th Int'l. Cong. Soil Sci. 4, 53-65.

151 Volk B.G. 1972 Everglades Histosol subsidence: 1. CO_2 evolution as affected by soil type, temperature, and moisture. Proc. Soil Crop Sci. Soc. Fla. 32, 132-135.

152 Volk B.G. and Loeppert R.H. 1982 Soil organic matter. *In* Handbook of Soils and Climate in Agriculture. Ed. V.J. Kilmer. CRC Press, Inc., Boca Raton, FL pp 211-268.

153 Wake L.V., Christopher R.K., Rickard A.D., Anderson J.E. and Ralph B.J. 1977 A thermodynamic assessment of possible substrates for sulfate reducing bacteria. Aust. J. Biol. Sci. 30, 155-172.

154 Warford A.L., Kosiur D.R. and Doose P.R. 1978 Methane production in Santa Barbara Basin sediments. Geo. Microbiol. J. 1, 117-137.

155 Waring S.A. and Bremner J.M. 1964 Ammonium production in soil under waterlogged conditions as an index of nitrogen availability. Nature 201, 951-952.

156 Weeraratna C.S. 1969 Absorption of manganese by rice under flooded and unflooded conditions. Plant and Soil 30, 121.

157 Whitefield M. 1969 Eh as an operational parameter in estuarine studies. Limnol. Oceanogr. 14, 547-558.

158 Widdel F. and Pfenning N. 1977 A new anaerobic sporing, acetateoxidizing, sulfate-reducing bacterium, *Desulfotomaculum* (emerd) *acetoxidans*. Arch Microbiol. 112, 119 122.

159 Wildung R.E., Garland T.R. and Buschbom R.L. 1975 The interdependent effects of soil temperature and water content on soil respiration rate in plant root decomposition in arid grassland soil. Soil Biol. Biochem 7, 373-378.

160 Williams C.H. 1967 Some factors affecting the mineralization of organic sulfur in soils. Plant and Soil 26, 205-223.

161 Williams S.R. and Gray T.R.G. 1974 Decomposition of litter on the soil surface. *In* Biology of Plant Litter Decomposition. Vol. 2. Eds. C.H. Dickinson and G.J. Pugh. pp 611-632. Acad. Press, NY, NY.

162 Williams W.A., Mikkelsen D.S., Mueller K.E. and Ruckman J.E. 1968 Nitrogen immobilization by rice straw incorporated in lowland rice production. Plant and Soil 28, 49-60.

163 Winfrey M.R. and Zeikus J.G. 1977 Effects of sulfate or carbon and electron flow during microbial methanogenesis in freshwater sediments. Appl. Environ. Microbiol. 33, 275 281.

164 Winfrey M.C., Nelson D.R., Klevickis S.C. and Zeikus J.G. 1976 Association of hydrogen metabolism with methanogenesis in Lake Mendota sediments. Appl. Environ. Micro-

biol. 33, 312-318.

165 Wolfe R.S. 1980 Respiration in methanogenic bacteria. *In* Diversity of Bacterial Respiratory Systems. Vol. 1. Ed. C.J. Knowles, CRC Press, Inc., Boca Raton, FL pp 161-186

166 Woolfolk C.A. and Whiteley H.R. 1962 Reduction of inorganic compounds by *Micrococcus lactilyticus*. J. Bacteriol. 84, 647-658.

167 Yoshida T. 1975 Microbial metabolism of flooded soils. *In* Soil Biochemistry Eds. E.A. Paul and A.D. McLaren pp 83-122, Marcel Dekker, Inc., NY, NY.

168 Zeikus J.G. 1977 The biology of methanogenic bacteria. Bacteriol. Rev. 41, 519-541.

Section III

Soil microorganisms, biofertilizers and biocontrol agents: their interactions with soil organic matter and effects on soil fertility

7. Soil microorganisms, soil organic matter and soil fertility

Y. HENIS

7.1. Introduction

In its broader sense, soil fertility may be defined as the fitness of the soil to support optimal plants growth. Since different plants may require different optimal growth conditions, there is no clear-cut concept of soil fertility. Fitness of soil to support plant growth depends on environmental, physical and chemical factors which are related to plants nutrition, and are intimately associated with the soil's organic matter through the activity of soil microorganisms.

Soil organic matter fractions found in any soil that supports plant growth may be arbitrarily divided into living organisms (plants, animals and microorganisms), nonliving fractions of animal and microbial origin undergoing different stages of decomposition, and soil organic matter, or humus, which comprises the final stage of this process. Soil humus is derived from plant, animal and microbial residues, as well as from products of microbial synthesis which occurred during its formation. Its slow mineralization by the soil microorganisms ensures a constant supply of essential elements to the growing plant. Soil microorganisms constitute a vast systematic array of bacteria, actinomycetes, fungi, protozoa, yeasts, algae, worms and insects. In general, the level of microorganisms in the soil is positively correlated with the level of organic matter, which is mainly associated with the upper 30-40 cm of the soil (2, 3, 23, 26, 32). Under optimal moisture and temperature conditions, a given soil will support a certain level of organisms of a typical composition. However, both the level and the composition will change rapidly as the conditions are changed. Such changes in condition may occur daily, and include drying and wetting, freezing and thawing, heating and cooling, and fumigation. The soil may be regarded as a living system composed of many individual creatures which respire (as is evident from O_2 uptake and CO_2 output), digest available nutrients such as glucose, release ammonia from amino acids and phosphate from nucleic acids, respond to metabolic poisons such as cyanide or azide, and release heat during organic matter decomposition.

Y. Chen and Y. Avnimelech (eds.), The Role of Organic Matter in Modern Agriculture.
ISBN 90-247-3360-X.
© *1986, Martinus Nijhoff Publishers, Dordrecht. Prii*

In the following section the role of microorganisms in soil fertility will be discussed in relation to the different fractions of organic matter mentioned above.

7.2. The living fraction (plants, animals, microorganisms)

As a rule, living tissues of plants and animals are free of microorganisms. However, plant roots support a large microbial population on their surface and in their soil environment. This rhizosphere population differs from the soil microbial population in composition, level and activity. This part of the soil microflora is under the influence of the root environment, into which the roots excrete a vast array of small organic molecules such as sugars, amino and carboxylic acids, as well as sloughed-off cells and moribund tissue. The rhizosphere population may improve plant growth, for example by producing plant hormones, acting as a buffer against plant pathogens, or improving uptake of microelements through the production of chelates. It may also, however, be responsible for detrimental activities such as loss of available N through denitrification, or oxidation of iron or manganese ions, thus creating deficiency symptoms in plants as a result of ion competition.

Excreta of living animals are incorporated into the soil organic matter through microbial decomposition. With regard to the living soil microorganisms, some are eaten up by soil protozoa, others are lysed by myxobacteria, and yet others are attacked by viruses or by bacterial predators such as *Bdellovibrio* (38). Mycoparasitic fungi, such as *Trichoderma* (29) and *Sporidesmium* (4), attack other fungi. Soil microorganisms also produce lytic enzymes such as proteases, glucanases and chitinases, but these are generally not active towards living organisms.

Soil microorganisms include bacteria, actinomycetes, fungi, algae, protozoa, and viruses (2, 26, 32). A typical composition is given in Table 7.1.

Table 7.1. Organic matter and microorganisms in arable soil (After E.W. Russell (32)).

Microorganisms	Number/g	Dry weight (kg/ha)
Bacteria & actinomycetes	3×10^9	700
Protozoa	3×10^4	35
Fungi	40 m Living mycelium/g	350
Other fauna		100-200
Total $\sim$		1200-1300
Organic matter		65-70 tons/ha (in top 15 cm)

In spite of the predatory and lytic processes, the soil microflora and micro-fauna are able to remain at a constant level provided that the conditions remain constant. This is because the rate of growth equals the death rate of the whole po-pulation, which is under a steady-state condition. Such a state is very similar to that described for a chemostat, in which a continuous culture is kept in a steady-state condition, *i.e.*, dx/dt=μx$-$Dx=O (33), where the growth rate (μ) of the soil microflora (x) is limited by the level and rate of decomposition of the soil organic matter, and the dilution rate (D) is determined by the predatory soil microfauna, mainly the protozoa. As in a chemostat, in which the level of the microorganisms at steady state is determined by the nutrients level in the entering solution, the level of the organic matter and its rate of degradation determines the flux of available nutrients entering the immediate microenvironment in the vicinity of the organic matter particles in the soil.

7.3. The dead fraction: Fresh organic matter

Bodies of dead plants, animals and microorganisms decompose quickly in the soil. However, both the rate of decomposition and its effect on plant growth de-pend greatly on the carbon to nitrogen ratio. The average ratio of carbon: nitro-gen: phosphorus: sulfur is 100:10:1:1 (2). Any degradation process in the soil de-pends on and is accompanied by microbial growth. Assuming that during their growth on organic matter soil organisms incorporate into their bodies some 30-40% of the organic matter dry weight, and release to the air some 60% of the de-composed carbonaceous plant material in the form of CO_2, and assuming that fresh plants contain 40% of their dry weight in the form of carbonaceous material, then the soil microflora needs 0.1 x 0.3-0.4 x 0.4 x 100 = 0.012-0.016 x 100 = 1.2-1.6% available nitrogen in order to decompose plant material without disturb-ing the nitrogen balance in the soil. Any material containing less than 1.2-1.6% N will cause nitrogen depletion, whereas material containing more than 1.6% N will increase the available N in the soil.

Apart from its effect on the N level through mineralization, fresh organic matter may affect the N level through denitrification, especially in waterlogged soils (2, 7, 32). It may also favor sulfate reduction, producing sulfides toxic to plants and corrosive to metals in soil (2). Finally, plant health may be affected by crop residues, either as carriers of pathogens or through their favorable effect on the saprophytic soil microorganisms which antagonize pathogens growth.

7.4. The natural soil organic matter: Humus

The highly stable colloidal soil organic fraction known as humus is a major contributor to the chemical and physical soil properties which are of greatest significance in crop production. It serves both as a soil conditioner and as a storage of available plant nutrients. Soil microorganisms play a major role in both its formation and its decomposition (25). Its relatively low rate of decomposition and its high nitrogen content (well above the critical level of 1.6%) ensure a steady supply of available N,P,K, and microelements to the growing plant as well as to the microflora active in humus degradation. The microorganisms active in its degradation are oligotrophic heterotrophs. It is not clear whether their build-up in the soil contributes to plant's nutrition as well. However, some of their products, *e.g.* polyuronides and polysaccharides, may improve the physical properties of the soil.

The beneficial effects of humus in improving soil structure are reflected in increased aeration, reduction in power requirement for plowing heavy soils, rapid germination of seeds and a reduced volume weight. Finally, some humus components may serve as chelates, improving cation uptake by plant roots, or may even show hormone-like activity (24).

Under steady state conditions, the rate of humus formation equals its rate of decomposition. Both processes are relatively slow under field conditions. For example, CO_2 evolution due to humus degradation in the field may be between 500 mg and 10.000 mg of CO_2 per m^2 per day, whereas under laboratory conditions it may be 5-50 mg of CO_2 per kg of soil per day. Under field conditions, a soil may loose 2-5 percent of the carbon present in its organic matter per annum (2). The effect of soil drying on humus decomposition and on nitrogen and phosphorus availability has been studied by Birch (6,7,8,9), who developed mathematical models for these processes. An analytical model of the microbial decomposition of soil organic matter has been developed by Smith (34). A carbon submodel after Smith is given in Fig. 7.1.

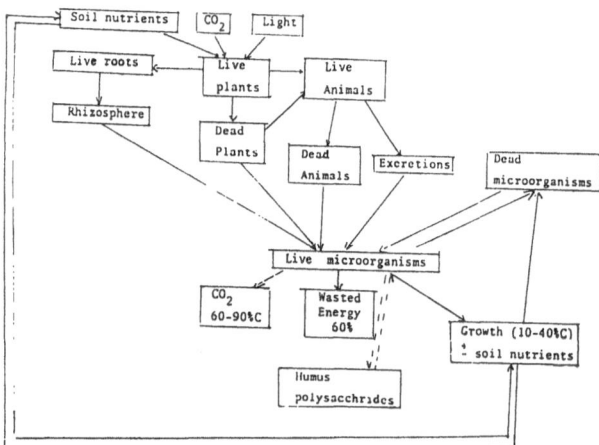

Fig. 7.1. Carbon submodel (After O.L. Smith (34)).

Heterotrophic activity in soil is a substrate-limited process, in which the microorganisms live under a constant state of starvation. The rate of humus degradation is far lower than the rate of utilization of its decomposition products by the plants and the soil. This is why no products of degradation or mineralization accumulate during this process. In some cases, however, breakdown products such as alcohols and organic acids may be detected.

7.5. The role of soil microorganisms in phosphorus availability to higher plants

Phosphorus is essential for both plants and microorganisms (24). It is found in soils in several forms, which include organic and inorganic, soluble and insoluble compounds (2, 15, 32). Of these, only the soluble inorganic phosphate is available to plants. However, when added to soil as a fertilizer, inorganic phosphate is soon fixed (14,15) or immobilized by being incorporated into the bodies of the soil microorganisms (2). These organisms can also affect phosphate availability by forming chelates (13), organic acids (1, 12, 13, 17, 36) or through the decomposition of phosphorus-containing organic matter (11, 35).

Many soil microorganisms, especially those colonizing the rhizosphere, are capable of solubilizing calcium phosphate, a common insoluble form of phosphate in soils (21, 30). Acid and chelates-producing rhizosphere organisms which use the root exudates and sloughed-off root tissue, increase phosphate solubility near the plant's root (17). They use calcium phosphate or apatite as a sole phosphorus source. However, to do this they must grow, multiply and produce acid, and for this they must be supplied with carbon, nitrogen and energy sources. On the other hand, some of the solubilized phosphate will be immobilized during the process. The products which lead to phosphate solubilization include both organic acids, produced by heterotrophic organisms, and inorganic acids such as nitric and sulfuric acids, produced by ammonia and sulfur oxidizing microorganisms (2).

Under anaerobic conditions (*e.g.* flooding) and in the presence of organic matter, microorganisms can increase phosphate availability indirectly, through the reduction of an accompanying element, thus changing the solubility of the phosphate-containing complex (2). Under highly unaerobic conditions and in the presence of available organic matter which serves as an electron donor, phosphate can be reduced to phosphite and phosphine (PH_3) which is highly volatile and flammable (2).

Organic phosphorus is found in soil in a number of compounds which include phytin (inositol hexaphosphate), lipids, nucleic acids and humic substances. The amount of organic phosphorus in soil is related to its level. The capacity of higher plants to utilize soil organic phosphorus is totally dependent on microbial activity (2, 11, 34). There is a highly significant correlation between nitrogen and phosphorus mineralization, the ratio between these two elements being N : P =

8-15:1 (2). However, because of the fixation of the mineralized phosphate and its binding in inorganic complexes, unlike nitrogen, phosphate may not become available to plants even after its mineralization. The capacity of plants to grow in phosphorus-poor soils is greatly inceased by ecto or endo-symbiotic (mycorrhizal) fungi (18), which are found in the roots of most plants and play an important role in phosphate uptake in phosphorus-deficient environments (16, 18).

7.6. Subsoil, humus and fertility

Both soil humus and microorganisms accumulate in the surface, rather then in the subsoil. Perhaps the most convincing evidence indicaing a correlation between organic matter, soil microorganisms and soil fertility, is the decline in soil fertility following the removal of the surface, humus-rich soil layers (3, 31). Application of inorganic fertilizers results in only a partial recovery of soil fertility. Best results were obtained with subsoil amended with organic manure. However, the recovery was slow and was due to a number of nutritional factors, mainly nitrogen. In addition, other factors such as soil aggregation, steady-rate release of plant nutrients through mineralization of organic matter and interaction between a number of factors could be involved (3). It may be concluded that soil organic matter contributes to soil fertility through the activity of the soil organisms, by supplying available nitrogen, phosphorus, potassium and microelements at a continuous and a steady rate, by improving the physical soil structure, and possibly by contributing growth factors and chelating agents which improve the microelements uptake.

7.7. Biological transformation of microbial residues in soil

The same organisms which bring about the decomposition of plant material and its transformation into humus, live in a perpetual cycle of growth, multiplication, death and decomposition (32). Their dead bodies are degraded either by autolysis or as a result of the activity of the living organisms. Because more organic matter is utilized for energy than for new cell synthesis, cell nitrogen and phosphorus are mineralized and become available for plants. The density of the cell components of the various microbial groups in soil have been discussed in detail (37).

Enhanced degradation of dead organisms by the remaining microorganisms partially explains the increase in soil fertility which usually follows physical and chemical treatments of the soil (6, 7, 8, 9, 20, 24). Other mechanisms involved are the direct erradication of plant pathogens (20), and the enhancement of their antagonists (27, 28, 29).

7.8. Nitrogen fixation and soil organic matter

Molecular nitrogen is a potential N-source for microorganisms which posses the nitrogenase system. Those asymbiotic, free-living procaryotes which live in the soil (*e.g. Azotobacter* sp., *Clostridium* sp.) will not fix molecular nitrogen in the presence of available nitrogen. Furthermore, these organisms require a carbon and energy source in the form of available carbohydrates to fix molecular nitrogen. Thus *Azotobacter* requires one g. glucose to fix 20 mg N, whereas *Clostridium* will fix 2-10 mg N per of carbohydrate consumed (2). This means that biological N-fixation in soil requires vast amount of available carbohydrates and low levels of combined, available nitrogen conditions both unpractical and unfavorable for plant growth. The amounts of molecular nitrogen fixed by the soil microorganisms do not exceed 2 lbs/acre/year (2). This may satisfy needs of natural flora but does not support agricultural crops.

7.9. Rhizosphere microflora, organic matter, and soil fertility

The rhizosphere microflora is dependent on the root environment, and is typical for every plant. It is more affected by the plant than by the soil. The microflora of the living plant differs from that of the soil which is not under the influence of the root (2). The rhizosphere population thrives on root exudates, root exretions, moribund and sloughed - off root tissue. Soil organic matter affect the rhizosphere indirectly, by favoring root growth and exudation. The rhizosphere population exerts both favorable and non-favorable effect on the plant growth. The balance, however, is more towards the favorable one. Non favorable effects include enhanced denitrification, competition for oxygen, oxidation of iron and manganese and supporting specific pathogens. Favorable effects include antagonism towards pathogens, phosphorus solubilization, symbiotic nitrogen fixation, production of chelating agents and growth factors.

7.10. Soil organic matter and plant diseases

Among means wed to augment microbial antagonists in soil, organic matter has received a special attention (5, 19, 27, 28, 30). Thus, the pathogen *Rhizoctonia solani* which causes damping off of many plants was reduced in soils amended with corn or oat tissue (27). Similarly mature oat tissue increased the number of microorganisms antagonistic to *Fusarium solani* in bean rhizosphere (28). Resistance of avocado groves to *Phytophthora cinnamomi* root rot in Australia was correlated with heavy application of animal manures (10). Similarly, chitin added to soil increased actinomycetes and fungi antagonistic to *R. solani* and to *Fusarium*

166

oxysporum, f.sp. pisi, the pea wilt pathogen, (causing an increase in the number of Fusarium - antagonistic organisms in the pea rhizosphere (22)). These observations clearly indicate that by using organic amendments, plants can be protected from pathogens. However, protection does not seem to be a full one and depends on environmental factors such as soil type, climatic conditions and presence of antagonists. More studies are required to turn these observations into practical application.

7.11. Conclusions

1. Both the formation and the mineralization of humus are essential for plant growth.
2. Both of these processes are brought about by soil microorganisms.
3. Rates of humus formation and mineralization depend upon physical and chemical soil conditions.
4. At any given time, only a small amount of free degradation products can be detected because the utilization rate is much faster than the degradation processes.
5. The availability of nitrogen to plants during the decomposition of fresh organic matter depends upon its N content.
6. Soil organic matter serves both as a soil conditioner and as a nutrient source, continuously releasing essential plant nutrients during mineralization.

7.12. References

1 Agnihorty, U.P. 1970 Solubilization of insoluble phosphates by some soil fungi isolated nursery seeds beds. Can J. Microbiol. 16, 877-880.
2 Alexander M. 1977 Introduction to Soil Microbiology. 2nd Edition. John Wiley & Sons, New York. 467 p.
3 Avnimelech Y. 1969 Decline in soil fertility after removal of top layers and its restoration. Report submitted to the Ministry of Agriculture. June 1969. (*In Hebrew*). 18 p.
4 Ayers W.A. and Adams P.B. 1981 Mycoparasitism and its application to biological control of plant diseases. *In* Biological Control in Plant Production. Ed., G.C. Papavizas. BARC Symposium No. 5. Allanheld, Osmun, London. 461p.
5 Baker K.F. and Cook J.R. 1974 Biological control of Plant Pathogens. W.H. Freeman and Co., San Francisco. 433 p.
6 Birch H.F. 1958 The effect of soil drying on humus decomposition and nitrogen availability. Plant and Soil 10, 9-31.
7 Birch H.F. 1959 Further observations on humus decomposition and nitrification. Plant and Soil 11, 262-286.
8 Birch H.F. 1960 Soil drying and soil fertility. Trop. Agric. Tv. 37, 3-10.
9 Birch H.F. 1960 Nitrification in soils after different periods of drying. Plant and Soil 12, 81-96.

10 Broadbent P. and Baker K.F. 1975 Soil suppressive to Phytophthora root rot in Eastern Australia. pp. 152-157. *In* Ed. G.W. Bruehl. Biology and Control of Soilborne Plant Pathogens. Amer. phytopathol. Soc. St. Paul, Minnesota.

11 Brown, M. 1974 Seed and root bacterization Annu. Rev. Phytopathol. 12, 181-197.

12 Dalton J.D., Russel G.C. and Sieling D.H. 1952 Effect of organic matter on phosphate availability, Soil Science 73, 173-181.

13 Duff R.P., Webley D.M. and Scott R.D. 1963 Solubilization of minerals and related materials by 2-ketogluconic acid producing bacteria. Soil Science 95, 105-114.

14 El-Sherif A.M. 1973 The relation between phosphatic fertilization and $CaCO_3$ content in highly calcarious soils of Egypt. Agrochimica 17, 546-553.

15 Emsley J. 1982 A fixation with phosphate. New Scientist 96, 915-918.

16 Gerderman J.W. 1968 Vesicular arbuscular mycorrhiza and plant growth. Annu. Rev. Phytopathol. 6, 397-418.

17 Gerretsen F.C. 1948 The influence of microorganisms on the phosphate uptake by the plant. Plant and Soil 1, 51-81.

18 Harley J.L. 1969. The Biology of Mycorrhiza. 2nd. ed Leonard Hill Ltd. London.

19 Hoitink H.A.J. 1980 Composted bark: A light weight growth medium with fungicidal properties. Plant Disease 64, 142-147.

20 Katan J. 1981 Solar heating (solarization) of soil for control of soilborne pests. Annu. Rev. Phytopathol. 19, 211-236.

21 Katznelson H. and Bose B. 1959 Metabolic activity and phosphate-dissolving capability of bacterial isolates from wheat roots, rhizosphere and non-rhizosphere soil. Can J. Microbiol. 5, 79-85.

22 Khalifa O. 1965 Biological control of Fusarium wilt of peas by organic soil amendments. Ann. Appl. Biol. 56, 129-37.

23 Kononova N.N. 1961 Soil Organic Matter. Pergamon Press, Oxford.

24 Larsen S. 1967 Soil phosphorus. Adv. Agron. 19, 151-210.

25 Lebedjantzev A.N. 1924 Drying of the soil as one of natural factors in maintaining soil fertility. Soil Sci. 18, 419-447.

26 McCalla T.N. 1959 Microorganisms and their activity with crop residues. Nebraska. Agric. Exp. Sta. Bull. No. 453.

27 Papavizas G.C. and Davey C.B. 1960 Rhizoctonia disease of bean as affected by decomposing green plant materials and associated microfloras. Phytopathology 50, 516-521.

28 Papavizas G.C. 1963 Microbial antagonism in bean rhizosphere as affected by oat straw and supplemented nitrogen. Phytopathology 53, 1430-35.

29 Papavizas G.C. and Lewis J.A. 1981 Introduction and augmentation of microbial antagonists for the control of soilborne plant pathogen. *In* Biological Control in Crop Production. Ed. G.C. Papavizas. BARC Symposium No. 5. Allanheld, Osmun, London. 461 p.

30 Raghu K. and MacRae I.C. 1966 Occurence of phosphate-dissolving micro-organisms in the rhizosphere of rice plants and in submerged soils. J. Appl. Bacteriol. 29, 582-586.

31 Reuss J.D. and Campbell R.E. 1961 Restoring productivity to leveled land. Soil Sci. Soc. Am. Proc. 25, 32-43.

32 Russell E.W. 1973 Soil Conditions and Plant Growth. 10th ed. Longmans, Green and Co., London. 849 p.

33 Slater J.H. 1979 Microbial population and community dynamics. *In* Microbial Ecology. Eds. J.M. Lynch and N.J. Poole. Blackwell Scientific Publications, Oxford. 266 p.

34 Smith O.L. 1979 An analytical model of the decomposition of soil organic matter. Soil Biol. Biochem. 11, 585-606.

35 Smith J.H. and Allison F.E. 1962 Phosphobacterin as a soil inoculant: Laboratory, greenhouse, and field evaluation. Technical Bull. No. 1263. ARS.USDA, Washington, D.C. 22 p.

168

36 Stevenson J.F. 1967 Organic acids in soil. *In* Soil Biochemistry Vol. Eds. A.D. McLaren and G.H. Peterson pp. 133-139. Marcel Dekker Inc. New York.

37 Webley D.M. and Jones D. 1971 Biological transformations of microbial residues in soil. *In* Soil Biochemistry Vol. 2 Eds. A.D. McLaren and J. Skujins. Marcel Dekker, Inc. New York, 527 pp.

38 Varon M. and Shilo M. 1980 Ecology of aquatic bdellovibrio. Advances in Aquatic Microbiology 2, 1-48.

8. The role of organic matter in the introduction of biofertilizers and biocontrol agents to soils

Y. HADAR

8.1. Introduction

Various groups of soil microorganisms have a potential use in agriculture.
These groups include the symbiotic nitrogen-fixing Rhizobia, free living bacteria
such as *Azotobacter* spp. and *Azospirillum* spp., vesicular-arbuscular and ecto-my-
corrhizal fungi, and biocontrol agents such as *Trichoderma* spp. . Together, these
organisms may be able to replace mineral fertilizers and chemical pesticides, thus
lowering costs and reducing pollution and environmental hazards.

However, as promising as these possibilities may seem in theory, only a few
micro-organisms have been used successfully in agriculture (7, 10, 36). Pessimism
as to the future of successful inoculation of soil with beneficial microorganisms is
demonstrated by Garrett's attitude towards biological control: "Such attempts to
boost the population of an antagonistic microorganism by inoculation alone have
been doomed to failure since their inception" (20). Similarly, Alexander (3) described failures to inoculate natural and already well-populated habitats with beneficial microorganisms. Only two types of inoculation were successful: plant pathogens inoculated into soil do cause disease, and *Rhizobium* added to leguminous
plants produces nodules. Indeed, *Rhizobium,* was the first commercial inoculant
produced during the early part of this century (55).

This pessimism as to the future of inoculation was based on ecological considerations, but was it founded in fact. Beneficial microorganisms are operating in
nature, providing plants with nitrogen, phosphorus or growth hormones, and pro-
tecting them from plant pathogens. The difficulty lies in manipulating these organisms so that they will act at economic satisfactory levels (7). This chapter describes the contribution of organic matter to the success of such inoculants.

8.2. Rhizobium

Rhizobium species are characterized by their ability to fix atmospheric nitrogen and to live symbiotically within the roots of leguminous plants. Rhizobia are aerobic Gram-negative rods, and are divided into two groups:(i) fast growers and (ii) slow growers. Their physiology and the mechanisms by which they recognize and infect their hosts has been reviewed extensively (8, 36, 51, 55, 69). The practice of inoculating seeds with artificial cultures of rhizobia dates from 1896 (55). Since that time, modern methods for growing rhizobia on a large scale and for the production of inoculant have been developed (8, 55, 56, 66). Moreover, the methods developed for rhizobia have served as models for the introduction of other beneficial microorganisms to soils.

Preparation of high-quality legume inoculants is dependent on the selection of suitable carrier material (55). The choice of a carrier should be based on its ability to support large populations of bacteria (10^9/g), provide nutritive medium for growth of rhizobia, and promote survival during distribution (55). To satisfy these requirements, the carrier should be free of toxins which might affect rhizobia, have a high content of organic matter, possess good absorption qualities and a high water holding capacity, be amenable to sterilization and processing, and be available locally at reasonable prices and in sufficient quantities.

Peat usually satisfies these requirements and is widely used as a rhizobia carrier. However, it should be examined carefully before use as peats are diverse in nature and may vary even within the same deposit (13). The rhizobia survive well in peat inoculants, but better results have been obtained with bacteria lyophilized in vegetable oil (32).

The association of *Rhizobium trifolii* with peat particles was examined by electron microscopy in order to understand the properties of peat that make it a favourable environment for the bacteria under adverse conditions (14). It was determined that micro-colonies developed both on the surface of the particles and in crevices. The bacteria are protected from desiccation within a fibrillar matrix, possibly an extracellular gam, which attaches them to the surface of the peat particles (14).

Table 8.1. Organic materials used as carriers for *Rizobium* inoculants.

Carrier	Reference
Corn cubs compost	11
Coir dust	19
Coal	12, 52, 60
Farm yard manure	61
Composted manure	61
Vegetable oil	32

In some countries, high quality peat is not available (61). Organic materials have been tested as alternatives to peat according to the characteristics discussed above. Some of these materials are listed in Table 8.1. Although peat seems to be the best of all the inoculant carriers tested, the ability of peat substitutes to promote both the survival of rhizobia and legume inoculation justifies their development as an alternative to peat.

8.3. Azotobacter and other free living bacteria

Azotobacter chroococcum, a Gram negative rod, is able to fix atmospheric nitrogen and produce plant growth regulators. However, western scientists have been sceptical (3, 6) about its usefulness to agricultural crops. In the Soviet-Union Azotobacterin, a fertilizing preparation containing *A. chroococcum* has been used since the 1930's (46). This is a peat preparation which was applied to soil or to seeds (6). However, only 34% of its applications resulted in an increase in crop yield, and those increases fell within the range of 8-12%, which was not statistically significant. When Azotobacterin was applied to heavily manured soils, it showed greater activity, thereby increasing tomato and corn yields by about 30% (46). Satisfactory results were acheived when *A. chroococcum* was used for growing vegetable seedlings in humus-peat feeding blocks made of a mixture of peat, saudust and cow manure (3:3:1.5). This mixture was amended with bacterial culture and the seedlings were transplanted to the field (46). It seems, therefore, that *Azotobacter* needs a sufficient amount of available organic matter in the soil in order to function, and thus it can develop only in fertile soils. Kumanda and Gaur (33) suggested that in such soils, minimal competition for nutrients between the *Azotobacter* and soil microflora resulted in greater benefit from bacterial inoculants. Improvement in *Azotobacter* performance in manured soil was observed by Ocampo *et al.* (50) Similarly, it was shown that addition of organic materials such as cellulose and rice straw to flooded soils increased the rate of atmospheric nitrogen fixation (54, 69). It was proposed that the anaerobic decomposition of organic matter leads to the formation of substrates which would be available to the N-fixing organisms (54).

Attemps to introduce other free-living beneficial bacteria to soil have been reported recently. An inoculant of *Azospirillum brasilense* was prepared by the method described for *Rhizobium,* using finely sieved sterilized peat. Yield increases were observed in field experiments with summer cereal crops in Israel (30, 51). Plant growth promoting *Pseudomonas* have been applied to potatoes and sugar beets with yield increases of 6-33% (62). The same investigators have applied these bacteria to seeds using organic polymers such as cellulose methyl-ether or xantangum (62, 63) finding high pseudomonads population in the rhizosphere of inoculated plants. Mendez-Castro and Alexander (43) were able to increase the number of *Pseudomonas* spp. on corn roots by adding glucose and mancozeb to the soil, thereby reducing competition. Mancozeb is known to decrease the number of bacteria in soil. These results emphasize the importance of available substrates for growth and survival of free-living bacteria in the rhizosphere.

8.4. Vesicular-arbuscular mycorrhizae

The vesicular-arbuscular (VA) mycorrhizal fungi are beneficial symbiotic microorganisms that increase growth and yield of most crop plants through improved phosphorus uptake, resistance to drought and salinity, and increased tolerance to pathogens (26, 27, 39, 44). The production of large quantities of high-quality pathogen-free inoculum for introduction into soil is however, problematic. Currently, since the fungi themselves cannot be cultured in the absence of a living root, the only way to produce suitable quantities of a VA mycorrhizal inoculum is on roots of susceptible host plants (25, 44, 45). Natural field soil containing mycorrhizal fungi has been used successfully as an inoculum. However, spores are not concentrated and large amounts of this material are required. Another method for producing mycorrhizal inoculum on plants in the greenhouse was described by Menge (44). Inoculated Sudan grass (*Sorghum vulgare* pers.) is grown in containers under conditions that will maximize VA mycorrhizal spore production. When spores are mature, plant tops are removed and roots, growth media and spores are ground up. Plants can be grown in peat instead of soil with lettuce serving as a host plant in nutrient film culture (27, 67). Inoculum produced this way weighs one tenth of the equivalent volume of soil-grown inoculum and is thus easier to transport (27). Composted plant residues were also evaluated as media for VA mycorrhizal inoculum (21, 22). Tests with maple, rhododendron and corn plants showed that maple compost contained substantial inoculum while corn compost was either devoid of mycorrhizal propagulas or inhospitable to mycorrhizal development. Chemical analyses showed that maple compost was low in extractable plant nutrients while corn compost was high in these nutrients (22).

Soil organic matter has a role in growth of VA mycorrhizal fungi in soil and in infection of plant roots (48). Pure sand and sand amended with peat or soil organic matter were inoculated with infected roots. After 125 days the inoculum was removed from the pots and seeds of white clover were sown. No infection was observed in the indicator plants grown in sand, but if the sand was amended with organic matter all plants became infected (28). These results emphasize the importance of organic matter in the process of plant infection with mycorrhizal fungi. It was also observed that the mycelium associated with infected roots often grew around decaying root fragments and other organic matter (31, 49). This is important since it restricts the adsorbing tissue to a localized nutrient rich zone. Thus, greater nutrient uptake will occur than if root and hyphal distribution were random (57). Mosse *et al.* suggest that soil organic matter affects the incidence of mycorrhizal infection indirectly through its effects on soil structure, water holding capacity and nutrient mineralization (48).

8.5. Ectomycorrhiza

The ectomycorrhizae include fungi belonging to many genera and are most common among forest and ornamental trees (39, 40). The growth of these fungal symbionts is stimulated by root exudates. Hyphae grow over the surface of host feeder roots and form a fungal mantle. The mantle replaces the root hairs with fungal strands, greatly increasing root surface absorbtive area. Hyphae then develop around root cortical cells, and form an interconnecting network known as the "Hartig net". Ectomycorrhizal fungi can be of major importance in forest regeneration (39). Inoculation of ectomycorrhizal fungi is most common under nursery conditions. Most ectomycorrhizal fungi produce sporophores that can be used for inoculation as well as soil inoculum. However the use of pure mycelial cultures of selected fungi for inoculation is the most biologically sound method, since pathogens and other contaminants are avoided (40, 41). Large scale nursery application of vegetative inoculum has been severely hampered by a shortage of inoculum. Moser (47) was the first to develop a method of growing vegetative inoculum of ectomycorrhizal fungi, using *Suillus plorans*. He grew the fungus in aerated liquid medium for 3 to 4 months and then transferred it to sterilized peat moss for another 2 to 4 months. He tried to apply agar mycelial inoculum or mycelial suspensions, but found them uneffective in the nursery (47). Different groups have successfully followed Moser's example with some modification in Argentina, Canada, Australia and the United States. The development of vegetative inoculum of *Pisolithus tinctorius* by Marx and co-workers is an excellent example. *P. tinctorius* was chosen by this group because it was considered valuable in reclamation of adverse sites with pine trees, it grew rapidly in pure culture, and it had a broad tree host range (40, 41).

A medium consisting of vermiculite and 5-10% peat by volume moistened with salts and glucose nutrient medium was excellent for growing mycelial inoculum. This inoculum was superior to basidiospore inoculum in infecting pine roots. Other organic substrates such as peanut hulls, corn cobs, or pine bark are not suitable because they release growth inhibitors during autoclaving (40). A commercial formulation of mycelial inoculum of *P. tinctorius,* which is grown on vermiculite peat medium, has been developed by Abbot Laboratories (40, 41) under the trademark MycoRize®. The use of peat in this formulation is important since it has a strong buffering capacity maintaining pH below 6. This was not possible with chemical buffers (42). The use of peat may have additional positive effects. *P. tinctorius* is able to produce fulvic and humic acids in pure culture, and vegetative growth of the fungus was stimulated by fulvic acid (64). In the culture medium, peat may be furnishing essential humic acids or their precursors to *P. tinctorius*, much as organic matter does in forest soil. Thus, peat may play a vital role not only in fungal growth but also in the production of effective inoculum (40). Soil organic matter has an impact on ectomycorrhizal development in forest soils (59). Soils with high organic matter content yielded larger numbers of active ectomycorrhizae than did low organic matter soils (24). This should be taken into consideration when making forest management decisions that may lead to a reduction in soil organic matter content. On the other hand, the presence of undecomposed green manure caused damage to pine seedlings, by eradicating mycorrhizal fungi (29).

8.6. Systems for biological control of soilborne plant pathogens

Biological control of soilborne plant pathogens by the application of antagonistic microorganisms to soil is a potential non-chemical means of plant disease control (10, 53). Among others, members of the genus *Trichoderma* are capable of hyperparasiting pathogenic fungi and have been very widely tested for control of many pathogenic fungi (4, 17, 18, 23, 34, 53, 58). Direct inoculation of antagonists to soil is not, however, economically feasible (53). Wells *et al.* were the first to report field control of *Sclerotium rolfsii* by the infestation of soil with *Trichoderma harzianum* grown on an autoclaved mixture of ryegrass and soil. Rather than to try to alter the soil microflora, they overwhelmed the infection court with *T. harzianum* and its food base, thus reducing disease incidence (68).

Since then, many workers have followed this method of growing the antagonist on organic substrate which serves as a delivery system and as a food base. Some examples are shown in Table 8.2. Wheat bran was used successfully as a

Table 8.2. Organic materials used for production and delivery of *Trichoderma* spp. to soil.

Carrier	Reference
Ryegrass seeds	68
Wheat bran	17, 18, 23
Wheat bran and peat mixture	58
Peat, soil and cellulose mixture	9
Wheat straw	2, 15
Barley grains	1
Diatomaceous earth and molasses	4
Sand and corn cob mixture	35, 53
Sand and wehat bran mixture	35

food base for *T. harzianum* (17, 18, 23). The wheat bran preparation proved superior to a directly applied *T. harzianum* conidial suspension in activity against *S. rolfsii* and in its ability to survive in the soil (17). The release of cell wall degrading enzymes is apparently the mechanism by which *T. harzianum* acts as an antagonist. High lytic activity was found in the wheat bran culture (16). During fungal growth, however, pH levels increased up to pH 8, probably due to the release of ammonia from proteins. Addition of peat (50% by volume) effectively stabilized the pH level at 5.5 without decreasing the *Trichoderma* population, and was beneficial in preventing bacterial contaminations (58). Moreover, survival of *Trichoderma* was better when grown on this mixture than on wheat bran alone. Another approach in which composted hardwood bark is used as a food base and carrier for antagonists is described in detail in a chapter by Hoitink and Kuter in this book. It seems that the use of a proper food base is a critical factor in determining the success or failure of directly applied soil biocontrol agents in controlling soil-born plant pathogens. Application of a food base with the biocontrol agent might overcome fungistasis and enable the introduced organism to grow, colonize the soil, and degrade propagules of plant pathogenic fungi.

Soil organic matter plays an important role in the activity of antagonists in naturally suppressive soils. A soil naturally suppressive to *Rhizoctonia solani* was found in Columbia, South America. The suppressiveness was attributed to a high organic matter level (35%) and to a high fungal population (10^6 propagules/g soil) of which 8×10^5/g soil were of the genus *Trichoderma*. A conducive soil examined contained only 10^2 *Trichoderma* propagules/g soil and 2% organic matter (9). In Australia, the soil in avocado groves and eucalyptus forests were supressive to *Phytophtora cinnamomi*. These soils were rich in organic matter and highly populated with antagonistic microganisms (5, 10, 38).

8.7. Conclusions

Many bacteria and fungi have beneficial effects on plant growth. This chapter describes some examples of the microorganism that fix nitrogen, mineralize organic forms of plant nutrients, and protect plants from pathogens. Indeed, all of these activities occur in nature, but the crucial problem is how to exploit these activities in modern agricultural practice (7, 36).

Soil is a highly populated environment (3, 36). The microbial population has the ability to maintain community stability and integrity in a variable environment, a state which Alexander termed by "homeostasis"(3). Thus alien microorganisms introduced into the soil die out rapidly. Such an organism may become established if the niche that it could occupy is not already filled (3). This concept may explain the success of inoculants of symbiotic microorganisms such as rhizobia and mycorrhitic fungi. These organisms colonize plant roots, a habitat that is newly available for microbial growth. However, successful introduction of free-living microorganisms is relatively rare. In many cases, no attempts were made to selectively favor growth of the introduced organism. It was assumed that the introduction of large amounts of inoculum would enable the microorganism to compete successfully with the indigenous microflora and establish itself in the rhizosphere. This chapter describes the role of organic matter in manipulating the biological balance to favor the introduced organism. This can be achieved by adding organic amendments to the soil, thus reducing nutrient competition between microorganisms, stimulating specific microbial groups, or directing plant roots to an area rich in microbial activity. Another approach involves introducing a beneficial microorganism within an artificially created niche. *e.g.* straw colonized by cellulolytic *Trichoderma* to yield the sugars, together with a nitrogen fixing *Azotobacter* (36, 37). In practice, this approach is dependent upon the development of a system for growth and delivery. This system should supply the microbe with a proper food base which, while unavailable to other microorganisms, will enable their establishment as well as interaction with soil microorganisms or plant roots. Thus achieving effective biological control or improving plant nutrition and growth.

8.8. References

1 Abd-El Moity T.H. and Shatla M.N. 1981 Biological control of white rot disease *Sclerotium* cepivorum of onion by *Trichoderma harzianum.* Phytopathol. Zeit. 100, 29-35.

2 Akhtar C.M. 1977 Biological control of some plant diseases laking genetic resistance of the host crops in Pakistan. Ann. N.Y. Acad. Sci. 287, 45-56.

3 Alexander M. 1971 Microbial Ecology. John Wiley, New York.

4 Backman P.A. and Rodriguez-Kabana R. 1975 A system for the growth and delivery of biological control agnets to the soil. Phytopathology 65, 819-821.

5 Broadbent P. and Baker K.F. 1975 Soils suppressive to *Phytophthora* root rot in Eastern Australia. pp. 152-157 *In* Biology and Control of Soil-Borne Plant Pathogens. Ed. G.W. Bruehl Ame. Phytopath. Soc. St. Paul, Minnesota.

6 Brown M.E. 1974 Seed and root bacterization. Annu. Rev. Phytopath. 14, 181-197.

7 Brown M.E. 1976 Microbial manipulation and plant performance. pp. 37-53 *In* Microbiology in Agriculture, Fisheries and Food. Eds. F.A. Skinner and J.G. Carr. Academic Press, London.

8 Burton J.C. 1982 Modern concept in legume inoculation pp. 105-114. *In* Biological Nitrogen Fixation Technology for Tropical Agriculture. Eds. P.H. Graham and S.C. Harris. Centero Internacional de Agricultura Tropical.

9 Chet I. and Baker R. 1981 Isolation and biocontrol potential to *Trichoderma hamatum* from soil naturally suppressive of *Rhizoctonia solani.* Phytopathology 71, 286-290.

10 Cook R.J. and Baker K.F. 1983 The nature and practice of biological control of plant pathogens. The Ame. Phytopath. Soc. St. Paul, Minnesota.

11 Corby H.D.L. 1976 A method of making a pure culture, peat type, legume inoculant, using a substitute for peat. pp. 169-173 *In* Symbiotic nitrogen fixation in plants Ed. P.S. Nutman Cambridge University Press, Cambridge.

12 Crawford S.L. and Berryhill D.L. 1983 Survival of *Rhizobium phaseoli* in coal based legume inoculants applied to seeds. Appl. Environ.Microbiol. 45, 703-705.

13 Date R.A. and Roughley R.J. 1977 Preparation of legume seed inoculants. 243-276 *In* A treatise on Dinitrogen Fixation IV, Agronomy and Ecology. Eds. R.W.F. Hardy and A.H. Gibson. Wiley New York.

14 Dart P.J. Roughley R.J. and Chandler M.R. 1969 Peat culture of *Rhizobium trifolii* an examination by electron microscopy J. Appl. Bacteriol. 32 352-357.

15 Davet P., Artigues M. and Martin C. 1981 Production en conditions non-aseptiques d'inoculum de *Trichoderma harzianum* pour des essais de lutte biologique. Agronomie 1, 933-936.

16 Elad Y., Chet I. and Henis Y. 1982 Degradation of plant pathogenic fungi by *Trichoderma harzianum* Can. J. Microbiol. 28, 719-725.

17 Elad Y., Chet I. and Katan J. 1980 *Trichoderma harzianum:* A biocontrol agent effective against *Sclerotium rolfsii* and *Rhizoctonia solani*. Phytopathology 70, 119-121.

18 Elad Y., Hadar Y., Hadar E., Chet I. and Henis Y. 1981 Biologial control of *Rhizoctonia solani* by *Trichoderma harzianum* in carnation, Plant Disease 65, 675-677.

19 Faizah A.W., Broughton W.J. and John C.K. 1980 Rhizobia in tropical legumes X. growth in coir dust soil compost. Soil Biol. Biochem. 12, 211-218.

20 Garrett S.D. 1956 Biology of root infecting fungi. Cambridge Univ. Press, London.

21 Guttay A.J.R. 1983 The interaction of fertilizers and vesicular arbuscular mycorrhizae in composted plant residues. J. Am. Soc. Hortic. Sci. 108, 222-224.

22 Guttay A.J.R. 1982 The growth of three woody plant species and the development of their mycorrhizae in three different plant composts. J. Am. Soc. Hortic. Sci. 107, 324-327.

23 Hadar Y., Chet I., and Henis Y. 1978 Biological control of *Rhizoctonia solani* damping off with wheat bran culture of *Trichoderma harzianum*. Phytopathology 69, 64-68.

24 Harvey A.E., Jurgensen M.F. and Larsen M.J. 1981 Organic reserves: Importance to ectomycorrhizae in forest soils. Forest Sci. 27, 442-445.

25 Hattingh M.J. 1975 Inoculation of Brazilian sour orange seed with an endomycorrhizal fungus. Phytopathology 65, 1013-1016.

26 Hayman D.S. 1983 The physiology of vesicular-arbuscular endomycorrhizal symbiosis. Can. J. Bot. 61, 944-963.

27 Hayman D.S. 1980 Mycorrhiza and crop production. Nature London 287, 487-488.

28 Hepper C.M. and Warner A. 1983 Role of organic matter in growth of a vesicular-arbuscular mycorrhizal fungus in soil. Trans. Br. Mycol. Soc. 81, 155-156.

29 Iyer S.G. 1980 Sorgum-sudan green manure; its effect on nursery stock. Plant and Soil 54, 151-162.

30 Kapulnik Y., Sarig S., Nur I., Okon Y., Kigel J. and Henis Y. 1981 Yield increases in summer cereal crops in Israeli field inoculated with *Azospirillum*. Expl. Agric. 17, 179-187.

31 Koske R.E., Sutton J.C. and Sheppard B.R. 1975 Ecology of Endogone in Lake Huron sand dunes. Can. J. Bot. 53, 87-93.

32 Kremer R.J. and Reterson H.L. 1983 Effect of carrier and temperature on survival of Rhizobium spp. in legume inocula: Development of an improved type of inoculant. Appl. Env. Microb. 45, 1790-1794.

33 Kundu B.S. and Gaur A.C. 1980 Establishment of nitrogen fixing and phosphate-solubilizing bacteria in rhizosphere and their effect on yield and nutrient uptake of wheat crop. Plant and Soil 57, 223-230.

34 Lewis J.A. and Papavizas G.C. 1980 Integrated control of *Rhizoctonia* fruit rot of cucumber. Phytopathology 70, 85-89.

35 Lewis J.A. and Papavizas G.C. 1983 Production of chlamidospores and conidia by *Trichoderma* spp. in liquid and solid growth media. Soil Biol. Biochem. 15, 351-357.

36 Lynch J.M. 1983 Soil Biotechnology. Blackwell Scientific Publications. Oxford

37 Lynch J.M. and Elliott L.F. 1983 Minimizing the potential phytotoxicity of Oxford wheat straw by microbial degradation. Soil Biol. Biochem. 15, 221-222.

38 Malajczuk N. 1979 Biological suppression of *Phytophthora cinnamomi* in eucalyptus and avocado in Australia. pp. 635-652 *In* Soil-Borne Plant Pathogens. Eds. B. Schippers and W. Gams. Academic Press, London.

39 Maronek D.M. 1981 Mycorrhizal fungi and their importance in horticultural crops production. Hortic. Rev. 3, 172-213.

40 Marx D.H. and Kenny D.S. 1982 Production of ectomycorrhizal fungus inoculum. pp. 131-146 *In* Methods and Principles of Mycorrhizal Research. Ed N.C. Schenck The Ame. Phytopathol. Soc., St. Paul. Minnesota.

41 Marx D.H., Ruehle J.L., Kenney D.S. Cordell C.E., Riffle J.W., Molina R.J., Pawuk W.H., Navratil S., Tinus R.W., and Goodwin R.W. 1982 Commercial vegetative inoculum of *Pisolithus tinctorius* and inoculation techniques for development of ectomycorrhizae on container grown tree seedling. Forest Sci. 28, 373-400.

42 Marx D.H. and Zak B. 1965 Effect of pH on mycorrhizal formation of slash pine in aseptic culture. Forest Sci. 11, 66-75.

43 Mendez-Castro F.A. and Alexander M. 1983 Method for establishing a bacterial inoculum on corn roots. Appl. Environ. Microbiol. 45, 248-254.

44 Menge J.A. 1983 Utilization of vesicular-arbuscular mycorrhizal fungi in agriculture. Can. J. Bot. 61, 1015-1024.

45 Menge J.A. and Timmer L.W. 1982 Procedures for inoculation of plants with vesicular-arbuscular mycorrhizae in laboratory, greenhouse and field. pp. 59-76. *In* Methods and Priciples of Mycorrhizal Research. Ed. N.C. Schenk. The Ame. Phytopath. Soc., St. Paul. Minnesota.

46 Mishustin E.N. 1970 The importance of non symbiotic nitrogen fixing micro-organisms in agriculture. Plant and Soil 32, 545-554.

47 Moser M. 1958 Die Kunstriche Mykorrhizaimpfung an forstpflanzen. 1. Erfahrungen bei der Reinkultur von Mykorrhizapilzen. Forstw. Cbl. 77, 32-40.

48 Mosse B. Stribley D.P. and Le Tcacon F. 1981 Ecology of mycorrhizae and mycorrhizal fungi. Adv. Microbiol. Ecol. 5, 137-210.

49 Nicolson T.H. 1959 Mycorrhiza in the Gramineae vesicular arbuscular endophytes, with special reference to the external phase. Trans. Br. Mycol. Soc. 42, 421-438.

50 Ocampo J.A., Barea J.M. and Montoyae E. 1975 Interaction between *Azotobacter* and phosphobacteria and their establishment in the rhizosphere as affected by soil fertility. Can. J. Microbiol. 21, 1160-1165.

51 Okon Y. and Hardy R.W. 1983 Developments in basic and applied biological nitrogen fixation. pp. 5-54. *In* Plant Physiology Vol. VIII Nitrogen Metabolism. Eds. F.C. Steward and R.G.S. Bidwell Academic Press N.Y.

52 Paczkowski N.W. Berryhill D.L. 1979 Survival of *Rhizobium phaseoli* in coal based legume inoculants. Appl. Environ. Microbiol. 38, 612-615.

53 Papavizas G.C. and Lewis J.A. 1981 Introduction and augmentation of microbial antagonists for the control of soilborne plant pathogens. pp. 305-322. *In* Biological Control in Crop Production. Ed. G.C. Papavizas. Allanheld, Osmum, London.

54 Rao U.R. 1978 Effect of carbon source on asymbiotic nitrogen fixation in a paddy soil. Soil Biol. Biochem. 10, 319-321.

55 Roughley R.J., Pulsford D.J. 1982 Production and control of legume inoculants. pp. 193-209. *In* Nitrogen Fixation in Legumes. Ed. J.M. Vincent Academic Press, Sydney.

56 Roughley R.J. and Vincent J.M. 1967 Growth and survival of *Rhizobium* spp. in peat culture.

J. App. Bacteriol. 30, 362-376.

57 St. John T.V. and Coleman D.C. 1983 The role of mycorrhizae in plant ecology. Can. J. Bot. 61, 1005-1014.

58 Sivan A., Elad Y. and Chet I. 1984 Biological control of *Phythium aphanidermatum* by a new isolate of *Trichoderma harzianum*. Phytopathology 74, 498-501.

59 Slankis V. 1974 Soil factors influencing formation of mycorrhizae. Annu. Rev. Phytopath. 437-457.

60 Strijdom B.W. and Deschodt C.C. 1976 Carriers of rhizobia and the effects of prior treatment on survival of rhizobia. pp. 151-168. *In* Symbiotic Nitrogen Fixation in Plants. Ed. P.S. Nutman. Cambridge University Press, Cambridge.

61 Subba Rao N.S. 1982 Biofertilizers. pp. 219-242. *In* Advances in Agricultural Microbiology. Ed. N.S. Subba Rao Butterworth Scientific, London.

62 Suslow T.V. 1982 Role of root colonizing bacteria in plant growth. pp. 187-223. *In* Phyto-pathogenic Procaryotes. Ed. M.S. Mount and G.H. Lacy Academic Press. New York.

63 Suslow T.V. and Schroth M.N. 1982 Rhizobacteria of sugar beets: Effects of seed application and root colonization on yield. Phytopathology 72, 199-206.

64 Tan K.H. and Nopanornbodi V. 1979 Fulvic acid and the growth of the ectomycorrhizal fungus *Pisolithus tinctorious*. Soil Biol. Biochem. 11, 651-653.

65 Vance C.P. 1983 *Rhizobium* infection and nodulation: a beneficial plant disease. Annu. Rev. Microbiol. 37, 399-424.

66 Van Schreven D.A. 1970 Some factors affecting growth and survival of *Rhizobium* spp. in soil-peat cultures. Plant and Soil 32, 113-130.

67 Warner A. Gee P. and Fyson A. 1981 The production of inoculum in nutrient film culture. Rothamsted Report, Part 1, 211.

68 Wells H.O., Bell D.K. and Jaworski C.A. 1972 Efficacy of *Trichoderma harzianum* as a biocontrol for *Sclerotium rolfsii*. Phytopathology 62, 442-447.

69 Yoneyama T., Lee K.K. and Yoshida T. 1977 Decomposition of rice residue in tropical soils IV. The effect of rice straw on nitrogen fixation by heterotrophic bacteria in some Phillippine soils. Soil Sci. Plant Nutr. 23, 287-295.

Section IV

Effects of soil organic matter and applied sewage sludge on soil structure and fertility

9. Soil organic matter extraction, fractionation, structure and effects on soil structure

M.H.B. HAYES

9.1. Introduction

Soil organic matter refers to the non-living heterogeneous mixture of organic components arising from the microbial and chemical transformations of organic debris (43). These transformations or the humification processes, give humus which has a degree of resistance to further degradation in the soil. The predominant components of humus are the humic substances, composed of fulvic acids, humic acids, and humin materials, and soil polysaccharides. Small amounts of protein-type or peptide compounds, nucleic acids, and lipids are invariably present in humus, but these, and especially the nucleic acids and protein-peptide structures which are not associated with and protected by the humic molecules, are thought to have a transient existence in the soil. Hence attention will focus here on the humic and polysaccharide substances as these are known to interact with soil inorganic colloids and are regarded as major contributors to the stabilization of naturally occurring aggregates in soil.

Humic substances are classified on the basis of their solubilities in aqueous acids and bases. *Humic acids* are precipitated from solution in aqueous alkali when the pH is adjusted to 1.0, *fulvic acids* remain in solution when these alkaline solutions are acidified, and *humin materials* are not solubilized in aqueous acid or base. In general it is thought that for the series humin, humic acids, and fulvic acids the carbon and nitrogen contents, and the molecular weights decrease, and the cation-exchange capacity (CEC) and the oxygen contents increase.

Soil structure refers to the association of particles in aggregates which gives rise to pores where air and water reside. Transmission pores have equivalent cylindrical diameters (e.c.d.) greater than 50 μm, and such pores allow free movement of air and the drainage of excess water (35), while storage pores, which retain water against gravity but release it to plant roots, have diameters of 0.5 to 50 μm. Fertile soils have appropriate distributions of transmission and of storage pores, and these distributions are regulated by the sizes of the aggregate structures. Microaggregates consist of associations of soil components less than 250 μm equivalent spherical diameter (e.s.d.) and these microaggregates and other soil components, such as sand grains, are bound in macroaggregates ⟩ 250 μm e.s.d., and the majority of these in well structured soils are in the range of 1-10 mm diameter (26, 27).

Farmers have always recognized the importance of organic matter for maintaining fertility and good soil structure. In the past they used crop rotation tillage practises which employed leys to conserve organic matter, and they have availed of the influences of freezing and thawing and utilized cultivation techniques to produce a good tilth, or a distribution of aggregates of a desired size range. Modern methods of growing the same crop year after year without introducing periodic leys lead to losses of soil organic matter, the breakdown of soil structure and the erosion of fine soil particles. Losses of fertility are exacerbated by soil compaction from heavy agricultural machinery.

The influences of earthworms (24), insects, burrowing animals, the pressure of plant roots, and mechanical mixing by ploughing and cultivating have long been recognized as means of bringing soil particles in contact so that aggregates are formed. So too has the contribution to aggregate formation of bridging by divalent and polyvalent cations between negatively charged soil colloids, but it is clear that aggregates formed in these ways are seldom stable when organic matter is absent. Although Schloesing (67) observed in 1874 that clays have a tendency to adhere in the presence of humus, the necessary background information and the instrumentation needed to study the extents and the mechanisms of the interactions between soil organic and inorganic colloids have become available gradually only during the past 50 years or so. There is, however, much that still needs to be studied about the role of humus materials in the formation and stabilization of soil aggregates.

Oxyhydroxide structures and carbonates of silicon and carbon can also exert strong influences in stabilizing soil structure. The view once held that oxyhydroxides coat clay particles and strongly influence the reactivities of the clay surfaces is losing favour. It is clear though that carbonates and the oxyhydroxides of iron and aluminium can act as interparticle cements in soil. When the precipitates extend between relatively few particles they can contribute to the stabilities of aggregates. In more extreme cases cementation or hardening can take place over an extended area of soil (35). Organic matter is not known to be involved in such cementation processes, but organic matter can of course interact with the charged oxyhydroxide structures.

Discussion will focus here on the extraction, fractionation, primary and tertiary structures of soil humic and polysaccharide molecules, and on interactions between these molecules and the soil inorganic colloidal constituents. On the basis of the information which is available about these structures and interactions, proposals will be put forward to suggest how humus materials are involved in the formation and in the stabilization of soil aggregates.

9.2. Structures of humus materials

In order to study the structure of any naturally occuring macromolecule it is necessary to isolate the material from its natural environment and to purify it. This is especially difficult in the case of humus because a large variety of compounds having similar structures is present in every batch in soil, and because these compounds are associated with the inorganic mineral colloids and other components of soils. There follows a brief outline of procedures used to isolate and to purify humus compounds and some deductions about structures will be made from data which have been obtained using degradative, spectroscopic, and physicohemical methods and measurements.

9.3. Extraction of humic substances

Achard (1), in 1786, was first to describe the isolation of humic materials in solution in aqueous sodium hydroxide, and the same basic solvent is still widely used today for the dissolution of humic substances in soils. Because humic molecules are polyelectrolytes, they remain insoluble in water in soil when the charges are neutralized by divalent and polyvalent cations, or by hydrogen ions in the cases of humic acids. When these ions are replaced by monovalent cations such as Li^+ Na^+, and K^+, which readily dissociate from the acidic humic structures, solvation of the polyanions takes place and the humic and fulvic acids dissolve in water Humins do not dissolve because some property, such as a lower charge density, a higher predominance of non-polar groups, higher extents of cross linking, or combinations of such properties, inhibit swelling and the extensive solvation needed to dissolve the macromolecules.

Best results for extraction under basic conditions are obtained when divalent and polyvalent cations are exchanged by H^+ ions. This is achieved by washing with 1 M HCl, then with water till the solution is chloride free. It is not necessary, however, to H^+ -exchange when the divalent and polyvalent cations are complexed, as was shown by Alexandrova (2) when she used sodium pyrophosphate to complex the calcium ions neutralizing the charges on humates in serosems and the humic components then dissolved in the sodium hydroxide in the mixture. However, as Bremner (7) showed, humic substances are oxidized when dissolved in base and artefacts are formed.

In their search for mild but efficient extractants, Bremner and Lees (8) showed that sodium pyrophosphate, adjusted to pH 7.0 with phosphoric acid, was the best of several neutral salt solutions and complexing agents tested. Replacement of the complexed ions by sodium allows solvation of the conjugate bases in the ionized polyelectrolytes to take place. It is unlikely that all of the divalent and polyvalent cations are exchanged in the neutral media, and the phenolic structures would not be ionized under these conditions. Hence only the most polar and highly charged and/or lower molecular weight humic components are dissolved in neutral sodium pyrophosphate solutions. Solution of H^+-exchanged humic materials in pyrophosphate is low except care is taken to prevent the medium becoming acid as the H^+ ions on the macromolecule are replaced by Na^+.

Organic solvents have not found favour for the extraction of humic substances. One reason is that they tend to be far less efficient than aqueous alkaline solutions, and another is the fact that it is difficult to recover the humic substances from the solvents. Hayes *et al.* (45) compared the efficiencies of a number of organic solvents for the extraction of the humic components from a sapric histosol and found that 2.5*M* aqueous diaminoethane was best. Its efficiency could, however, be attributed to the alkalinity of the solution because anhydrous diaminoethane was a very poor solvent for humic materials.

Hayes (42) has reviewed the properties of good organic solvents for the extraction of soil humic substances. His data show that the best of the solvents have an electrostatic factor (EF) value greater than 140, a base parameter or pK_{HB}(a measure of the ability of the solvent to act as an acceptor in hydrogen bonding processes) value greater than 2, and polar (δp), hydrogen bonding (δh), and proton acceptor (δb) solubility parameters (6) of or greater than 6, 5, and 5, respectively. Water satisfies these criteria, but it is a poor solvent for unionized H^+-exchanged humic substances because it is highly associated through hydrogen bonding and is unable to break the hydrogen bonds linking the macromolecules.

Some of the dipolar aprotic solvents satisfy the criteria for good solvents for humic substances. Dimethylformamide, dimethylsulfoxide or DMSO containing up to 6 per cent of the total volume of concentrated HCl has been shown by Law and Hayes (51) and Fagbenro *et al.* (29) to compare favourably with 0.1 or 0.5*M* NaOH for the extraction of humic substances from a sapric histosol and from two tropical soils. Addition of acid dispensed with the need to exchange the cations for H^+ ions. Divalent and polyvalent cations neutralizing the negative charges on the humic polyelectrolytes were solvated in DMSO. Anions solvate with difficulty only, in this solvent and so ion exchange to give the undissociated H^+-exchanged form is needed for efficient extraction to take place. DMSO is capable of disrupting the inter- and intramolecular hydrogen bonding characteristic of H^+-exchanged humic substances.

9.4. Extraction of soil polysaccharides

Polysaccharides have been extracted from soils in hot water using a soxhlet procedure (21), 98 per cent methanoic acid under reflux conditions (64), 0.3*M* H_2SO_4 (30), dipolar aprotic solvents (45), and 0.5*M* NaOH (74). Highest yields have been obtained by extraction with aqueous solutions of sodium hydroxide, and Swincer *et al.* (74) have shown that yields of polysaccharide in the basic solution extract can be significantly increased by pretreating the soil with $1M$ HCl or HF. The same authors have found that acetylation of the residual materials after extraction in aqueous sodium hydroxide, using acetic anhydride and concentrated H_2SO_4 at 60°C for 2 hours and extracting in chloroform, gave an increase in carbohydrate material of 23 per cent. The amounts of carbohydrate materials extracted in the sequence $1M$ HCl, $0.5M$ NaOH, and in chloroform following acetylation amounted to 57 to 74 per cent of the carboxylate estimated to be present in hydrolysates after soil samples were pretreated in 72 per H_2SO_4 followed by dilution of the acid to $0.5M$ concentration, and then refluxing.

Reviews dealing with the extraction of soil polysaccharides have been provided by Mehta *et al.* (54), Swincer *et al.* (76), Greenland and Oades (38), Hayes and Swift (43), Cheshire (15), and Stevenson (72).

9.5. Fractionation of humic substances

Swift (73) has presented an extensive account of the principles involved and the methods used to fractionate soil humic substances. He has stated that one of the main reasons for carrying out a fractionation is to determine the range of the variations for properties such as molecular weight, functional group content, elemental composition, charge density, *etc.* The techniques for fractionation utilise differences in solubilities, molecular weight, charge, adsorption characteristics, and in their abilities to complex with metal ions.

Fractionation on the basis of differences in solubilities at different pH values has not proved to be as successful for humic acids as might be expected (46), and this can probably be attributed to the fact that many components will be coprecipitated at any pH value. Theng *et al.* (81) have shown that useful fractionation can be obtained by salting out procedures, using ammonium sulphate at pH 7.0. Again, salting out would not be expected to produce clear cut fractions because the boundaries between components are indistinct and coprecipitation is likely to take place.

Dubach *et al.*(25) and others have shown that metal ions can be used to obtain some degree of fractional precipitation of humic acids. The technique might use gradually increasing concentrations of the metal ion, or changing the identity of the ion.

Gel chromatography has been the most successful of the fractionation procedures used till this time. Swift and Posner (74) have discussed some of the problems encountered with this procedure, and in particular that of adsorption by the gel. Cameron *et al.* (10) have successfully used the technique to isolate fractions ranging in molecular weights from 2.6×10^3 to 1.36×10^6, and they showed that the most abundant component of the mixture which they worked with had a molecular weight of the order of 100×10^3.

Nowadays a variety of membranes are available which have discrete pore sizes and selections of these can be used to fractionate humic substances on the basis of molecular size differences. It is important to choose membranes which do not adsorb or reject humic macromolecules. Cameron *et al.* (10) achieved successful fractionation using membrane filtration procedures.

Centrifugation, or more accurately ultracentrifugation provides an appropriate fractionation procedure, although the procedures are time consuming and require expensive instrumentation. The technique is especially useful for measurements of molecular weight values.

Although humic substances isolated by means of some of the procedures mentioned above may be relatively homogeneous with regard to molecular weight, they are likely to be polydisperse with respect to charge density. Electrophoresis provides the classical procedure for fractionating macromolecules on the basis of charge density differences, and recent developments in the general procedure, which include isoelectric focusing and isotachophoresis, have given promising fractionations of humic substances (58).

9.6. Fractionation of soil polysaccharides

The polysaccharide materials isolated from soils are likely to be contaminated with humic acids except acidic solvents are used. Polyclar-AT, a poly(vinyl pyrrolidone) resin product was introduced by Swincer *et al.* (75) to adsorb the brown humic contaminants and the polysaccharides were eluted in the carrying solvent. More recently Amberlite XAD-8 (from Rohm and Hass), a macroreticular, nonionic acrylic ester hydrophobic macromolecular resin, is being used to separate polysaccharides from the brown fulvic acids (82) which are adsorbed by the resin. The sorbed components may be removed by washing through with dilute base, or even with mildly alkaline buffer (51) solutions.

Gel chromatography provides an appropriate procedure to fractionate soil polysaccharides on the basis of molecular size differences (5, 30). Again, components which are homogeneous with regard to molecular sizes are likely to be highly charge heterogeneous. Fractionation on the basis of charge density heterogeneity can be achieved by using anion exchange processes, employing resins, or anion exchange Sephadex and anion exchange cellulose preparations (5, 30, 44).

9.7. Primary structures of humic substances

Primary structures refer to the component molecules or building blocks which compose the macromolecules. The usual procedures for determining primary structures involve degradation of the macromolecules and the identification of the components in the degradation digests. Such degradation and identification processes are relatively simple when the primary structures are held together by labile bonds, such as the glycosidic linkages of polysaccharides and the peptide bonds of proteins. In the cases of humic substances it would appear that the 'backbone' consists of difficult to cleave carbon to carbon bonds and, possibly, ether linkages. For these reasons high inputs of energy are needed to cleave the structures to their component molecules. Advantage is taken of the presence of functional groups which lower the activation energy needed to cleave carbon to carbon bonds. In cases where such groups are absent, the amounts of energy needed to cleave the carbon to carbon bonds would yield products such as carbon dioxide and water which are meaningless for structural assignments. Hayes and Swift (43) and Hayes (41) have described in considerable detail the kinds of products obtained in a variety of different degradation reactions, and they have considered a number of the reaction mechanisms which are likely to operate in the degradation digests.

Haworth and his colleagues (3, 16, 17) showed that nearly half of the masses of humic acid materials could be accounted for as hydrolizable materials, such as sugars and amino acids, when the macromolecules were boiled in $6M$ HCl. Zinc dust distillation at 550°C of the residue gave sufficient yields of compounds such as naphthalene, anthracene, phenanthrene, 1, 2-bezanthracene, chrysine, pyrene, and coronene to suggest that the core or backbone of humic acids is composed of fused aromatic structures, and they suggested that the remaining components of the macromolecules consist of saccharides, peptides, phenols and complexed metals attached to the cores. This thesis was weakened when it was shown that model substances, such as those which could be released into the digest (*e.g.* 3,5-dihydroxybenzenecarboxylic acid), or present as contaminants in humic samples gave rise to similar digest products.

The polycyclic core or backbone concept may be considered to be substantiated by the identification of benzene polycarboxylic acids in alkaline permanganate digests (68, 69). Such acids would be generated from the oxidation of fused aromatic structures and from aliphatic substituents attached to the aromatic nuclei. The same acids were identified in the alkaline cupric oxide digests of humic substances, and the reaction conditions in such digests would not be likely to degrade fused aromatic structures. It is therefore unproven that humic acids contain fused aromatic structures. Aliphatic substituents on the aromatic structures would also give benzenepolycarboxylic acid structures in oxidative degradation processes, but there is also the possibility that the polycarboxylic acids arose from carbonylation reactions in the alkaline digests. We await results from model studies which will show whether or not carbonylation is feasible under these conditions.

Substantial amounts of a variety of aliphatic dicarboxylic acids have been identified in the alkaline permanganate and alkaline cupric oxide digests. These could arise from the cleavage of double bonds separated by different numbers of carbon to carbon single bonds. Tri, and tetracarboxylic acids have also been identified, but these could have been formed from oxidizable branch groups (such as primary alcohols, or olefinic bonds) on the unsaturated hydrocarbon structures.

Degradation reactions in sodium suphide (10%) solutions at elevated temperatures (56) (250°C), and in digests where the humic substances were refluxed in phenol (47), have provided strong evidence of attachments of aliphatic substituents to single ring aromatic structures. Although the digest products from these degradation reactions provided evidence for hydroxysubstituents, there were no indications of benzenepolycarboxylic acid structures, or of sufficient numbers of oxidizable substituents to give four, five, or six carboxyl groups on one aromatic hydrocarbon component.

Reduction during refluxing in sodium amalgam has given a variety of phenolic structures, none of which had more than a single carboxyl substituent (43, 65). From our knowledge of the mechanisms of degradation in this medium, the products identified suggest that many were released from aromatic ether linkages.

Jackson *et al.* (47) and Colclough (23) have adapted the phenol and paratoluenesulphonic acid reflux technique, introduced by coal scientists, to degrade humic acids to ether soluble components amounting to up to 60% of the weights of the starting humic acids. Interpretations based on the structures identified and on model studies indicate that substantial amounts of aliphatic structures link aromatic hydrocarbons in the humic macromolecules.

None of the degradation processes used until the present time are able to cleave all of the humic macromolecules to identifiable digest products. In fact it is rare to find that more than 35 to 40 per cent of the masses of the starting materials can be isolated as ether- or ethylacetate-soluble products. Furthermore, any degradation procedure which is highly energetic, or which involves oxidation processes is likely to bring about substantial changes in the primary structures before and after they are released into the digests from the macromolecules. Therefore consideration must be given to the changes in structures which are likely to take place or to give rise to the compounds identified. This requires a knowledge of the reaction mechanisms which can operate so that reasonable deductions can be made of the types of molecules which are the parent structures of the compounds identified.

All of the degradation procedures mentioned here do not, of course, cleave the same linkages. This has caused Hayes and Swift (43) to suggest that reactions using an appropriate sequence of reagents and conditions might give increased degradations of humic macromolecules. It would be appropriate to start with hydrolysis under anaerobic conditions, then to subject the insoluble residue to refluxing in sodium amalgam, followed perhaps by refluxing the non-degraded residue with phenol, and the residues from this process might then be subjected to degradation with Na_2S at elevated temperature in an atmosphere of hydrogen. Finally, the residues from the foregoing reagents and reactions might be methylated and degraded in alkaline media containing permanganate or cupric oxide.

Some limited information about humic primary structures is provided by spectroscopic analysis procedures. Infrared spectra tend to be featureless, but give evidence for -OH stretch at around 3400 cm^{-1}, carbonyl stretch at 1720 cm^{-1}, and the carboxylate bands at 1610 and 1318 cm^{-1} are generally well defined. Electron spin resonance (ESR) spectroscopy shows that the free radicals in humic substances range from 10^{16} to 10^{18} spins g^{-1}, although we still do not know where the radicals are located or the extents to which they influence reactions of humic substances.

Nuclear magnetic resonance (NMR) spectroscopy has provided more information than the other spectroscopy procedures about the composition of humic substances. The development of cross polarization magic angle spinning (CPMAS) ^{13}C NMR is allowing reasonably well defined spectra to be obtained for solid humic samples. The data obtained by Wilson and his colleqgues for instance (90, 91, 92) indicate that humic substances are less aromatic than might be predicted from consideration only of the products identified in degradation digests. However, an aromatic carbon fraction (f_a) of the order of 0.2 to 0.35 can be considered to be relatively high when account is taken of the fact that hydrolysable components (consisting largely of sugars and with some amino acid residues) can compose 30 to 45 per cent of the masses of humic acids. The predominant chemical shifts reported provide strong evidence for polyethylene $(CH_2)_n$ groups, for 0-alkyl groups characteristic of carbohydrates, for carboxylic and phenolic acids, and of course for aromaticity. However, despite the emphasis which is being placed on the contributions which advanced NMR techniques are making to our knowledge of humic structures, it is well to remember that the instruments which are available at this time cannot provide conclusive identification of any of the component molecules in the humic structures, or about the linkages which join individual pairs of molecules together. Solution and solid state NMR instrumentation provide, however, the basis of very useful procedures for investigating similarities and differences between humic samples from varied environments, and between fractions of humic substances isolated from the same batch.

9.8. Secondary and tertiary structures of humic substances

Biopolymers whose synthesis is genetically controlled have a definite sequence in the order of the binding together of their primary structures. Because there is no evidence for genetic control in the synthesis of humic substances, and because the macromolecules may well arise from several condensation reactions, it is possible that few molecules in any batch are exactly the same. For that reason there is little point in trying to make accurate determinations of humic secondary structures. Appropriate predictions of these can be made if the primary structures are known and if the shapes and sizes, or tertiary structures, are understood.

Humic substances are amorphous in the solid state and thus studies of the sizes and shapes of the molecules are carried out in solution. Because these substances are polyelectrolytes they can be expected to obey the theories applied to charged macromolecules in solution.

Ultracentrifugation has provided the most reliable information about the sizes, shapes, and solution conformations of humic macromolecules. In their classic study Cameron et al. (10) carefully fractionated humic acids using gel chromatography and pressure filtration through graded porosity membranes, and then estimated, from equilibrium ultracentrifugation measurements, molecular weights for the samples ranging from 2.6×10^3 to 1.36×10^6. Ultracentrifugation also provided data for frictional ratio (f/f_{min}) values, where f is the frictional coefficient of the molecule under consideration and f_{min} is that of a condensed sphere (containing no solvent occupying the same volume).

Data by Cameron et al. (10) showed that values for the frictional ratios varied with the molecular weights of the humic acids (see also Hayes and Swift (43) p. 278 and Stevenson (72) p. 291), and these could be fitted to a linear relationship $f/f_{min} = 0.3M^{1/6}$, where M is the molecular weight value, which would be followed if the humic molecules had random coil or oblate ellipsoid (condensed discusshaped) conformations in solution. There was an excellent fit of the data (for the plot of f/f_{min} versus molecular weight) for the calculated straight line relationship for molecular weight values up to about 3×10^5 and the authors considered that the shapes of the macromolecules would be more in keeping with the random coil than with the oblate ellipsoid structures. The points on the plot were seen to deviate from linearity for the higher molecular weight humic samples, and this was interpreted to indicate more extensive branching in such instances. These studies used tris (or 2-amino-2hydroxy-1,3-propanediol) buffer, pH 9.2 for the fractionation and centrifugation experiments, and not all of the humic acid materials were completely dissolved in it. It may be assumed that the insoluble components had lower charge densities and/or higher molecular weight values, and/or higher degrees of cross linking than the components which did dissolve.

9.9 General conclusions from studies of humic structures

Humic substances, as defined from the classic fractionation procedures, may contain from 30 to 50 per cent hydrolyzable materials, which are mostly sugars and with some amino acids from peptide or protein structures. The non-hydrolyzable 'backbone' may be 30-50 per cent aromatic, composed of single ring aromatic structures, many of which have hydroxyl, and/or methoxyl substituents with or without carboxyl and/or hydrocarbon substituents. Linkages between the component molecules would include aliphatic polymethylene structures linking the aromatic nuclei, and it would appear that many polymethylene structures contain some olefinic groups. It is highly likely that ether linkages, including aromatic-aromatic, aromatic-aliphatic, and aliphatic-aliphatic, contribute to the structures, and there is good evidence to indicate that carboxyl groups, which may be attached to aliphatic and to aromatic structures, quinones, aliphatic hydroxyl, and carbaldehyde functional groups contribute to the structures.

There is overwhelming evidence to suggest that humic acids have random coil-type conformations in solution. The strands would bend randomly with respect to time and space to give a shape which is roughly spherical. There would be a gaussian distribution of mass within the sphere. Solvent would be free to perfuse throughout the whole molecule, although it would be most restricted towards the centre where the mass density is greatest.

When NH_4^+, or monovalent metal cations neutralizing the negative charges on the humic polyelectrolytes are replaced by divalent and polyvalent cations, or by H^+ ions, the macromolecules shrink, water is excluded, and eventually precipitation to sols or gels takes place. H^+-exchanged humic acids are insoluble in water because the carboxyl groups are not sufficiently dissociated to carboxylate anions which are readily solvated. Replacement wit divalent and polyvalent cations allows a single cation to satisfy two or more charges and to from bridges within and/ or between strands.

As drying proceeds, the strands are brought closer together enabling van der Waals attractive forces between the strands to supplement the bridging and hydrogen bonding effects. Thus humic molecules which have random coil conformations in solution would become close packed, condensed, amorphous structures when dried. Because dried humic materials are difficult to rewet, it would appear that the more polar groups are orientated towards the interior and that the more hydrophobic structures are exposed to the outside.

Little is known about the nature of nitrogen and sulphur in the 'backbone' of humic structures. Reference was made to amino acids in peptides, and these could be held to the 'backbone' as Schiff bases, and as carbon to nitrogen bonds α- to the carbonyl groups of quinones. Although significant amounts of heterocyclic structures have not been isolated from among the degradation products in humic digests, it would appear likely that some such compounds are contained in the macromolecules. Such structures would be expected should Maillard or browning-type reactions (involving interactions of amino acids with sugars) contribute to the synthesis of humic macromolecules. Sulphur would, of course, be a component of cysteine, cystine and methionine, but it could also be present in thiol, thioether, and heterocyclic sulphur compounds in the humic macromolecules.

9.10. Structures of soil polysaccharides

Hydrolysis of soils (15) generally yield the hexose sugars glucose, galactose, and mannose, the pentoses xylose and arabinose and sometimes traces of ribose, the 6-deoxy sugars (in which the $-CH_2OH$ groups in C-6 are replaced by $-CH_3$) fucose (or 6-deoxy-L-galactose), and rhamnose (or 6-deoxy-L-mannose), the amino sugars glucosamine (or 2-amino-2-deoxy-D-glucose) and galactosamine (or 2-amino-2-deoxy-D-galactose), the uronic acids (where the $-CH_2OH$ groups on C-6 are replaced by $-CO_2H$) glucuronic acid and galacturonic acid, trace amounts of the methylated sugars 4-O-methyl-D-galactose, 2-O-menthyl-L-rhamnose, 2-O-methylxylose, and 2-O-methylarabinose, the amino sugar muramic acid (or 3-O-carboxyethyl-D-glucosamine), and the sugar alcohols mannitol and inositol. In nature the amino sugars are in the acetylated forms, but the acetyl groups are released in the hydrolysis process.

As yet no one has claimed to have isolated a polysaccharide from soil which satisfies all of the criteria of homogeneity, although the composition of some of those described by Hayes et al.(44) and by Clapp et al. (21) may well have been close to being homogeneous macromolecules. Determinations of the secondary and tertiary structures of polysaccharides is relatively easy when homogeneous structures are available to work with, but no one has yet had sufficient confidence in their isolates to carry out extensive structural studies. In general we know from Cheshire (15) that soil polysaccharides are laevorotatory, which suggests that β-linkages between the sugar units predominate. A predominance of β-glycoside linkages would indicate a preference for linear helical structures. Where α-glycosidic linkages predominate, the polysaccharides would assume more globular (as in starch) or random coil structures . The β-configuration and the linear helix-type structure allows intimate contact between the component sugars of the polysaccharides and flattened surfaces such as those of clays and inorganic soil adsorbents.

9.11. Interactions of humus materials with soil inorganic components

In mineral soils most of the organic matter is intimately associated with the inorganic colloidal components (34, 72, 77, 85). The most usual way to study the nature of humus mineral complexes is to interact humus components with well characterized clays and oxyhydroxides. Because the soil environment consists of a heterogeneous mixture of organic and inorganic materials and living organisms, it is very difficult to divise laboratory experiments which might simulate the complicated mixtures of components and the physical, chemical and biological interactions which are involved in the formation of a stable soil aggregate. Thus attention will be focused here on laboratory studies of the binding of humic substances and polysaccharides to clays.

Reference will be made to the importance of such bindings to oxyhydroxides, although few studies have been made in this area, and attention will be given to the ways in which organo-mineral complexes can be involved in the formation and stabilization of aggregates.

9.12. Humic substances - clay interactions

Clays have net negative charges, as do humic substances in solution. Hence the overall effect would be one of repulsion if sodium humate polyanions were interacted with a monovalent cation-exchanged clay, and this is inherent from the small uptake obtained when Chaney and Swift (12) mixed humic acids in solutions at pH 7 with Na^+-and Ca^{2+}-exchanged dispersed soil samples. However, nearly all of the humic materials were sorbed when the pH was 3.5. At this pH the humic acids were still in solution, but it is obvious that most of the carboxylate and all of the phenolic structures would be protonated. Under such circumstances the decreased negative charge in the humic macromolecules would allow the random coil structures to shrink and some inter- and intramolecular hydrogen bonding could take place. Uncharged stretches of the humic macromolecules could sorb to the silicate surfaces by weak hydrogen bonding, and by van der Waal's forces, for example. In the cases of Ca^{2+} and of other divalent and polyvalent exchanged clays and soils, sorption mechanisms could also involve coordination of carboxylate groups to the cation, or more likely to the primary hydration shell of the cation. This principle is illustrated in the studies of Theng and Scharpenseel (80).

Figure 9.1 presents a selection from the isotherms obtained by Theng and Scharpenseel (80) for the sorption of sodium humate preparations with various homoionically exchanged montmorillonite clays. Sorption for Na^+ and the other monovalent cation-exchanged clays was small, but the amounts sorbed increased as the valence of the exchangeable cations increased. The ionic potential (or the ratio of the valency of the cation to its radius) was linearly related (with a positive slope) to the logarithm of the slope of the isotherms in the cases of sorption by the Ba^{2+}, Ca^{2+}, Zn^{2+}, Co^{2+} and La^{3+} -exchanged clays, and this was considered to be good evidence for cation bridging between the humate adsorbate and clay adsorbent. This kind of linear relationship was found also for sorption by the monovalent cation-exchanged clays, but the slope was negative. The positive slope was interpreted as indicative of outer-sphere coordination complexes (water-bridge between the adsorptive and the cation), and the negative slope suggested inner-sphere complexes, or direct coordination of the adsorptive to the cation.

Uptake of humic acids by the Fe^{3+}- and Al^{3+}-exchanged clays was less than would be predicted from considerations of the ionic potentials of the cations, and

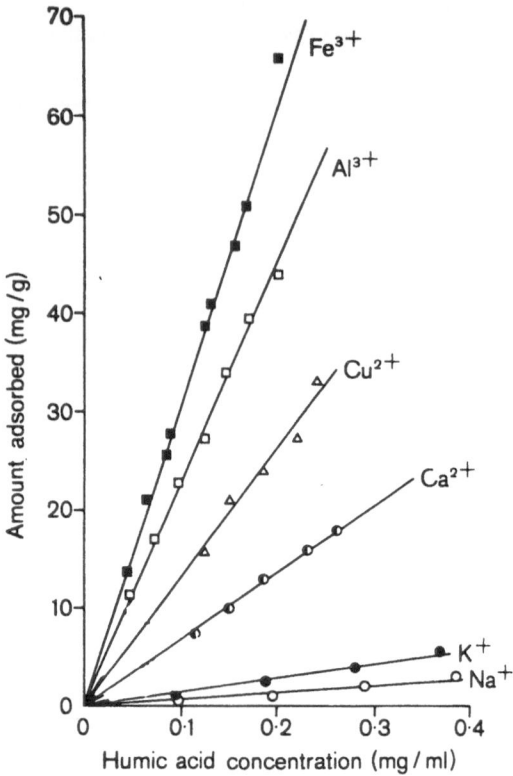

Fig. 9.1. Isotherms at pH 7 for sorption of humic acids by clays exchanged with different cations (from Theng and Scharpenseel (80)).

this was thought to result from the formation of polyhydroxy cations of lesser charge at the clay surfaces.

Theng (78) obtained similar linear isotherms for the sorption of fulvic acids at pH 7, by homoionic-exchanged monomorillonites, and Kodama and Schnitzer (49, 50), among others, have also shown how the resident cations neutralizing the charges on the clays influence fulvate sorption. In a further study Kodama and Schnitzer (50) showed that fulvic acids complexed Cu^{2+} ions, and the ability to complex increased with pH indicating the involvement of ionized fulvic groups in the complexation mechanisms. However, monovalent ion-exchanged clays would not sorb fulvate anions. At low pH values, where the fulvic acids were not dissociated but were, of course, still water soluble, adsorption was observed even between the clay layers.

Linear or constant partition (C-type) isotherms indicate, according to Giles *et al.* (33) that the numbers of binding sites remain constant as sorption proceeds. This would mean that when one site is filled another is made available, and in general this would suggest that the sorbed molecules expand or open in some way the sorbent structures to allow the sorptive molecules to penetrate to interior sorption sites in the sorbent structure. Interlayer sorption by expanding layer clays would provide one process which could explain linear isotherms. However, the C-type shape was obtained even when no interlayer sorption was detected. In order to explain the phenomenon Theng (79) considered that the humic molecules generated new sorption sites by pushing clay domains apart and allowing the sorptive molecules access to the pores formed from the turbostratic arrangements of the domains.

The initial parts of the isotherms obtained by Evans and Russel (28) for sorption of humic acids by H^+- and Ca^{2+}-exchanged montmorillonite and kaolinite clays were also linear. They showed that sorption decreased as the pH of the medium was raised from 3 to 6, and the decrease was especially marked for the pH range 3.5 to 4.5. There is the possibility, of course, that precipitation and not sorption would take place at the low pH values. However, in the cases of the experiments by Chaney and Swift (12), for example, the humic acids at pH 3.5 were not precipitated when centrifuged at the speed needed to sediment the clay-humic complexes. Any suspicion that precipitation resulting from complexation by cations neutralizing the charges on the clays could take place can be removed by equilibrating the humic solution at the appropriate pH with the concentration of the cation associated with the clay and observing whether or not a precipitate settles during centrifugation.

Results obtained by Chassin *et al.* (14) for sorption of Na^+-humates and fulvates by Al^{3+}- and Ca^{2+}-montmorillonite gave L-shaped isotherms with plateaus at equilibrium solution concentrations between 0.8 and 1.2 mg cm^{-3}. The humic substances were dissolved in pH 13 sodium hydroxide solutions and these were dialysed against distilled water to remove the excess base. Fractionation on the basis of molecular weight differences was achieved by use of gel chromatography, and it was observed that the lower molecular weight fulvates were preferentially sorbed.

9.13. Soil polysaccharide-clay interactions

There are few studies which have looked at the sorption of soil polysaccharides by clays. Finch *et al.* (30) took a mixture of polysaccharides extracted in $0.3M$ H_2SO_4 from a sapric histosol and they reacted these with a freshly prepared H^+-montmorillonite. The clay was passed through a H^+-exchanged resin and immediately frozen and subjected to freeze drying. Samples of the freeze dried clays

198

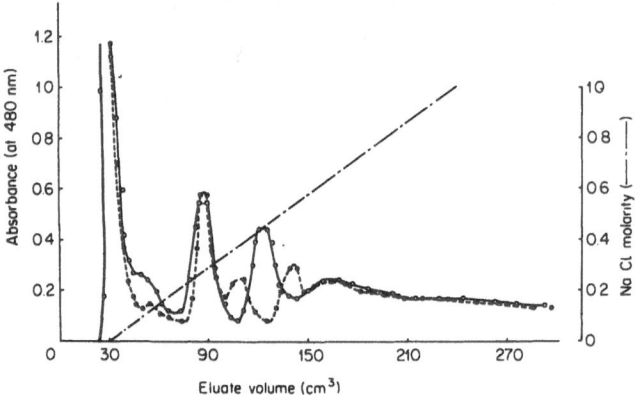

Fig. 9.2. Anion-exchange chromatography on DEAE-G50 Sephadex in pH 6.0 phosphate buffer of a soil polysaccharide mixture before (-o-o-) and after (-x-x-) reaction with H[+]-exchanged montimorillonite (from Finch *et al.* (30)).

were added to the mixtures of polysaccharides in solution, and after equilibration and centrifugation the supernatant was eluted through a column of anion exchange DEAE -G50 Sephadex using a 0 to $1M$ NaCl gradient. For the control, the same concentration of polysaccharide solution which had not been equilibrated with the clay was passed through an identical column.

Figure 9.2. shows the elution pattern obtained for the polysaccharide mixtures. It is evident that fractionation of components could be obtained on the basis of charge density differences, and there was clear evidence for the preferential sorption by the clay of one of the polysaccharide components. Barker *et al.* (5) showed that this component had one uronic acid unit for every six to seven sugar residues. No proof was obtained to indicate whether or not the uronic acids were involved in the sorption processes. The carboxyl groups were not dissociated at the pH of the medium and so these could have taken part in hydrogen bonding to the siloxane surface. Some Al^{3+} ions could have been released from the clay lattice in the acid medium, and if so coordination between these ions and the carboxyl groups would have contributed to the sorption process. Parfitt (62) has provided infrared evidence for coordination between the carboxyl groups of poly galacturonic acid and Al^{3+} ions in Al^{3+}-exchanged montomorillonite.

There is no information available about the configuration of the glycosidic linkages in the different polysaccharides in the components shown in Figure 9.2. The importance of the configurations of such linkages is illustrated in the studies of Olness and Clapp (57). They showed that substantially more of the commercial polyglucose dextran, Polytran, was adsorbed by Na^+-montmorillonite than the polysaccharide B-512F synthesized by a strain of *Leuconostoc mesenteroides*. Polytran has a molecular weight of the order of 2×10^6 and with 75 per cent of the glucose units linked β- $(1 \rightarrow 3)$ and 25 per cent linked β- $(1 \rightarrow 6)$, whereas in the case of B-512F 95 per cent of the linkages are α- $(1 \rightarrow 6)$ and 5 per cent are α- $(1 \rightarrow 3)$. Reference was made above to the fact that sugar molecules having β-glycosidic linkage configurations can make more intimate contact, than those having α-configurations, with flattened sorbent surfaces, such as those of clays.

9.14. Humus-oxyhydroxide interactions

The role of oxyhydroxides in binding humus components in soils is very imperfectly understood, and systematic studies in this area are relatively new. At pH values where oxyhydroxides are positively charged, binding by ion exchange of humic polyanions could be expected. Turchenek and Oades (85) have shown, however, that highly crystalline iron oxides have little affinities for humic substances, but this might be explained by the relatively small surface areas exposed by such crystalline materials. According to Schwertmann *et al.* (71) there are substantial amounts of amorphous iron hydroxides in soils whose crystallizations might be inhibited by humus materials. The mechanisms of such inhibitions are unknown, as are the extents of associations between the organic and inorganic amorphous materials.

Parfitt *et al.*(63) have shown that humic and fulvic acids are adsorbed by goethite gibbsite, and imogolite and their data suggest that ligand exchange processes are involved in the binding. The work of Broadbent *et al.* (9) and Tokudome and Kanno (84) show that large amounts of humic materials are bound in Andosols derived from volcanic ash, and which contain allophane and imogolite as the predominating clays. Wada (87) has drawn attention to the important role of these minerals in holding humus in Andosols, but he has also stressed that there is a very good linear correlation between Al and the humus contents of such soils whether or not allophane is present.

9.15. Humus and soil aggregates

Soils which are lacking in structure, and in which the component particles are randomly dispersed in the soil solution, will not be aggregated by additions of soluble humus materials. Components of the humus may well be adsorbed by the soil constituents, but mixing of the soil and amendments, as well as adsorption, would be needed for aggregates to form. Addition of calcium ions, and of other divalent and polyvalent cations, and mixing and drying would promote improvement of structure, but the aggregates formed in this way generally have limited stabilities under the impact of rain drops, or when subjected to mechanical perturbations when the soil moisture contents are at field capacity or higher. The presence of humus materials promotes good soil structure, and the discussion which follows will try to explain some of the reasons why and how humus helps to stabilize soil aggregates.

In order to stabilize an aggregate by addition of any amendment material it is necessary for that material to penetrate into the aggregate and to bind together two or more of the component structures in the aggregate. In the cases of organic molecules in solution it is important that these should diffuse to internal sorption sites and be adsorbed, or in some way bound to neighbouring soil components. It is evident that the larger the molecule the more likely it is to be able to bridge across two or more soil particles.

The most effective molecules for stabilizing soil aggregates would have linear or linear helix conformations in solution. Such conformations would span the longest distances, and even relatively short macromolecules could be effective binders (if enough were present) of soil inorganic colloids. Molecules with random coil structures would span shorter distances than linear molecules with the same molecular weights. However, when a random coil structure anchors to two or more soil components it could provide several strands bridging the gap between the components. Guidi et al. (39) and Paglai et al. (61) have shown that molecular size influenced the pore sizes and the stabilities of the aggregates in some of the soils which they investigated. In general, they found that the efficiencies of the polyglucose dextrans (from Pharmacia, Sweden) decreased in the order T-2000 ⟩ T-500 ⟩ T-10. The predominant linkages were α- $(1 \to 6)$ and with some α-$(1 \to 3)$, and the molecular weights were 1,960,000, 476,000 and 10,300 for the T-2000, T-500 and T-10 materials, respectively. Such structures would have random coil-type solution conformations. Soluble cellulose, with its β- $(1 \to 4)$ glucose linkages and linear helix structure would be expected to be a better stabilizer of aggregates. Page (59, 60) has demonstrated in laboratory and field studies that aggregates are well protected by cellulose xanthate, and Harrison (40) has shown that CS_2 is given off as the xanthate is sorbed by clay. Thus the sorbate would consist of cellulose molecules held by the inorganic colloids.

The organic molecules in soil which are most likely to be water soluble and mobile are fulvic acids and polysaccharides. Polysaccharides would arise as exudates from plant roots or as exocellular microbial polysaccharides, or contained in bacterial or fungal cell walls. Many of these polysaccharides have the properties of gums, are highly viscous, and will generally stay at the location where they were exuded or secreted. Recent studies, using scanning electron microscopy (SEM) and transmission electron microscopy (TEM) suggest that finely divided clays when migrating in soil become entrapped by the gums, providing a degree of protection for the polysaccharide materials against the hydrolysis of soil enzymes (31, 32, 48). In this way the gums and mucilages would act as nuclei for the formation of microaggregates and aggregates. The entrapment of the clays in the gums might represent a physical entrapment, and not adsorption in the classical sense. Water soluble polysaccharides would be free to move with the soil solution. These could enter aggregates and be adsorbed by clays linking particles and even domaines.

Waksman and Martin (88) introduced the concept of microbial products for promoting good soil structure, and later Martin (52, 53) showed that the most effective slimy bacterial products were polysaccharides. During the 1950's good soil structure was statistically correlated with microbial gums and polysaccharides (20, 66). However, some doubts were cast on this stabilization role of polysaccharides when Mehta *et al.* (54) showed in 1960 that the crumb structure of a Swiss braunerde resisted degradation after treatment with periodate. Greenland *et al.* (36, 37) considered that the unusual stability of the soil crumbs after the periodate treatment might be attributed to mechanical binding by fungal hyphae and myceliae. Also, the soil contained free calcium carbonate, and Clapp and Emerson (22) have shown that such soils can retain their structure when treated with periodate.

Recent work by Cheshire *et al.* (19) has shown that long periods of treatment with periodate are needed to degrade substantial amounts of carbohydrates in soil.

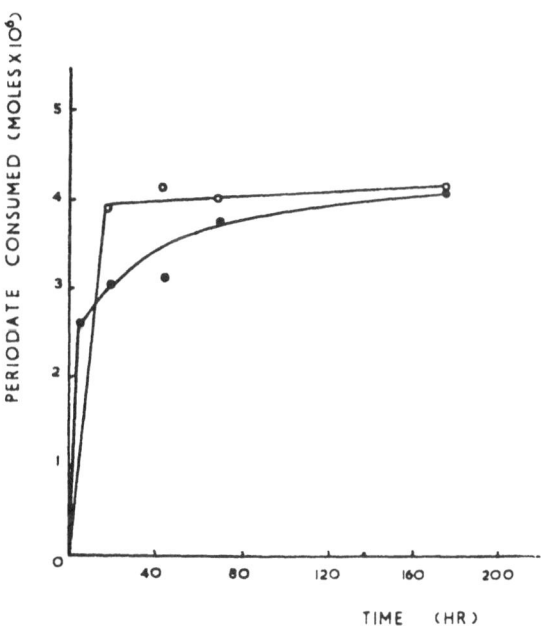

Fig. 9.3. Uptake of periodate by a soil polysaccharide in the presence (- - -) and in the absence (-o-o-) of Na^+-exchanged kaolinite.

This is understandable when the polysaccharides are adsorbed by the soil components, or are in some way rendered inaccessible to the periodate. Finch *et al.* (30) have shown, for example, that adsorption of a soil polysaccharide by Na^+-exchanged kaolinite followed the periodate oxidation of the macromolecule. Figure 9.3. indicates that oxidation of the polysaccharide was the same after one week of reaction with periodate whether or not clay was in the medium. However, degradation was much slower in the presence of the clay, and it can be assumed that sorption of the clay provided a steric constraint to the formation of the five-membered cyclic intermediate required for cleavage with periodate.

Some of the effects of plant roots on soil structure appear to contradict the concept that root mucigels are important for promoting good soil structure. Micro-aggregates are broken down in the rhizosphere of some plant roots, and one explanation offered for this is that root exudates chelate iron and aluminium ions bridging humus components to the inorganic colloids. However, the disruptive effects as the roots grow and expand must be considered in this context, and also the influences which the secretions might have on the existing organic substances stabilizing the aggregates might be important.

Edwards and Bremner (26, 27) have represented soil micro-aggregate structures as

$$[(C - P - OM)_x]_y$$

where C is clay, P is a polyvalent metal cation, OM is humified soil organic matter, and x and y are whole numbers whose magnitudes are determined by the sizes of the clay particles. If the polyvalent metal ions are exchanged for sodium ions the micro-aggregates can be disrupted by mild shaking. The polyvalent cation bridge is the most essential part of this mechanism, and the fact that soils can be dispersed by reacting with complexing agents, such as pyrophosphate, provides some proof of this.

Chaney and Swift (11) found significant correlations between soil aggregate stabilities and the humus materials extracted in pyrophosphate and in pyrophosphate followed by aqueous sodium hydroxide. Correlations were better for the pair of solvents and this caused them to suggest that the less oxidized and higher molecular weight substances were more strongly bound to the inorganic colloids than the more highly oxidized and lower molecular weight substances extracted in pyrophosphate alone.

Turchenek and Oades (86) have shown that the clay-humus associations can vary from soil to soil, and their data indicate significant differences between the humic materials associated with the various clay fractions even in the same soil.

Fractionation of the different soil components was achieved by sedimentation and density gradient techniques, and it was found that the more aliphatic type humic substances were preferentially associated with the heavier fine clays, and there was some evidence to suggest that the materials associated with the lighter and coarser fractions were more highly aromatic. The association of Ca, Fe, and Mn ions with the lighter fractions suggested to the authors that cation bridges were involved in binding the more oxidized humic substances to the clays, and they considered that physical adsorption forces might be important for binding the more highly aliphatic materials. This type of research is new, but it promises a very fruitful approach to the study of clay-humic interactions. The fact that there appears to be some selectivity on the part of clays for humic substances is important, and the uses of separation techniques, coupled with CPMAS ^{13}C NMR would be expected to provide information about the extents of the differences within and between soils.

Drying would appear to be very important for the formation of stable aggregates. There are no data to suggest that stable aggregates are formed by mixing aqueous suspensions of soil inorganic and humic colloids. Drying brings the adsorbent and adsorbate species into intimate contact and the associations formed have significant resistance to disruptive forces. Chassin et al. (13) have given a clear demonstration of the effects of drying. They prepared Ca^{2+}-humate - Ca^{2+}-montmorillomite complexes by dehydrating the mixture at pH 7.0 and 40°C, and by flocculating the components in $1M$ $CaCl_2$ and then drying at 40°C. Where flocculation was not involved the humate coated the clays as the water was removed. Flocculation prior to drying caused the humate particles to associate in spherical structures and these associated with the clay surfaces as the suspension was dried. The complexes formed between the clays and the humates were strong, and the uptake of water by the clays, the porosities of the aggregates, and the surface areas of the particles were strongly influenced by the amounts of organic matter in the clay-humate associations, and by the method of preparation used.

The manner in which humic acids and humin materials interact in the field with the inorganic soil colloids in still unresolved. It is known that humic substances in stable soil aggregates have a very long half life, and this is considered to be attributable to the fact that the sorbed materials are in some way protected from enzymatic attack. When the aggregates are broken, as happens during cultivation practices in tillage operations, the humic substances lose their protection and degradation takes place, though at a considerably slower rate than that of fresh plant and animal residues.

Although humic acids have not been shown to penetrate between the layers of expanding clays in laboratory studies, there are reports of intercalated humic acids in clays in the field. One possible explanation for this would be the chemical synthesis of the macromolecules between the layers, with the clay surfaces and the cations neutralizing their charges acting as catalysts. Another explanation might be that at some time an appropriate pH existed in the soil to dissolve some humic acids and these were then able to slowly diffuse between the clay layers. However, clay interlayer humic acids appear to be rare, and for the most part these humic substances are held on the external surfaces of clays and between domaines. It would be appropriate for humic substances to bridge neighbouring domaines and to be protected by these domaines. It is conceivable that the clays and humic substances came together in moist conditions and that the associations were strangthened on drying. The initial bonding would be through divalent and polyvalent cation bridges, but the close contact on drying would enhance the participation of secondary forces in the sorption process.

Mention was made of the probable random coil structure of humic substances in solution, and it was postulated that on drying the coils shrink to give solid or gel structures. At the molecular level the macromolecules would adhere through 'bridging' cations, van der Waals forces *etc.* This intermolecular humic affinity would help to provide a short chain of humic molecules holding clay particles and domaines together in micro-aggregates.

It is unlikely that humic molecules can provide the 'long and strong bridges' between the components in macro-aggregates having predominances of macro-pores. Even if they could they would be exposed to attack by micro organisms in the pores. Some synthetic polymers and cellulose xanthate appear to be able to bridge these gaps and to stabilize the aggregates in most soils (59, 60). However, it is aften considered that many soils owe their stabilities to the mechanical binding effects of grass roots and to fungal hyphae and myceliae. It is well to remember that the roots and myceliae must adhere to the components of the aggregates in order to confer such stability. The gums, mucilages, and other polysaccharide structures secreted by roots and by the microfauna would be appropriate adhesives.

9.16. References

1 Achard F.K. 1786 Untersuchung des Torfs. Grell's Chem. Ann. 2, 391.

2 Alexandrova L.N. 1960 The use of sodium pyrophosphate for isolating free humic substances and their organic mineral compounds from the soil. Soviet Soil Sci. 2, 190-197.

3 Atherton N.M., Cranwell P.A., Floyd A.J. and Haworth R.D. 1967 Humic acid - I ESR spectra of humic acids. Tetrahedron 23, 1653-1667.

4 Bailey G.W. and White J.L. 1970 Factors influencing the adsorption, desorption and movement of pesticides in soil. Residue Rev. 32, 29-92.

5 Barker S.A., Hayes M.H.B., Simmonds R.G. and Stacey M. 1967 Studies on soil polysaccharides I. Carbohyd. Res. 5, 13-24.

6 Barton A.F.M. 1975 Solubility parameters. Chem. Rev. 75, 731-753.

7 Bremner J.M. 1950 Some observations on the oxidation of soil organic matter in the presence of alkali. J. Soil Sci. 1, 198-204.

8 Bremner J.M. and Lees H. 1949 Studies on soil organic matter II. The extraction of organic matter from soil by neutral reagents. J. Agric. Sci. 39, 274-279.

9 Broadbent F.E., Jackman R.H. and McNicoll J. 1964 Mineralization of carbon and nitrogen in some New Zealand allophanic soils. Soil Sci. 98, 118-128.

10 Cameron R.S., Thornton B.K., Swift R.S. and Posner A.M. 1972 Molecular weight and shape of humic acid from sedimentation and diffusion measurements on fractionated extracts. J. Soil Sci. 23, 394-408.

11 Chaney K. and Swift R.S. 1984 The influence of organic matter on aggregate stability of some British soils. J. Soil Sci. 35, 223-230.

12 Chaney K. and Swift R.S. 1985 The effect of polysaccharide and humic substances on the stability of reformed aggregates. *Submitted* J. Soil Science.

13 Chassin P., Le Berre B. and Nakaya N. 1978 Influence des substances humiques sur les proprietes des argiles IV. Hydration des associations montmorillonite acides humiques. Clay Min. 13, 1-15.

14 Chassin P., Nakaya N. and Le Berre B. 1977 Influence des substances humiques sur les proprietes des argiles II. Adsorption des acides humiques et fulviques par la montmorillonite. Clay Min. 12, 261-271.

15 Cheshire M.V. 1979 Nature and Origin of Carbohydrates in Soils. Academic Press, London and New York.

16 Cheshire M.V., Cranwell P.A., Falshaw C.P., Floyd A.J. and Haworth R.D. 1967 Humic acid - II. Structure of humic acids. Tetrahedron 23, 1669-1682.

17 Cheshire M.V., Cranwell P.A. and Haworth R.D. 1968 Humic acid - III. Tetrahedron 24, 5155-5167.

18 Cheshire M.V. and Hayes M.H.B. 1985 Composition, origins, structures, and reactivities of soil polysaccharides. *In* Soil Colloids and Their Associations in Soil Aggregates. Eds. M.F.L. De Boodt, M.H.B. Hayes and A. Herbillon. Plenum. *In* preparation.

19 Cheshire M.V., Sparling G.P. and Mundie C.M. 1983 Effect of periodate treatment of soil on carbonhydrate constituents and soil aggregation. J. Soil Sci. 34, 105-112.

20 Chesters G., Attoe O.J. and Allen O.N. 1957 Soil aggregation in relation to various soil constituents. Soil Sci. Soc. Amer. Proc. 21, 272-277.

21 Clapp C.E., Dawson J.E. and Hayes M.H.B. 1979 Composition and properties of a purified polysaccharide isolated from an organic soil. pp. 153-167. *In* Proc. Intern. Symp. Peat in Agriculture and Horticulture. Bet Dagan, Israel.

22 Clapp C.E. and Emerson W.W. 1965. The effect of periodate oxidation on the strength of soil crumbs. I. Qualitative studies II. Quantative studies. Soil Sci. Soc. Am. Proc. 29, 127-134.

23 Colclough P. 1980 Degradation of humic acids with phenol and ptoluenesulphonic acid. Internal Report, Chemistry Department, University of Birmingham.

24 Darwin C. 1881 and 1927 The formation of vegetable mould through the action of worms with observations on their habits. Faber and Faber, London.

25 Dubach P., Mehta N.C. and Deuel H. 1961 Extraktion von Huminstoffen aus dem B-Horizonteines Podsols mit ADTE. Z. Pflanzenernahr Düng. Bodenk. 95, 119-123.

26 Edwards A.P. and Bremner J.M. 1967 Dispersion of soil particles by sonic vibration. J. Soil Sci. 18, 47-63.

27 Edwards A.P. and Bremner J.M. 1967 Microaggregates in soils J. Soil. Sci. 18, 64-73.

28 Evans L.T. and Russell E.W. 1959 The adsorption of humic and fulvic acids by clays. J. Soil Sci. 10, 119-132.

29 Fagbenro J., Hayes M.H.B., Law I.A. and Agboola A.A. 1985 Extraction of soil organic matter and humic substances from two Nigerian soils using three solvent mixtures. *In press.* International Humic Substances Society, 2nd Intern. Conf. (Birmingham).

30 Finch P., Hayes M.H.B. and Stacey M. 1967 Studies on soil polysaccharides and their interactions with clay preparations. Int. Soil Sci. Soc., Trans. II and IV Comm., 1966, Aberdeen. pp. 19-32.

31 Foster R.C. 1981 Polysaccharides in soil fabrics. Science 214, 665-667.

32 Foster R.C. 1982 The fine structure of epidermal cell mucilages of roots. New. Phytol. 727-740.

33 Giles C.H., Smith D. and Huitson A. 1974 A general treatment and classification of the solute adsorption isotherm. I. Theoretical. J. Colloid Interface Sci. 47, 755-765.

34 Greenland D.J. 1965 Interaction between clays and organic compounds in soils. I and II. Soils Fertil. 28, 415-425 and 521-532.

35 Greenland D.J. and Hayes M.H.B. 1978 Soils and soil chemistry. In The Chemistry of Soil Constituents. Eds. D.J. Greenland and M.H.B. Hayes. Wiley, Chichester and New York. pp. 1-27.

36 Greenland D.J., Lindstrom G.R. and Quirk J.P. 1962 Organic materials which stabilize natural soil aggregates. Soil Sci. Soc. Am. Proc. 26, 366-371.

37 Greenland D.J., Lindstrom G.R. and Quirk J.P. 1961 Role of polysaccharides in stabilization of natural soil aggregates. Nature London 191, 1283-1284.

38 Greenland D.J. and Oades J.M. 1975 Saccharides. In Soil Components Vol 1. Ed. J.E. Gieseking. Springer Verlag, Berlin. pp. 213-261.

39 Guidi G., Pagliai M., Petruzzelli G. and Aringhieri R. 1978 Changes in some physical properties of clay soils induced by dextrans. Z. Pflanzenernaehrung Bodenkd. 141, 367-377.

40 Harrison R. 1982 A study of some montmorillonite-organic complexes. Ph. D. Thesis, University of Birmingham.

41 Hayes M.H.B. 1985 Structures of humic substances. In Organic Matter and Rice. The International Rice Research Institute, Los Banoo, The Philippines. pp. 93-115.

42 Hayes M.H.B. 1985 Extraction of humic substances from soil. In Humic Substances in Soil, Sediment and Water. Eds. G. Aiken et al. Wiley, New York. pp 329-362.

43 Hayes M.H.B. and Swift R.S. 1978 The chemistry of soil organic colloids. pp. 179-320. In The Chemistry of Soil Constituents. Eds. D.J. Greenland and M.H.B. Hayes. Wiley, Chichester and New York.

44 Hayes M.H.B., Stacey M. and Swift R.S. 1975 Techniques for fractionating soil polysaccharides. Trans. 10th Intern. Cong. Soil Sci. Moscow. Suppl. Vol. pp. 75-81.

45 Hayes M.H.B., Swift R.S., Wardle R.E. and Brown J.K. 1975 Humic materials from an organic soil: A comparison of extractants and of properties of extracts. Geoderma 13, 231-245.

46 Hobson R.P. and Page H.J. 1932 Studies of the carbon and nitrogen cycle in soils. J. Agric. Sci. 22, 497-515.

47 Jackson M.P., Swift R.S., Posner A.M. and Knox J.R. 1972 Phenolic degradation of humic acid. Soil Sci. 14, 75-78.

48 Kilbertus G. 1980 Etudes des microhabitats concenus dans les agregats du sol. Leur relation avic la biomass bacterienne et la taille des procargotes presents. Rev. Ecol. Biol. Sol 17, 543-557.

49 Kodama H. and Schnitzer M. 1968 Effects of interlayer cations on the adsorption of a soil humic compound by montmorillonite. Soil Sci. 106, 73-74.

50 Kodama H. and Schnitzer M. 1974 Further investigations on fulvic acid-Cu^{2+} montmorillonite interactions. Clays Clay Min. 22, 107-110.

51 Law I.A. and Hayes M.H.B. 1985 Extraction of humic substances from soil using acidified dimethyl sulphoxide. In press. International Humic Substances Society, 2nd Intern. Conf. Birmingham.

52 Martin J.P. 1945 Microorganisms and soil aggregation. I. Origin and nature of some of the

aggregating substances. Soil. Sci. 59, 163-174.

53 Martin J.P. 1946 Microorganisms and soil aggregation. II. Influence of bacterial polysaccharides on soil structure. Soil Sci. 61, 157-166.

54 Mehta N.C., Streuli H., Muller M. and Deuel H. 1960 Role of polysaccharides in soil aggregation. J. Sci. Food Agric. 11, 40-47.

55 Oades J.M. 1978 Mucilages at root surfaces. J. Soil Sci. 29, 1-16.

56 O'Callaghan M.R. 1980 Some Studies in Soil Chemistry. Ph. D. Thesis, University of Birmingham.

57 Olness A. and Clapp C.E. 1975 Influence of polysaccharide structure on dextran adsorption by montmorillonite. Soil Biol. Biochem. 7, 113-118.

58 Orioli G.A. and Curvetto N.R. 1982 Evaluation of extractants for soil humic substances. Plant and Soil 66, 353-361.

59 Page E.R. 1980 Cellulose xanthate as a soil conditioner: Field trials. J. Sci. Fd. Agric. 31, 718-723.

60 Page E.R. 1980 Cellulose xanthate as a soil conditioner: Laboratory experiments. J. Sci. Fd. Agric. 31, 1-6.

61 Pagliai M., Guidi G. and La Marca M. 1980 Macro- and micromorphometric investigation on soil-dextran interactions. J. Soil Sci. 31, 493-504.

62 Parfitt R.L. 1972 Adsorption of charged sugars by montmorillonite. Soil Sci. 113, 417-421.

63 Parfitt R.L., Fraser A.R. and Farmer V.C. 1977 Adsorption on hydrous oxides. III Fulvic and humic acid on goethite, gibbsite and imogolite. J. Soil Sci. 28, 289-296.

64 Parsons J.W. and Tinsley J. 1961 Chemical studies of polysaccharide material in soils and composts based on extraction with anhydrous formic acid. Soil Sci. 92, 46-53.

65 Piper T.J. and Posner A.M. 1972 Sodium amalgam reduction of humic acid. I. Evaluation of the method; II. Application of the method. Soil Biol. Biochem. 4, 513-531.

66 Rennie D.A., Truog E. and Allen O.N. 1954 Soil aggregation as influenced by microbial gums, level of fertility, and kind of crop. Soil Sci. Soc. Am. Proc. 18, 399-403.

67 Schloesing Th. 1874 Determination de l'argile dans la terre arable. C.R. Acad. Sci. Paris 78, 1276-1279.

68 Schnitzer M. 1978 Humic substances: chemistry and reactions. pp 1-64. *In* Soil Organic Matter Eds. M. Schnitzer and S.U. Khan. Elsevier, Amsterdam and New York.

69 Schnitzer M. and Khan S.U. 1972 Humic Substances in the Environment. Marcel Dekker, New York.

70 Schnitzer M. and Kodama H. 1977 Reactions of minerals with soil humic substances. *In* Minerals in Soil Environments. Eds. J.B. Dixon and S.B. Weed. Soil Science Soc. Amer., Madison. pp. 741-770.

71 Schwertmann U., Fischer W.R. and Papendorf H. 1968 The influence of organic compounds on the formation of iron oxides. Trans. 9th Intern. Congress Soil Sci. Adelaide 1, 645-654.

72 Stevenson F.J. 1982 Humus Chemistry. Genesis, Composition, Reactions. Wiley, New York.

73 Swift R.S. 1985 Fractionation of soil humic substances. *In* Humic Substances in Soil, Sediment and Water. Eds. I.G. Aiken *et al* Wiley, New York. pp 387-408.

74 Swift R.S. and Posner A.M. 1971 Gel chromatography of humic acid. J. Soil. Sci. 22, 237-249.

75 Swincer G.D. Oades J. M. and Greenland D.J. 1968 Studies on soil polysaccharides. I. The isolation of polysaccharides from soil. II. The composition and properties of polysaccharides in soils under pasture and under fallow wheat rotation. Aust. J. Soil. Res. 6, 211-239.

76 Swincer G.D. Oades J.M. and Greenland D.J. 1969 Extraction, characterization, and significance of soil polysaccharides. Advan. Agron. 21, 195-235.

77 Tate K.R. and Theng B.K.G. 1980 Organic matter and its interactions with inorganic soil

208

constituents. In Soils with Variable Charge. Ed. B.K.G. Theng. N.Z. Soc. of Soil Sci. Lower Hutt. pp. 225-249.

78 Theng B.K.G. 1976 Interactions between montmorillonite and fulvic acid. Geoderma 15, 243-251.

79 Theng B.K.G. 1979 Formation and properties of clay-polymer complexes. Elsevier, Amsterdam, New York.

80 Theng B.K.G. and Scharpenseel H.W. 1975 The adsorption of ^{14}C-labelled humic acid by montmorillonite. Proc. Intern. Clay Conf., Mexico City. pp. 643-653.

81 Theng B.K.G., Wake J.R.H. and Posner A.M. 1968 Fractional precipitation of soil humic acid by ammonium sulphate. Plant and Soil 29, 305-316.

82 Thurman E.M. and Malcolm R.L. 1983 Structural study of humic substances: New approaches and methods. pp. 1-23. *In* Aquatic and Terrestrial Humic Materials. Eds. R.F. Christman and E.T. Gjessing. Ann Arbor Science, Michigan.

83 Tisdall J.M. and Oades J.M. 1982 Organic matter and water stable aggregates in soils. J. Soil Sci. 33, 141-163.

84 Tokudome S. and Kanno I 1968 Nature of the humus in some Japanese soils. Trans. 9th Intern. Congress Soil Sci. Adelaide 3, 161-173.

85 Turchenek L.W. and Oades J.M. 1978 Organo-mineral particles in soils. *In* Modification of Soil Structure. Eds. W.W. Emerson, R.D. Bond and A.R. Dexter. Wiley, Chichester and New York, pp. 137-144.

86 Turchenek L.W. and Oades J.M. 1979 Fractionation of organo-mineral complexes by sedimentation and density techniques. Geoderma 21, 311-343.

87 Wada K. 1977 Allophane and imogolite. *In* Minerals in Soil Environments. Eds. J.B. Dixon and S.B. Weed. Soil Science Socity of America, Madison. pp. 603-638.

88 Waksman S.A. and Martin J.P. 1939 The role fo microorganisms in the conservation of the soil. Science 90, 304-305.

89 Whitehead D.C. and Tinsley J. 1964 Extraction of soil organic matter with dimethyl-formamide. Soil Sci. 97, 34-42.

90 Wilson M.A. 1984 Soil organic matter maps by nuclear magnetic resonance. J. Soil Sci. 35, 209-215.

91 Wilson M.A., Barron P.F. and Goh K.M. 1981 Differences in structure of organic matter in two soils as demonstrated by ^{13}C-cross polarization nuclear magnetic resonance spectroscopy with magic angle spinning. Geoderma 26, 323-327.

92 Wilson M.A., Heng S., Goh K.M., Pugmire R.J. and Grant D.M. 1983 Studies of litter and acid insoluble soil organic matter fractions using ^{13}C-cross polarization nuclear magnetic resonance spectroscopy with magic angle spinning. J. Soil Sci. 34, 83-97.

10. Sewage sludge organic matter and soil properties.

C.E. CLAPP, S.A. STARK, D.E. CLAY and W.E. LARSON

10.1. Introduction

Organic matter from sewage sludge can exert significant influence on the physical, chemical and biological properties of soils. Organic matter in general contributes greatly to soil's productive capacity. Organic matter incorporated into the soil surface can affect its structure, as denoted by porosity, aggregation, and bulk density, as well as causing an impact as expressed in terms of content and transmission of water, air and heat, and of soil strength. Nutrients are mineralized during organic matter decomposition; C, N, and cation exchange capacity increase following organic matter additions. Other soil chemical properties such as pH, electrical conductivity, and redox potential are changed. The soil biosystem can be altered by addition of new energy sources for organisms, reflected by changes in micro- and macro-biological populations, which in turn influence synthesis and decomposition of microbially-produced soil humic substances, nutrient availability, interactions with soil inorganic components, and other exchanges with soil physical and biochemical properties.

Sewage sludges and other municipal wastes can benefit soil organic matter relationships in the overall soil-plant-water biosphere. Large quantities of these waste materials have traditionally been incinerated, dumped in oceans, or deposited in landfills. With the recent interest in decreasing air and water pollution, agricultural scientists are advocating utilization of these wastes on cropped land rather than just disposal of them. The composition of waste materials must be determined prior to land application, though, to avoid potentially hazardous high levels of trace metals and toxic organic compounds. In general, for any organic waste such as sewage sludges, soil and crop management practices (53) will largely control the nature of the chemical and physical changes which occur.

While it is commonly recognized that organic matter imparts a desirable physical condition to soils, as well as altering chemical and biological relationships, the mechanisms by which it reacts have not been well documented. Likewise, the impact of organic matter on desirable soil physical, chemical and biological properties is not well understood.

Y. Chen and Y. Avnimelech (eds.), The Role of Organic Matter in Modern Agriculture.
ISBN 90-247-3360-X.
© *1986, Martinus Nijhoff Publishers, Dordrecht.*

It is the purpose of this paper to summarize and discuss the influences of organic matter from sewage sludge on the physical, chemical and biological properties of soils.

10.2. Effect of sewage sludge organic matter on soil physical properties

Literature reviews investigating changes in soil physical properties due to sewage sludge addition are available (24, 25). For the most part, these reviews have discussed changes in soil physical properties in a general sense. The objective of this review is to provide more of the specific details from studies available in the literature. Information is provided from experiments conducted in countries around the world, including the United States (19, 20, 28, 35, 37, 50, 54, 58), Norway (92), Switzerland (22), Hong Kong (97), Italy (26, 27, 67, 68, 69), France (65), Great Britain (29), and Canada (94).

10.2.1. Bulk density

Bulk density (BD), defined as the weight of oven-dry soil per unit volume, is an indicator of the soil's physical condition. BD is usually related to a soil's porosity, texture, hydraulic conductivity, aggregation, compaction, and organic matter content.

Several researchers have investigated the effect of sewage sludge addition on changes in BD of mineral soils (Table 10.1). In general, soil BD values decreased as the rate of sewage sludge application increased. The reduction in soil BD may last for as long as 48 months, if relatively high applications of sludge are used (29).

Soil BD values were reduced by sludge addition regardless of soil texture or sludge type used. The reduction of soil BD appeared to be influenced primarily by the rate of sewage sludge organic matter addition. Based on data from 23 sources, utilizing 7 waste types, 21 soil types ranging from coarse sand to clay loam, and study periods ranging from 1 to 85 years, the following highly significant ($r^2 = 0.69$) relationship between soil BD and organic-C addition was found (43):

$$\Delta BD = 3.99 + 6.62\ (\Delta C), \qquad [1]$$

where ΔBD (%) = the waste-treated soil BD minus the control soil BD, divided by the control soil BD times 100, and ΔC (%) = treated soil organic-C content minus the control soil C content. Reduction in soil BD was probably due not only to the dilutional effect of adding less dense organic matter to the more dense mineral matter, but also to increased soil aggregation.

Table 10.1. Changes in soil bulk density resulting from sewage sludge addition

Reference	Soil type	Sludge type	Sludge application Rate (Mg/ha/yr)	Time (yr)	Bulk density (Mg/m³)	Measurements taken after last addition (mo)	Remarks
28	Hubbard cos Udorthentic Haploboroll)	anaerobic	0	2	1.43	~6	-rototill
			112		1.37		-values estimated
			225		1.24		from figures
			450		1.03		-vegetable crops
29	silt loam	undigested	0	1	1.48	36	-grass crops
			27		1.39		-values are averages
			53		1.36		of two sludge types
			80		1.13		
			107		1.07		
			133		0.98		
29	sandy loam	undigested	0	1	1.43	36	”
			27		1.26		
			53		1.25		
			80		1.26		
			107		1.23		
			133		1.19		

Table 10.1. Continued.

Reference	Soil type	Sludge type	Sludge application		Bulk density (Mg/m³)	Measurements taken after last addition (mo)	Remarks
			Rate (Mg/ha/yr)	Time (yr)			
29	clay loam	undigested	0	1	1.64	36	"
			107		1.52		
			133		1.54		
29	boulder clay	digested	0	1	1.78	48	-grass crops
			35		1.79		-rototill
			70		1.75		
			140		1.70		
			280		1.46		
29	boulder clay	undigested	0	1	1.78	48	"
			20		1.76		
			160		1.51		
29	calcareous loam	digested	0	1	1.13	24	"
			19		1.10		
			38		1.11		
			76		1.09		
			152		0.99		

Table 10.1. Continued.

Reference	Soil type	Sludge type	Sludge application Rate (Mg/ha/yr)	Time (yr)	Bulk density (Mg/m^3)	Measurements taken after last addition (mo)	Remarks
29	calcareous loam	digested	125 250	2	0.87 0.72	12	"
37	Arkport sl (Psammentic Hapludalf)	compost	0 50 200 500	1	1.38 1.22 0.80 0.69	9	-3 to 6 cm sample -rototill
37	Langford sil (Typic Fragiochrept)	compost	0 50 200 500	1	1.26 1.18 0.94 0.68	9	"
50	Celina sil (Aquic Hapludalf)	anaerobic	0 22 56 90	1	1.35 1.27 1.29 1.21	12	-0 to 5 cm sample -rototill -no crop
50	Blount sil (Aeric Ochraqualf)	anaerobic	0 56	1	1.18 1.03	12	"

Table 10.1. Continued.

Reference	Soil type	Sludge type	Sludge application		Bulk density (Mg/m³)	Measurements taken after last addition (mo)	Remarks
			Rate (Mg/ha/yr)	Time (yr)			
50	Tracy sl (Ultic Hapludalf)	anaerobic	0 56	1	1.23 1.13	12	"
54	Cheshire 1 (Entic Haplorthod)	anaerobic (A)	0 41 82	1	1.13 1.09 1.07	~24	-vegetable crops -soil frames
54	Cheshire 1	anaerobic (B)	0 65 130	1	1.13 1.05 0.99	~24	"
58	Sango sil (Glossic Fragiudult)	compost	0 23 41 82 164	2	1.37 1.32 1.27 1.22 1.12	~ 5	-sorghum
94	Guelph 1 (Typic Hapludalph)	anaerobic	0 8.3	2*	1.38 1.32	33	-10 to 15 cm sample -disk and harrow

* 2 years between additions.

10.2.2. Aggregation and aggregate stability

Aggregation, or the binding together of individual soil particles, gives rise to what is known as soil structure. Typically, a well-structured soil (*i.e.*, high level of strongly-bound aggregates) has greater resistance to the forces of erosion and has improved air-water relationships. In general, hydraulic conductivity, infiltration rate, air diffusivity, surface drainage, and ease of root penetration will increase with increasing aggregation.

Improving or increasing aggregation is more desirable on finer textured soils such as silt loams, clay loams, and clays. A fine textured soil will behave much like a coarse one with respect to water infiltration and drainage, if its clay and silt particles are bound together into aggregates. Modern-day farming techniques, such as conventional tillage, row cropping, and complete vegetation removal on some soils, can decrease the degree of aggregation at the surface and on more sensitive soils, can completely destroy surface soil structure. The need to increase soil aggregation in many situations is apparent, however, the aggregates formed must be resistant to degradation by the forces of water (*e.g.* rain drop and irrigation impact) and tillage operations.

The addition of sewage sludge to soil has been found to be an effective method to not only increase total aggregation, but also increase the proportion of water-stable aggregates (Table 10.2.). Three years after a single sewage sludge addition of 50 Mg/ha to a heavy clay soil, the soil's percentage of water-stable aggregates (92) more than doubled. In an incubation study (19) lasting 6 months, the percentage of stable aggregates were greater in the sludge-treated soil, (28-35%), when compared to the untreated silt loam soil (17%).

The effect of sewage sludge addition on water-stable aggregate formation is less on soils with existing high levels of aggregate stability. After five annual additions (22) of sewage sludge totalling 61 Mg/ha to a clay soil with high aggregate stability, and cropped to a grass/clover mix, aggregate stability increased by only 2%

The type of crop grown or rotational sequence modified the effect sewage sludge had on aggregate stability. Using various rotational sequences (grass/clover, corn, and wheat) on three different soil types (clay, sandy loam and loamy sand), and five annual sludge additions, it was found that the longer the time period from the last grass/clover crop in the rotation, the less effect sludge had on soil aggregation (22). When sludge was applied to permanent grassland plots and plots in crop rotation annually for a five-year period, sludge had a favorable effect only on the plots in rotation (wheat-corn-beets-wheat-corn).

Data summarized in Table 10.2. suggest that sludge is more beneficial as a soil conditioner if applied to fine-textured soils used for row crop production. Sewage sludge can also be useful in mineland reclamation. Ten months after relatively high applications of sewage sludge to a silty clay loam mine spoil, water-stable ag-

Table 10.2. Changes in water-stable aggregation of soil resulting from sewage sludge addition

Reference	Soil type	Sludge type	Sludge application		Measurement		Measurements taken after last addition (mo)	Remarks
			Rate (Mg/ha/yr)	Time (yr)	Type	Results (%)		
19	Beltsville sil (Typic Fragiudult)	digested, pH 7.5 digested, pH 11.5 raw	0 75 75 75	1	water-stable aggregates	16.6 33.3 28.4 34.5	6	-incubation study -25°C -subsoil
22	clay	not given	0 12	5	aggregate stability	78.1 80.2	~5	-values estimated from figures -25 Mg/ha organic matter added -grass/clover mix
22	clay	not given	0 10 38	5	aggregate stability	51 55 59	~5	-wheat/corn/beet rotation -no effect on meadow soils
22	sandy loam	not given	0 4 8 12	5	aggregate stability	65.4 68.3 70.2 74.7	~5	-lysimeter study -grass crop

217

Table 10.2. Continued

Reference	Soil type	Sludge type	Sludge application Rate (Mg/ha/yr)	Sludge application Time (yr)	Measurement Type	Results (%)	Measurements taken after last addition (mo)	Remarks
26,68	sandy loam (Typic Psammaquent)	aerobic	0	2	water stability index (WSI)*	4.8	1	-low & high rates added equal amounts of organic-C
			68			8.3		-organic-C content assumed to be equal for each sludge type
			203			8.4		-corn crop
		anaerobic	49			11.7		
			147			13.2		
		composted aerobic	78			8.3		
			234			9.9		
		composted anaerobic	74			5.9		
			222			7.9		
35	silty clay loam mine spoil	digested	0	1	water-stable aggregates > 0.25 mm 10 min. wet sieving	12	10	-grass crops
			224			24		-values estimated from figures
			448			33		
			672			39		
			896			42		
50	Celina sil	anaerobic	0	1	mean weight diameter**	mm 0.31	12	-0 to 5 cm sample
			22			0.89		-no crop
			56			0.74		-rototill
			90			0.69		

Table 10.2. Continued

Reference	Soil type	Sludge type	Sludge application		Measurement		Measurements taken after last addition (mo)	Remarks
			Rate (Mg/ha/yr)	Time (yr)	Type	Results		
						mm		
50	Blount sil	anaerobic	0	1	mean weight diameter	0.32	12	"
			56			0.94		
50	Tracy sl	anaerobic	0	1	mean weight diameter	0.88	12	"
			56			2.44		
						(%)		
54	Hartford ls (Entic Haplorthod)	anaerobic (A)	0	1	aggregates >1 mm	1.3	9	-greenhouse study -values are averaged for acid and neutral soils
			46			1.4		
			91			2.9		
			183			3.5		
54	Hartford ls	anaerobic (B)	0	1	aggregates >1 mm	1.3	9	"
			25			2.5		
			51			3.6		
			102			6.3		
54	Cheshire 1	anaerobic (A)	0	1	aggregates >1 mm	5.3	~24	-soil frames
			41			8.0		
			82			9.4		

Table 10.2. Continued

Reference	Soil type	Sludge type	Sludge application Rate (Mg/ha/yr)	Time (yr)	Measurement Type	Results (%)	Measurements taken after last addition (mo)	Remarks
54	Cheshire 1	anaerobic (B)	0 65 130	1	aggregates >1 mm	5.3 6.6 6.2	~24	-soil frames
65	loamy clay	lime/FeCl$_3$ treated	0 35 70	2***	water-stable aggregates without pretreatment	19.7 19.4 22.2	48	
65	loamy clay	lime/FeCl$_3$ treated	0 10	1	water-stable aggregates	19.7 22.7	6	
92	clay	not given	0 50 100	1	% water-stable aggregates (2-0.6 mm)	5 13 18	36	
94	Guelph 1	anaerobic	0 8.3	2****	water-stable aggregates (1-2 mm)	51.0 53.3	33	

* WSI= 100 (1-A/B); A & B are weights of 1-2 mm aggregates passing 0.25 mm sieve after 5 and 60 min of wet sieving, respectively.
** Diameter (mm) of water-stable aggregates from wet-sieving 2-8 mm aggregates for 10 min.
*** 3 sludge applications.
**** 2 yr between additions.

gregates greater than 0.25 mm in diameter increased from 12 to 42% for a 900 Mg/ha addition (35).

The type of sewage sludge applied may also have an effect on aggregate stability. When equivalent rates (organic-C basis) of aerobic- and anaerobic-digested sludges, both composted and not composted, were applied to a sandy loam soil of poor aggregate stability, the stabilizing effect of the uncomposted anaerobic sludge was always greater and longer lasting than the other sludge treatments (26, 68). This finding was later followed by a study in which the same aerobic and a-naerobic sludges were extracted to obtain fractions of (1) fats, waxes, and oils, (2) resins, and (3) water-soluble polysaccharides (27). The three fractions were added separately to 1- to 2-mm diameter aggregates of two soil types. Water stability of soil aggregates were increased markedly by both the fat/wax/oil (ether-soluble) fraction and the resin (alcohol-soluble) fraction. The ether-soluble fraction, being more hydrophobic, had a greater effect on water stability than the alcohol- or water-soluble fractions. The anaerobic sludge had one-third more ether-soluble organic matter compared to the aerobic sludge.

10.2.3. Porosity and pore size distribution

Porosity, or the total soil volume occupied by pore space, is inversely proportional to a soil's bulk density. As with bulk density, porosity is generally viewed as an index of other soil physical properties such as tilth, compaction, air-water relationships, and root penetration. Another important related soil physical property is that of pore size distribution. A recent classification scheme (23) considers pores smaller than 50 μm in diameter to be storage pores. Pores larger than this size are generally drained at field capacity. Storage pores are important water and nutrient reservoirs for plants and microorganisms. Pore diameters ranging from 50 to 500 μm are considered transmission pores. These pores are important for the movement of water and the exchange of gases through soil. Transmission pores are important for root penetration. Most roots need pores in the range of 100 to 200 μm in which to grow (69). Lastly, pores greater than 500 μm in diameter are termed fissures. A high percentage of fissures is usually considered an index of poor soil structure (68).

Several studies have investigated the effect of sewage sludge addition on soil porosity and pore size distribution (Table 10.3.). The addition of sewage sludge increased total porosity (26, 67, 68, 69), large pore space (50), or non-capillary pore space (54). These increases occurred on various soil types, ranging from sandy loams to silt loams, and were found using several different types of sewage sludge. Increases in total soil porosity were evident for as long as 12 months after sewage sludge addition (67). However, increases in porosity appeared to be more apparent in the same cropping season during which sludge was applied. Porosity decreased

Table 10.3. Changes in soil porosity resulting from sewage sludge addition

Reference	Soil type	Sludge type	Rate (Mg/ha/yr)	Time (yr)	Type	Results 2 mo	6 mo (%)	12 mo	Measurement taken after last addition (mo)	Remarks
50	Celina sil	anaerobic	0	1	large pore space*	7.5	19.6	6.8	2,6, & 12	-rototill
			22			17.8	21.2	12.6		-0 to 5 cm sample
			56			14.0	17.3	10.1		
			90			21.4	25.4	14.2		
50	Blount sil	anaerobic	0	1	"	19.8	19.8	17.6	2,6, & 12	"
			56			32.7	28.7	19.9		
50	Tracy sl	anaerobic	0	1	"	27.9	28.1	20.6	2,6, & 12	"
			56			29.5	33.3	22.9		
54	Cheshire 1	anaerobic (A)	0	1	% non-capillary pore space**		25.3		~24	-soil frames
			41				27.7			
			82				28.5			
54	Cheshire 1	anaerobic (B)	0	1	"		25.3		~24	"
			65				30.0			
			130				29.8			

Table 10.3. Continued

Reference	Soil type	Sludge type	Rate (Mg/ha/yr)	Time (yr)	Type	Results 4 mo	12 mo	Measurement taken after last addition (mo)	Remarks
						(%)			
67	sandy loam	aerobic	0	2	% of total area occupied by pores***	16.0	9.4	4 & 12	-see Table 10.2
			68			47.1	24.6		
			203			49.3	44.3		
		anaerobic	49			36.2	27.9		
			147			38.3	31.8		
		composted	78			39.5	22.9		
		aerobic	234			49.5	26.0		
		composted	74			35.6	30.9		
		anaerobic	222			50.1	43.2		

Reference	Soil type	Sludge type	Rate (Mg/ha/yr)	Time (yr)	Type	<50	50-500	>500	Measurement taken after last addition (mo)	Remarks
						(µm)				
						(%)				
67	sandy loam	aerobic	0	2	% of total pore area***	20	41	39	12	-values estimated from figures
			68			38	50	12		
			203			22	68	10		
		anaerobic	49			21	62	17		
			147			23	54	23		
		composted	78			31	50	19		
		aerobic	234			25	63	12		
		composted	74			35	50	15		
		anaerobic	222			30	55	15		

Table 10.3. Continued

Reference	Soil type	Sludge type	Sludge application		Measurement			Measurement taken after last addition (mo)	Remarks
			Rate (Mg/ha/yr)	Time (yr)	Type	Results (%)	(no./cm²)		
69	sandy loam (Typic Psamm-aquent)	aerobic	0	1	% of total area occupied by pores*** and no of pores per cm²	14.6	22	<1	-rates were equivalent to 50 Mg/ha of manure on an organic-C basis
		anaerobic	68			27.2	34		
		composted aerobic	49			29.6	36		
			78			31.4	37		
		composted anaerobic	74			36.9	43		

* % of total soil volume drained at 50 cm water suction.

** % of large pores drained at unspecified suction.

*** from thin sections.

after the winter months, but was still greater than soil receiving no sewage sludge (50, 67). Depending on the original physical condition of the soil, sewage sludge may have no significant effect on porosity (22).

Pore size distribution was observed to shift from a relatively high percentage of fissures before sludge addition to greater percentages of storage and transmission pores after addition (67), which are more important from an agronomic point of view (24).

10.2.4. Hydraulic conductivity

Several studies have determined the maximum rate of water movement through soils, or saturated hydraulic conductivity (sat. K) (Table 10.4.). In all cases, sat. K was increased significantly by the addition of sewage sludge. Increases in sat. K were generally greater for finer textured soil. When relatively low quantities of a digested sludge (1 to 2 cm) were added to an eroded soil of unknown texture, sat. K increased more than 8-fold, 20 days after sludge application (97). Sat. K values continued to increase for a few weeks, but then decreased. This decline was thought to be due to clogging of pores by microbial decomposition products. When moderate rates of sewage sludge were added to a loamy clay soil, sat. K remained twice that of the untreated soil for approximately one year (65).

It is presumed that soil application of a sludge high in lime may have a greater influence on sat. K due to the flocculating capabilities of Ca^{++} on soil colloids. However, when equal quantities of the same sludge (one high in lime, the other low) were applied to a silt loam soil, the opposite result occurred (19). The sat. K of a soil receiving the low pH, digested sludge treatment was approximately 56% greater than that receiving the high pH, digested sludge treatment. Both treatments, however, had sat. K values no different than the control after 79 days of incubation. In this same study, digested sludges had a greater ability in increasing sat. K than undigested or raw sludges. The decreased ability of raw sludge to increase sat. K may be due to its unstabilized nature. Raw sludges have organic nutrients that are more easily mineralized by soil microorganisms as compared to digested sludges. This fact may cause rapid growth of fungal mycelium and other microorganisms which could result in clogged soil pores and surface crusting, thereby restricting water movement.

In contrast to the finding that sludge addition at least temporarily increased sat. K is the fact that sludge addition at the same time decreased unsaturated hydraulic conductivity (unsat. K). At any given water content, unsat. K decreased with increasing sewage sludge additions to a coarse sand soil (28).

Table 10.4. Changes in saturated hydraulic conductivity of soil resulting from sewage sludge addition

Reference	Soil type	Sludge type	Sludge application Rate (Mg/ha/yr)	Time (yr)	Saturated hydraulic conductivity (cm/h)	Measurement taken after last addition (mo)	Remarks
19	Beltsville sil	digested, pH 7.5 digested, pH 11.5 raw	0 75 75 75	1	4.0 19.5 12.5 6.5	1	-values estimated from figures -incubation study -subsoil
28	Hubbard cos	anaerobic	0 112 225 450	2	108 115 122 127	6	
37	sand	compost	0 500 2,500	1	14.3 18.9 42.1	1	-laboratory mixtures
65	loamy clay	lime/FeCl$_3$ treated	0 35 70	2*	10.3 21.7 19.7	12	
97	eroded	digested	0 1 cm** 2 cm	1	20 d 40 d 60 d 3 3 3 25 31 8 29 90 60	20d, 40d, 60d	-column study -0 to 8 cm sample -values estimated from figures

* 3 sludge applications
** liquid sludge in cm.

10.2.5. Moisture retention

The addition of sewage sludge to soil increases water retention at both field capacity (-33 kPa) and wilting point (-1500 kPa). This effect is summarized in Table 10.5. The increase in water retention (weight basis) at various matric potentials of sludge-treated soils is probably due to the increase in total porosity, storage pores space (< 50-μm diameter), and the water absorption capacity of organic matter.

The greatest percentage increases in water retention at both field capacity (FC) and wilting point (WP), were for treatments on coarser textured soils or using higher application rates (Table 10.5.). Application of 56 Mg/ha of anaerobic sludge to a Blount silt loam caused an increase in both FC and WP of 14.9 and 14.7% respectively; whereas, the same application rate on a sandy loam soil increase these parameters by 17.1 and 51.7% respectively (50).

One review of the available literature (43) found that approximately 80% of the observed variations in percentage increases in water retention at both FC and WP, could be explained by soil texture (%sand) and increases in organic-C. This analysis indicated that with organic waste application to fine-textured soils, increases in water retention at FC is greater than at WP. This effect is probably the result of increased aggregation, producing a greater number of larger size pores which would not drain under gravity (-33 kPa). While, for coarse-textured soils, the percentage of sand present in the soils produced a larger increase in water retention at WP than at FC, perhaps due to an increase in number of smaller pores not draining at -1500 kPa.

Plant-available waterholding capacity (AWC), or the difference between moisture retained at FC and WP (weight basis), increased with increasing sludge application rates for medium and fine-textured soils (Table 7.5.). Two studies, one on coarse sand (28) and the other on sandy loam (50), found no significant increase in AWC even at an application rate as high as 450 Mg/ha/yr.

When moisture retention is computed on a volume basis, increases may not be as dramatic or even non-existent. FC values (weight basis) ranged from 30.1 to 47.3% for the control and 250 Mg/ha/yr treatment, respectively (29), however, when AWC values were converted to a volume basis, the control FC was equal to the 250 Mg/ha value, 34.0 mL/cm^3. The lack of difference was due to the crease in BD caused by sludge addition.

10.3. Effect of sewage sludge organic matter on soil chemical properties

To better understand the effect of sludge organic matter on soil chemical properties it is important to consider that sewage sludge is produced by several different technical processes and applied to different soils under different climatic con-

227

Table 10.5. Changes in water retention of soil at various matric tensions resulting from sewage sludge addition

Reference	Soil type	Sludge type	Sludge application Rate (Mg/ha/yr)	Time (yr)	SAT	FC	WP	AWC	Measurement taken after last addition (mo)	Remarks
19	Beltsville sil		0	1		13.8	6.0	7.8	6	-incubation study -25°C -subsoil
		digested pH 7.5	75			16.4	7.9	8.5		
		digested pH 11.5	75			16.3	7.8	8.5		
		raw	75			16.9	7.6	9.3		
20	Woodstown sil (Aquic Hapludult)	anaerobic composted	0	1				12.5		
			240					14.5		
			240					18.5		
28	Hubbard cos	anaerobic	0	2	43.3	9.2	5.5	3.7	~6	-FC measured at 10 kPa
			112		46.6	11.6	7.6	4.0		
			250		54.3	14.3	11.4	2.9		
			450		58.8	21.3	17.3	4.0		

Water retention (% by weight): SAT, FC, WP, AWC

Table 10.5. Continued

Reference	Soil type	Sludge type	Sludge application Rate (Mg/ha/yr)	Time (yr)	Water retention (% by weight)				Measurement taken after last addition (mo)	Remarks
					SAT	FC	WP	AWC		
29	calcareous loam	digested	0	1		30.1			24	
			19			32.2				
			38			32.4				
			76			35.2				
			152			39.4				
29	calcareous loam	digested	125	2		43.1			12	
			250			47.3				
35	silty clay loam mine spoil	digested	0	1	55	27	12	15	10	-values estimated from figures
			224		60	31	14	17		
			448		66	36	17	19		
			896		75	44	23	21		
50	Celina sil	anaerobic	0	1		17.3	6.0	11.3	6	-0 to 5 cm sample
			22			20.3	6.8	13.5		
			56			21.3	7.5	13.8		
			90			21.9	7.4	14.5		

Table 10.5. Continued

Reference	Soil type	Sludge type	Sludge application Rate (Mg/ha/yr)	Time (yr)	Water retention (% by weight) SAT	FC	WP	AWC	Measurement taken after last addition (mo)	Remarks
50	Blount silt loam	anaerobic	0 56	1		20.1 23.1	9.5 10.9	10.6 12.2	6	”
50	Tracy sil	anaerobic	0 56	1		14.5 17.5	6.0 9.1	8.5 8.4	6	”
54	Cheshire 1	anaerobic (A)	0 17	1		22.8 24.6			∼6	-soil frames
54	Cheshire 1	anaerobic (B)	0 35	1		22.8 27.6			∼6	”
54	Cheshire 1	anaerobic (A)	0 41 82	1		24.3 25.1 25.8			∼24	-soil frames
54	Cheshire 1	anaerobic (B)	0 65 130	1		24.3 27.2 29.9			∼24	”

Table 10.5. Continued

Reference	Soil type	Sludge type	Sludge application Rate (Mg/ha/yr)	Time (yr)	Water retention (% by weight)				Measurement taken after last addition (mo)	Remarks
					SAT	FC	WP	AWC		
65	loamy clay	lime/FeCl$_3$ treated	0	2*		33.2	12.7	20.5	~ 12	
			35			34.6	13.8	20.8		
			70			41.9	16.6	25.3		
94	Guelph l	anaerobic	0	2**		21.7	10.9	10.9	33	
			8.3			21.2	10.6	10.6		

* 3 sludge applications
** 2 yr between additions

ditions. Moreover, at the end of the digestion process different kinds and amounts of inorganic compounds (15, 86) are associated with sludge organic matter (Table 10.6.). Given these considerations, variation in soil response to sludge application is understandable.

Table 10.6. Range and median of major constituents in sewage sludge*

Component	Range		Median
	Low	High	
		(g/kg)	
Organic C	65	480	304
Total N	‹1	176	33
Total P	‹1	143	23
Total S	6	15	11
K	0.2	26	3
Na	0.1	31	2
Ca	1	250	39
Mg	0.3	20	4
Fe	‹1	153	11
Al	1	135	4
		(mg/kg)	
NH_4^+-N	5	67,600	920
NO_3^--N	2	4,900	140
Zn	101	27,800	1,740
Cu	84	10,400	850
Cr	10	99,000	890
Pb	13	19,700	500
Mn	18	7,100	260
Hg	1	10,600	5
Ni	2	3,520	82
Cd	3	3,410	16
Co	1	18	4
B	4	760	33

* Adapted from Sommers (86)

10.3.1. Carbon

Unlike fresh plant and animal residues that have been incorporated into the soil, most sewage sludges have been through a biological treatment, where partial decomposition and stabilization have occurred. Therefore, the rate of decomposition in soil may be slower than most fresh organic residues, resulting in longer lasting increases in the levels of soil organic matter (35, 54) (Table 10.7.) and shifts in the composition of the soil organic fractions. Anaerobically-digested sludge contains relatively large amounts of stable organic compounds, consisting of resistant microbial tissue, lignin, cellulose, lipids, organic-N compounds, and humic-like materials (59). Sewage sludge also contains more ether- and alcohol-soluble fractions than plant material as determined by chemical fractionation (79). Moreover, because waxes, fats, oils, and resins decompose slowly in soils, their contribution to the total organic matter tended to increase with time after sludge application. Changes in IR spectra indicated that microbial decomposition of anaerobic sewage sludge in a Baywood sand (Entic Haploxeroll) under laboratory conditions (83), involved the disappearance of polysaccharides, proteinaceous material, and sulfate compounds with the emergence of carboxylates and nitrates. These reactions were slowed by decreasing the pH and saturating the soil with water. Sludge application for 6 years to a Blount silt loam (36) resulted in a shift in the composition of organic compounds towards that of sludge. Carbohydrate-C decreased from 18 to 10%and oil and grease fractions increased from 1.7 to 11.9%of the total organic-C in the control and sludge-treated plots, respectively.

The rate of organic matter decomposition influences many physical, chemical, and biological processes in the soil. Specific rates of decomposition of sewage sludge have ranged from 2 to over 60%per year (1, 36, 62, 78). Soil temperature is a major factor directly influencing the rate of sewage sludge decomposition (89). Sludge decomposition in a column study (62) was not influenced by soil moisture in a sandy soil, but decompostion rate was reduced in a saturated silt loam soil, and was almost completely stopped in a saturated clay soil.

10.3.2. Nitrogen

The N content of sewage sludge depends on the degree and type of processing, and can range from less than 1 to greater than 170 g/kg (Table 10.6.). The primary inorganic form is NH_4^+-N, representing approximately 30% of the total N contained in anaerobic sludge (15). Sewage sludge addition to soils can markedly increase the total soil N content (Table 10.7.).

Transformations of N in sewage wastes applied to soil encompass the whole spectrum of the N cycle. Nitrification of NH_4^+-N to NO_3^--N is important because NO_3^--N may leach and contaminate the ground water, or may be subject to deni-

Table 10.7. Changes in soil carbon and nitrogen contents resulting from sewage sludge addition

Reference	Soil type	Sludge type	Sludge application Rate (Mg/ha/yr)	Time (yr)	Total C (g/kg)	Total N (g/kg)	Time from last application (mo)
20	Woodstown sil	anaerobic	0	1	10.0	0.60	18
			160		15.0	1.70	
			240		20.0	1.90	
31	Hubbard sl	anaerobic	0	3	10.1	0.98	60
			40		13.3	1.18	
			79		15.2	1.52	
			155		22.8	2.50	
31	Hubbard sl	aerobic	81	3	12.4	1.16	60
31	Hubbard sl	waste-activated	164	3	20.1	1.72	60
36	Blount sil	anaerobic	0	6	9.5	1.00	9
			50		22.9	2.12	
42	Plano sil (Typic Argiudoll)	anaerobic	0	3	-	1.63	4
			20		-	2.48	4
			20		-	2.03	12

Table 10.7. Continued

Reference	Soil type	Sludge type	Sludge application		Total C (g/kg)	Total N (g/kg)	Time from last application (mo)
			Rate (Mg/ha/yr)	Time (yr)			
42	Warsaw sl (Typic Argiudoll)	anaerobic	0	3	-	1.04	4
			20		-	1.56	4
			20		-	1.57	12
50	Celina sil	anaerobic	0	1	9.9	-	2
			56		23.3	-	2
			56		20.7	-	12
50	Blount sil	anaerobic	0	1	12.3	-	2
			56		26.0	-	2
			56		25.2	-	12
50	Tracy sl	anaerobic	0	1	14.3	-	2
			56		35.6	-	2
			56		19.4	-	12
63	Teel sil (Fluvaquentic Eutrochrept)	anaerobic	0	1	-	1.80	1
			160		-	4.90	1
			160		-	3.78	6

Table 10.7. Continued

Reference	Soil type	Sludge type	Sludge application		Total C (g/kg)	Total N (g/kg)	Time from last application (mo)
			Rate (Mg/ha/yr)	Time (yr)			
63	Teel sil	aerobic	160	1	-	14.20	1
			160		-	7.33	6
77	Orangeburg fsl (Typic Paleulult)	digested	0	6	-	-	
			56*		3.2	-	6
77	Troup fsl (Grossarenic Paleulult)	digested	0	6	-	-	
			56*		5.6	-	6
88	Hubbard sl	anaerobic	0	3	10.5	0.92	12
			40		22.6	2.28	
			79		34.2	3.62	
			155		58.3	6.08	
88	Hubbard sl	aerobic	81	3	23.6	2.64	12
88	Hubbard sl	waste-activated	164	3	46.6	4.12	12
88	Hubbard sl	anaerobic	0	1	9.6	0.79	48
			63		13.6	1.30	
			125		17.0	1.74	
			199		21.6	2.23	

* Values reported are for increases over 0 rate (not given).

trification. Dewatering sludge prior to land application can result in volatilization of a large percentage of the NH_3, with the remaining NH_4^+-N being nitrified. Rate of sludge application influenced the rate of nitrification (76, 80), with rapid nitrification at low sludge rates, whereas at high rates, initial inhibition was followed by rapid nitrification.

Sewage sludge may supply a large portion of the N required for crop growth (18, 35, 54, 81, 84). Mineralized-N from sludge-derived organic matter may become made available for plant uptake for several years after application. An aerobic incubation study conducted in Minnesota (88) reported that cumulative N mineralization was approximately linearly related to the square root of time. In another study (55), up to 55% of the organic-N added in the sludge mineralized in the first year, while 14 to 25% and 36 to 61% of the organic-N applied from anaerobically and aerobically-digested sludge mineralized during a 13-week incubation period, respectively (56). For ^{15}N-labeled sludge (89), 40% of the N was mineralized after 24 weeks of incubation at 21°C. Four to 48% of the sludge organic-N was mineralized in a 16-week incubation study (80), and after an 18-week incubation 60% of the sludge-applied N was mineralized (44). Other workers in England (76) however, did not observe mineralization of sludge organic-N in a laboratory incubation experiment.

The ability of sludge to supply N for crop growth decreases as the sludge decomposes. Sludge application in California (75) released 35, 10, 6, and 5% of the residual organic-N during the first through fourth year after application, respectively. An annual decay series of 20, 15, 6, 4, and 2% was found in Wisconsin (40) following sludge application. In Minnesota (31), the potential of the sludge-derived organic matter to mineralize N was greatly reduced, five years after sludge application, suggesting that the fertilizer-N requirement should be determined by using both the sludge decay series and the potential of the original soil organic matter to mineralize N. A considerable amount of N was released from the sludge two years after application on a Woodstown silt loam (20), but N mineralization decreased with further time. Sewage sludge may also produce a "priming" effect in native soil organic matter decomposition rates, thus releasing extra N for plant growth (81).

Sewage sludge applications to soils can influence the C/N ratio. Sludge additions for six years to a Blount silt loam (36) increased the C/N ratio from 9.4 to 10.8, measured two years after the last application. Since the C/N ratio in the sludge was 7.8, organic-N initially was lost from the system faster than organic-C. However, on a strip-mined spoil (35), the C/N ratio was decreased from 10.9 to 10.3.

Denitrification may reduce NO_3^--N concentration. High sludge application to a Warsaw sandy loam (80), incubated under aerobic conditions resulted in the loss of inorganic- and/or organic-derived N. This fact, coupled with the observed decreases in NH_4^+-N, indicated that N was lost from the system by volatilization or denitrification, or that the inorganic-N was immobilized or fixed and not released back into the inorganic-N fractions. In a pot study (74) using ^{15}N-labeled fertilizer, 7.8 and 13.1% of the applied N was lost from control and sludge-treated pots, respectively. This loss was attributed to denitrification.

10.3.3. pH

Application of treated sewage sludge to soil may result in alterations in the soil pH. The magnitude of the pH change depends on many soil properties, including texture, buffering capacity, and length of time after the last sludge application. Modification of soil pH is important also because trace metals become more plant-available as the pH decreases, and microbial activity decreases as the pH decreases below 6 or increases above 7. Several authors have reported that soil pH decreased following sludge application (35, 45, 54). Sludge application (40 Mg/ha) to a Woodstown silt loam (initial pH 5.7) gave soil pH values of 5.7, 4.9, and 5.1 after one, six and twenty months, respectively (20). The decrease in pH between one and six months was probably caused by the production of organic acids during sludge decomposition and/or by NH_4^+-N nitrification, while the slight rise in pH between six and twenty months may indicate that the soil-sludge system has stabilized. In Indiana (50), pH changes were not observed when sludge was added to three silt loam soils. Other workers (51, 97) have observed increases in soil pH when initial pH was low and the sludge contained large quantities of lime. Applications of limed sewage sludge at 60 Mg/ha/yr for five years to a Port Byron silt loam (Typic Hapludoll) in Minnesota (18) increased the pH in the 0 - 15 cm soil depth from 5.3 to 7.6, and in the 15 - 30 cm soil depth from 6.1 to 7.3. The increase in pH was attributed to the lime in the sludge.

10.3.4. Cation exchange capacity

The influence of sewage sludge on the cation exchange capacity (CEC) depends on soil texture, initial soil CEC, length of time from the last application, and nature of the sewage sludge. Increases in the soil organic matter may increase the CEC (20, 50, 54, 63) and thus its capacity to retain nutrients. In Maryland (20), the CEC of an untreated soil ranged from 0.055 to 0.064 meq/g soil, but increased with increasing rates of sludge application (240 Mg/ha/yr) to 0.154 meq/g soil. However, after two years, the CEC had decreased to 0.087 meq/g soil. Much of the increased CEC has been attributed to the high exchange capacity of the sludge-derived organic matter. The subsequent decrease in CEC may be due to the breakdown of the sludge organic matter. In an Indiana study (50), organic matter derived from sludge had a CEC of 2.5 meq/g of organic matter, and six months after sludge application the increase in soil CEC per unit increase in organic-C averaged 5.8 meq/g organic-C for three soils. Based on these increases, and assuming that sludge organic matter contains 50% C, the calculated CEC for the sludge organic fraction averaged 3.0 meq/g.

10.3.5. Electrical conductivity

Sewage sludge contains large quantities of salts (15) (Table 10.6.) which may increase the soil solution electrical conductivity (EC). Increasing the EC increases the osmotic potential of the soil solution, which may reduce the ability of plants to absorb water at high water suctions. Application of 60 Mg/ha/yr for five years of limed sewage sludge to a Port Byron silt loam in Minnesota (18) increased the EC in the 0 - 15 cm soil depth from 0.32 to 0.46 dS/m, and in the 15 - 30 cm soil depth, from 0.30 to 0.43 dS/m. Similarly, application of 164 Mg/ha/yr of a primary-settled-waste-activated sludge to a Hubbard sandy loam in Minnesota (88) over three years, increased the EC from 0.08 to 1.40 dS/m. With no further sludge addition, the EC had dropped to 0.49 dS/m four years later (31), but increased in the control plots from 0.08 to 0.30 dS/m. The increase in EC in the control plots was probably caused by salts contained in N fertilizer. In an Illinois study (35), the EC increased from 2.2 dS/m in the control plots to 6.6 dS/m in the sludge-treated plots (900 Mg/ha). A sludge application rate of 240 Mg/ha/yr in Maryland (20) initially increased the EC from 0.41 to 5.50 dS/m, however by the end of the growing season, the EC had dropped to 3.90 dS/m. By the next growing season, the EC was reduced to 1.13 dS/m, a level tolerated by most plants. The increase in EC was primarily due to high concentrations of Mg, Ca, and Cl, while the decrease in EC was due to the leaching of soluble salts out of the root zone.

10.3.6. Phosphorus

Phosphorus supplied by sludge can vary in concentration by two orders of magnitude (Table 10.6.). The amount of P associated with the organic and the inorganic fractions is highly variable (15, 86). Sludge application can provide substantial amounts of P for crop growth (3, 4, 35) and increased P concentration in the soil (Table 10.8.). Soils can fix substantial amounts of P, which may limit its vertical movement even when large quantities of sludge-containing P are applied (21, 42). Increasing the soil organic matter, however, may decrease the ability of the soil to sorb P. Addition of organic acids to kaolinite, gibbsite, and goethite (66) reduced the amount of P sorbed, because the acids were adsorbed by ligand exchange on the mineral surfaces, and competed with P for adsorption sites. Following sludge application in Switzerland (21) on a soil with a high P-adsorption capacity, pronounced accumulation of P occurred in the 0 - 15 cm soil layer, producing no P-leaching. However, in Belgium (30) there was a loss of 1575 kg P_2O_5/ha from the surface 25 cm of sludge-treated soil. This loss was attributed to leaching or to overestimation of the P content in the applied sludge. Work in Illinois (33) also showed that P moved down the soil profile. Changes in soil pH can influence P solubility, with P becoming less soluble at pH values less than 5.5 or greater than 7.5.

Table 10.8. Changes in soil phosphorus content resulting from sewage sludge addition

Reference	Soil type	Sludge type	Sludge application Rate (Mg/ha/yr)	Time (yr)	Soil depth (cm)	Extracting method	measured P (mg/kg)	Time from last application (mo)
20	Woodstown sil	anaerobic	0 160 240	1	-	0.025N HCL & 0.03N NH$_4$F	10 196 253	1
20	Woodstown sil	anaerobic	0 160 240	1	-	"	37 180 200	18
42	Plano sil	anaerobic	0 20	3	0-15	0.025N HCL & 0.03N NH$_4$F	39 214	10
42	Plano sil	anaerobic	0 20	3	0-15	"	28 159	20
42	Warsaw sl	anaerobic	0 20	3	0-15	0.025N HCL & 0.03N NH$_4$F	39 166	10
42	Warsaw sl	anaerobic	0 20	3	0-15	"	44 140	20

Table 10.8. Continued

Reference	Soil type	Sludge type	Sludge application		Soil depth (cm)	Extracting method	measured P (mg/kg)	Time from last application (mo)
			Rate (Mg/ha/yr)	Time (yr)				
46	Cecil scl (Typic Hapludult)	anaerobic	0 133	2	0-15	0.05N HCL & 0.025N H$_2$SO$_4$	12 18	1
46	Cecil scl	anaerobic	0 133	2	15-30	"	1 2	1
63	Teel sil	anaerobic	0 160	1	0-2.5	Total P	400 6300	1
63	Teel sil	aerobic	0 160	1	0-2.5	Total P	400 5900	1
70	dredge spoil sil	anaerobic	0* 100	1	0-20	Total P	856 1150	48
70	dredge spoil sil	anaerobic	0* 100	1	20-40	Total P	443 640	48
70	dredge spoil sil	anaerobic	0* 100	1	40-60	Total P	387 1230	48
77	Orangeburg fsl	digested-liquid	0 56	6	0-150	Total P	96 190	12

* Total P measured in fertilized control.

The rate of P movement through soils treated with sludge depends upon many variables, including soil texture, content of Al, Fe, and Ca, and the redox potential. After soil columns treated with sewage sludge became anaerobic, the P concentration of the leachate decreased (57). It was postulated that the lower P concentration of the leachate may have been caused partially by the formation of Fe phosphates at the bottom of the columns when the anaerobic solution came into contact with the air. It has also been shown (72) that the rate of phosphate sorption and release by aerobic and anaerobic soils was dependent upon the P concentration of the soil. The difference in behavior of P under aerobic and anaerobic conditions was attributed to a change in the ferric oxyhydroxide compounds formed during soil reduction. The gel-like reduced ferrous compounds release P when the solution P levels are low, and sorb P when solution P levels are high. High concentrations of Ca, Fe, and Al can reduce the soluble P levels by forming complexes with P.

Much of the applied P can remain in a form unavailable for plant uptake. A Georgia study (46) showed that 5 cm of sludge supplied 2340 kg P/ha, only 88 kg P/ha was removed by the crop and 1160 kg P/ha remained in the soil crust. Sludge did not significantly increase extractable soil P levels, therefore, the unrecovered 1090 kg P/ha (47%) was retained in unavailable forms. This result suggested that high levels of P in the surface horizon may lead to surface water pollution problems if sediments are eroded. Other work (5) has shown that P in runoff water was highly variable and depended on the sludge application rate and rainfall intensity.

10.3.7. Metals

Metal concentrations in sewage sludges are highly variable (Table 10.6.). Large quantities of metals contained in sludge may be associated with the organic fraction (85), therefore, the decomposition of organic-C may control their release. The presence of excessive levels of oxidizable sludge organic matter may limit availability of some metals (35). High amounts of Fe and P supplied by the sludge may also limit the availability of some metals through formation of sparingly soluble precipitates (12). However, because sludge contains many different chelating compounds, some metals may actually be increased in mobility (71). The chelating tendency (52) of metals in descending order was Cu-Ni-Co-Fe, with Mn and Zn showing little reactivity. It also has been reported (32) that plant Cd uptake decreased with increasing CEC, and that soil organic matter was not observed to decrease Cd uptake more than its contribution to the CEC. Decreasing the soil pH by sludge application increased the levels of water-soluble Mn more than the amount of Mn supplied by the sludge (45). Application of sludge in Florida (77) caused movement of Ca, Mg, and K into the 90 - 120 and 120 - 150 cm soil zones, due to leaching by the liquid in the sludge and rainfall. Extractable Fe, Zn, and Al increased to the 90 cm depth. Extractable Cu remained in the top 30 cm of the soil.

10.3.8. Redox potential

Anaerobic conditions are typified by low O_2 content and high CO_2 concentrations, and can exist under local conditions found in small aggregates or generally when the soil profile is water-saturated. The organic matter contained in sewage sludge may increase aggregation and the number of small pores in a soil (69), which may result in an increase in local anaerobic conditions. However, anaerobic conditions may also be increased because sewage sludge contains large amounts of organic-C, and upon mineralization, O_2 is utilized and CO_2 is produced. Soils are considered anaerobic at redox potentials (Eh) of less than +350 mv at pH 7. The major electron donors in soils are fresh plant residues and soil organic matter. However, all chemical elements can transfer electrons and thus, change their oxidation state. The stability of one form over other forms depends largely on the electron availability, and if the O_2 supply is insufficient, soil microorganisms are forced to utilize progressively weaker electron acceptors such as NO_3^-, MnO_2, FeOOH, SO_4^{2-}, and H^+. These conditions can result in increased denitrification (100), reduced nitrification, N mineralization, and decomposition and formation of phytotoxic compounds such as NO_2^-, H_2S, CH_4, and C_2H_2 (14). Following a sludge addition of 240 Mg/ha to a Woodstown silt loam (20), the Eh decreased rapidly to -100 mv and reached -200 mv in 40 days. The CO_2 measurements at 20-40 cm depths were increased from 0.3 to 1.7% in the untreated check soil to 3.6 to 10% in treated plots, respectively.

In water-logged paddy soils, organic N compounds can act as the electron acceptor (38). In an incubation study (95), it was observed that the addition of sewage water low in C to soil, resulted in a redox potential of 200 mv. However, when C was added to the sewage water, the redox potential dropped to -200 mv, indicating that the oxidation states of elements other than N were changing.

10.4. Effect of sewage sludge organic matter on soil biological properties

The close relationship between organic matter and soil organisms has been shown by the abundance of research and published results. However, there are few data or publications on the effects, either beneficial or detrimental, of sewage sludge organic matter on soil organisms. These effects on soil micro- and macro-flora and -fauna often can not be separated from the primary effects on soil physical and chemical properties. The biological component of soil which includes bacteria, actinomycetes, fungi, algae and soil micro- and macrofauna makes significant contributions to waste recycling by decomposing waste organic compounds; eliminating pathogenic microorganisms; involvement in the N, P, and S cycles; and by influencing the solubility and mobility of inorganic ions in soil (60). We will attempt to cover briefly the direct interactions involving sewage sludge organic matter and microorganisms, macrofauna and higher plants.

10.4.1. Microorganisms

Considering that the soil microflora, especially bacteria and fungi, can directly influence soil physical and chemical properties, *e.g.*, aggregation and plant nutrient status, it is surprising that little attention has been given to these organisms in waste management. By the same token, microbial activities such as enzyme processes in sewage sludge-treated soils have also received little attention.

Application of sewage sludge increased microbial numbers in three widely different Ohio soils (61). Numbers of bacteria and actinomycetes were generally directly related to sludge loading rate. Maximum populations were found after one month incubation and decreased after three and six month incubations. The fungal population increased but not as much as the other two groups of organisms. In this study (61), 354 bacterial isolates were characterized from two soils treated with anaerobically-digested sludge. Among the more significant findings was the change in bacterial population from one dominated by gram-positive bacteria in untreated soil to one where gram-negative bacteria made up 50% or more of the isolates from the sludge-treated soils. Other changes included a reduction in number of spore formers, a decrease in average cell size and an increase in colony pigmentation. Bacteria isolated from the sludge-treated soil also differed physiologically from the normal soil microflora as indicated by a faster relative growth rate, better growth at $5°C$ but less at $35°C$, and tolerance of a higher concentration of NaCl. Biochemically these isolates also had more catalase and cytochrome oxidase activity, were better able to utilize citrate, but were less able to hydrolyze starch or produce acid from carbohydrates. Lastly, the sludge soil isolates were generally more resistant to antibiotics.

An Illinois study (91) on a silt loam soil, comparing fertilized control plots with soils treated with up to 370 Mg/ha of sludge over 6 years, provided a rather complete view of microbial interactions (Table 10.9.). No clear-cut evidence of restrictions in microbial populations and their activities was observed. Although total bacterial, fungal, and actinomycete populations were never less than the control in the highest sludge-treated plots, a significant increase in populations of these major groups was found only once during three sampling dates. Percentage of denitrifiers, and protease and amylase activities were increased as a result of sludge application, while invertase and urease activities were unaffected by sludge

Table 10.9. Microorganism populations and activities on continuous maize plots after six years of sewage sludge addition Adapted from Varanka *et al.* (91)

| | Sludge applied | | |
Determination	0	370 Mg/ha	LSD
Bacteria (x 10^6/g)	50.2	54.8	NS***
Actinomycetes (x 10^6/g)	2.5	15.4	9.4**
Fungi (x 10^6/g)	0.12	0.24	0.10**
Denitrifiers (%)	32.4	38.0	NS
Azotobacter (%)	82.0	26.2	29.7*
CO_2 (mg/72h/10g)	0.57	1.12	0.40**
Dehydrogenase μg TPF/24h/g)	355	1629	844*
Invertase (nM gl/h/g)	39	41	NS
Amylase (nM malt/h/g)	2.1	2.3	NS
Cellulase (nM gl/h/g)	7	6	NS
Urease (units)	126	187	NS
Protease (units)	1.10	2.73	0.80**

* Significant at 5% level.
** Significant at 1% level.
*** NS = not significant.

treatments. An increase in dehydrogenase activity was found only at the last sampling date, while *Azotobacter* population and cellulase activity were each decreased at only one date.

In a five-year Swiss experiment (87) on a sandy loam soil with two rates of a-naerobically-stabilized sludge and a fertilized control, sludge application resulted in increased populations of heterotrophic soil microorganisms (aerobic bacteria, actinomycetes, yeasts, hyphal fungi), but decreased autotrophic algal populations. Sludge also increased mineralization processes (respiration, ammonification) and biological activiteis (catalase, alkaline phosphatase). Sludge application had a greater effect on increasing the soil microbial counts and their activities in a grass-land soil than in an arable soil. The authors pointed out that from the long-term point of view, it must be considered that heavy metals and other toxic substances in sewage sludge could pose a danger to the diversity of microorganisms, to the biochemical processes of soils, and to the microbial decomposing potential to eliminate foreign organic substances. On the other hand, the addition of an energy source as well as a diverse population of microorganisms in sewage sludge, may broaden the spectrum of soil microbial activity and increase the soils' potential to degrade contaminating organics.

A composted raw sewage sludge study in Italy (11) provided evidence that apart from some disturbances to mineralization processes, the application of composted sludge at high rates did not appear to endanger the microbiological activities assayed in four soils. The effects were more evident on microbial activities than on numbers of responsible organisms, suggesting a chemical and/or physical influence on microbial enzymes.

Research on changes in rhizosphere microbial populations resulting from sewage sludge application has been carried out by two groups in Italy (73, 90). One group (73) compared liquid and composted sewage sludges in a field experiment, counting several types of bacteria, fungi and actinomycetes, and collecting maize root samples for rhizosphere and mycorrhiza analyses. They suggested that the rhizosphere effect was more important for microbial growth at the soil-root interface than any other external effect, including treatments with organic and inorganic fertilizers. The other study (90) involved pot trials with maize and field trials with wheat and maize for different rates and types of sewage sludge. Measurements were made of microbial (bacteria, fungi, and actinomycetes) growth and oxygen consumption, protease activity, nitrification and denitrification, acid phosphatase activity and hormone production in soil and rhizosphere. All activities and microbial development were found to be highest in the rhizosphere and dependent on sludge amounts.

10.4.2. Macrofauna

Soil macrofauna such as earthworms (*Eisenia foetida*), collembola, mites and nematodes can markedly alter both the magnitude and pathways of sludge organic matter decomposition by their feeding activities. Unfortunately, there is little information on the effect of these animals on organic matter decomposition in sewage sludge-treated soils. In a New York study (64) of earthworm activity in activated sludge-treated soils within laboratory microcosms, three roles in altering C mineralization were attributed to earthworms: accelerating decomposition and stabilization of sludge in the soil, mechanically mixing sludge with soil, and changing the distribution of organic and mineral matter. In addition, labile constituents of sludge were converted to earthworm biomass and respiration. The accelerated rate of sludge stabilization by the earthworm biomass also reduced odors and pathogens. The results of an earlier study (39) were confirmed in that biomass of earthworms increased only when sludge was added to soil. Another New York group (93), however, has suggested that Cd accumulation by earthworms may pose a health risk to humans through concentrating Cd in garden soils where sludge has been applied, and may endanger birds feeding on the earthworms.

Sewage sludge applications distinctly reduced microarthropod populations in a field study (99), compared with fertilized and unfertilized controls. Numbers of collembola and mites, in four out of five genera tested, decreased following sludge application. It was not clear which sludge components were responsible, however toxic anaerobic decomposition products were suspected by the loss of toxicity after soil incorporation.

10.4.3. Plants

It has been shown (54, 96) that seed germination can be inhibited if planting is made directly into soil amended with fresh sewage sludge. However, these effects lasted only a few days and were dependent on rate of sludge addition. Germination can be inhibited by high levels of ammonia which may be present in sludge or be produced during sludge decomposition. Salts and organic compounds such as ethylene oxide (98) may also be involved. Although germination inhibition was greatest in soils freshly amended with sludge (82, 96), probably due to high initial rates of decomposition and more rapid evolution of volatile compounds, delayed planting or storing sludge before application reduced or eliminated these inhibiting effects. Most evidence for inhibition, however, was based on laboratory or greenhouse experiments, while field studies (47) have shown no significant germination differences between control and sludge-amended soils. Under disturbed soil conditions, enhancement of germination was observed (6) in dredged river sand, probably resulting from improved moisture conditions of the sludge-treated soil.

Root growth and development in soil, as evidenced by ease or resistance to penetration, can be related to soil strength, bulk density, pore sizes, and organic matter levels. Addition of sewage sludge to soil decreased root penetration resistance (2), resulting in improved root: shoot ratios compared with soil receiving inorganic fertilizer. In a soil column study (49), root growth of winter wheat was greatly influenced by raw sludge placed at different depths. In a grassland soil (29), root weights in the 0 - 15 cm depth increased from 1.4 to 2.9 mg/cm^3 from an application of 133 Mg/ha of undigested sludge, but decreased in the 15-30 cm depth.

Land application of sewage sludge has not been limited to agricultural crops. Organic matter in sludge has been considered extremely important in reclaiming mine spoils (2, 33, 35), and revegetating other problem areas such as highway banks and dredged sediments (6, 70). Another important aspect of sludge utilization is its role in forest fertilization (10, 48). Differences which must be considered in forest soils compared with agricultural soils include: a pH typically below 5.5 where mobilization of metals may cause problems, less frequent and higher sludge applications can be made (up to 80 Mg DM/ha at 5- to 10-year intervals), and slopes of up to 30% can be used effectively. It should be possible to manage sludge applications on forested land in a sound and environmentally safe manner

Table 10.10 Dry matter yields of crops from soils treated with sewage sludge compared with fertilized controls in field experiments

Reference	Crop	Sludge application Rate (Mg/ha/yr)	Time (yr)	Control (Mg/ha/yr)	Yield Sludge (Mg/ha/yr)	Increase (%)
7	Maize* fodder grain	116	4	14.9 6.5	20.5 9.8	38 51
8	Maize** fodder grain	15	7	16.8 7.9	17.4 8.6	4 9
8	Reed canarygrass**	18	7	9.8	11.2	14
9	Maize*** fodder grain	116	5	17.6 9.2	19 9.6	8 4
16	Potato	450	1	1.8****	6.7	272
17	Snap bean	450	3	5.2	16.4	215
41	Maize fodder grain	60	1	6.2**** 3.1****	11.9 7.1	92 129

* Means of 3 years.
** Means of 7 years.
*** Means of 5 years, no additional sludge.
**** Unfertilized.

with consequent dramatic increase in growth rates.

Increased crop yields from sewage sludge application have been well documented (25, 34) (Table 10.10.). It has been suggested (13) that the most interesting effects of organic manures, among them sewage sludge, was not the direct (or short-term) fertilizer effects, but the so called organic matter effect, *i.e.*, an increase in soil productivity which could not be explained by mineral nutrients alone . Digested sludge applied at up to 150 Mg/ha on a light calcareous loam increased the yields of five different crops (25), but did not increase wheat grain yield (Table 10.11.). These increases over fertilized control plots were attributed to an increase in soil moisture, resulting from higher organic matter levels, as well as from the slow release of N and P from sludge.

Table 10.11 Crop yields in sewage sludge-treated and untreated soils Adapted from Guidi and Hall (25)

Crop	No sludge*	Sludge**	Increase
		(Mg/ha)	(%)
Wheat-grain	2.72	2.74	0.7
Potato	7.63	9.02	18
Cabbage	4.90	8.29	69
Red Beet	3.02	5.21	73
Lettuce	0.33	0.68	106
Rye-grass	1.01	2.85	182

* All treatments received adequate NPK.
** Mean of liquid digested sludge applied at 18-150 Mg DM/ha.

10.5. References

1 Agbin N.N., Sebey B.R. and Markstrom D.C. 1977 Land application of sewage sludge: V. Carbon dioxide production as influenced by sewage sludge and wood waste mixtures. J. Environ. Qual. 6, 446-451.

2 Ayerst J.M. 1978 The effect of compaction of coal shale on the revegetation of spoil heaps Reclamation Review 1, 27-30.

3 Bear F.E. and Prince A.L. 1955 Agricultural value of sewage sludge. New Jersey Agric. Exp. Sta. Bull. 733.

4 Braids O.C., Ardakoni M.S. and Molina J.A.E. 1970 Liquid digested sewage sludge gave field crops necessary nutrients. Ill. Res. 12, 6-7.

5 Brink N. 1971 Runoff from fertilizing with digested sludge. Publ. no. 7. Agricultural School, Uppsala, Sweden.

6 Chawla V.K., Yip J. and Cohen D.B. 1976 Recycling of liquid digested sewage sludge on dredged river sand. Environ. Canada Report No. EPS 4-WP-76-3.

7 Clapp C.E., Dowdy R.H., Larson W.E., Linden D.R. and Stark S.A. 1975 Liquid sewage

sludge as a fertilizer and soil amendment. p. D.1-D.17. *In* Utilization of Sewage Wastes on Land. Research progress report. USDA-ARS, University of Minnesota, St. Paul MN.

8 Clapp C.E., Larson W.E., Dowdy R.H., Linden D.R., Marten G.C. and Duncomb D.R. 1983 Utilization of municipal sewage sludge and wastewater effluent on agricultural land in Minnesota. pp 259-292. Proceedings of 2nd Intern. Symp. on Peat and Organic Matter in Agric. and Hort., Volcani Center, ARO, Bet Dagan, Israel.

9 Clapp C.E., Larson W.E., Harding S.A. and Titrud G.O. 1980 Sewage sludge as a fertilizer and soil amendment. pp. D.1-D.23. *In* Utilization of Sewage Wastes on Land. Research progress report. USDA-ARS, University of Minnesota, St. Paul MN.

10 Cole D.W., Henry C.L., Schiess P. and Zasoski R.J. 1983 The role of forests in sludge and wastewater utilization programs. pp. 125-143. *In* Workshop Proceedings: Utilization of Municipal Wastewater and Sludge on Land. Eds. A.L. Page *et al.* Univ. of California, Riverside CA.

11 Coppola S. 1983 Soil microbial activities as affected by application of composted sewage sludge. pp. 170-195. *In* The Influence of Sewage Sludge Application on Physical and Biological Properties of Soils. Eds. G. Catroux, P. L'Hermite and E. Suess D. Reidel Publ. Co., Dordrecht, Holland.

12 Cunningham J.D., Keeney D.R. and Ryan J.A. 1975 Yield and metal composition of corn and rye grown on sewage sludge-amended soil. J. Environ. Qual. 4, 448-454.

13 De Haan S. 1983 General comments on the organic value of sludge. pp. 62-63. *In* The Influence of Sewage Sludge Application on Physical and Biological Properties of Soils. Eds. G. Catroux, P. L'Hermite and E. Suess, D. Reidel Publ. Co., Dordrecht, Holland.

14 DeLaune R.D., Reddy C.N. and Patrick W.H. Jr. 1981 Organic matter decomposition in in soil as influenced by pH and redox conditions. Soil. Biol. Biochem. 13, 533-534.

15 Dowdy R.H., Larson R.E. and Epstein E. 1976 Sewage sludge and effluent use in agriculture. pp. 118-153. Proc. Land Application of Waste Material Conf., Soil Conserv. Soc. Am., Ankeny IA.

16 Dowdy R.H. and Larson W.E. 1975 The availability of sludge-borne metals to various vegetable crops. J. Environ. Qual. 4, 278-282.

17 Dowdy R.H., Larson W.E., Titrud J.M. and Latterell J.J. 1978 Growth and metal uptake of snap beans grown on sewage sludge-amended soil: A four-year field study. J. Environ Qual. 7, 252-257.

18 Duncomb D. R., Clapp C.E., Dowdy R.H., Larson W.E. and Polta R.C. 1983 The effect of limed sewage sludge on soil properties and maize production. pp. F.1-F.12. *In* Utilization of Sewage Wastes on Land. Research progress report. USDA-ARS, University of Minnesota, St. Paul MN.

19 Epstein E. 1975 Effect of sewage sludge on some soil physical properties. J. Environ. Qual. 4, 139-142.

20 Epstein E., Taylor J.M. and Chaney R.L. 1976 Effects of sewage sludge and sludge compost applied to soil on some soil physical and chemical properties. J. Environ. Qual. 5, 422-426.

21 Furrer O.J. 1981 Accumulation and leaching of phosphorus as influenced by sludge application. pp. 235-240. *In* Phosphorus in Sewage Sludge and Animal Waste Slurries. Eds. T.W.G. Hucker and G. Catroux. D. Reidel Publ. Co., Dordrecht, Holland.

22 Furrer O.J. and Stauffer W. 1983 Influence of sewage sludge application on physical properties of soils and its contribution to the humus balance. pp. 65-74. *In* The Influence of Sewage Sludge Application on Physical and Biological Properties of Soils. Eds. G. Catroux, P. L'Hermite and E. Suess. D. Reidel Publ. Co., Dordrecht, Holland.

23 Greenland D.J. 1977 Soil damage by intensive arable cultivation: temporary or permanent? Philos. Trans. Roy. Soc. London Ser. B. 281, 193-208.

250

24 Guidi G. 1981 Relationships between organic matter of sewage sludge and physio-chemical properties of soil. pp. 530-544. *In* Characterization, Treatment, and Use of Sewage Sludge. Eds. P. L'Hermite and H. Ott. D. Reidel Publ. Co., Dordrecht, Holland.

25 Guidi G. and Hall J.E. 1984 Effects of sewage sludge on the physical and chemical properties of soils. pp 295-395. *In* Processing and Use of Sewage Sludge. Eds. P. L'Hermite and H. Ott. D. Reidel Publ. Co., Dordrecht, Holland. (In press).

26 Guidi G., Pagliai M. and Giachetti M. 1983 Modifications of some physical and chemical soil properties following sludge and compost applications. pp. 122-130. *In* The Influence of Sewage Sludge Application on Physical and Biological Properties of Soils. Eds. G. Catroux P. L'Hermite and E. Suess. D. Reidel Publ. Co., Dordrecht, Holland.

27 Guidi G., Petruzzelli G., Giachetti M. and Levi-Minzi R. 1983 Effect of three fractions extracted from an aerobic and an anaerobic sewage sludge on the water stability and surface area of soil aggregates. Soil Sci. 136, 158-165.

28 Gupta S.C., Dowdy R.H. and Larson W.E. 1977 Hydraulic and thermal properties of a sandy soil as influenced by incorporation of sewage sludge. Soil Sci. Soc. Am. J. 41, 601-605.

29 Hall J.E. and Coker E.G. 1983 Some effects of sewage sludge on soil physical conditions and plant growth. pp. 43-60. *In* The Influence of Sewage Sludge Application on Physical and Biological Properties of Soils. Eds. G. Catroux, P. L'Hermite and E. Suess. D. Reidel Publ. Co., Dordrecht, Holland.

30 Hanotiaux G., Heck J.P., Rocher M., Barideau L. and Marlier-Geets O. 1981 Evolution of phosphorus in sewage sludge after its application to soil. pp. 399-411. *In* Phosphorus in Sewage Sludge and Animal Waste Slurries. Eds. T.W.G. Hucker and G. Catroux. D. Reidel Publ. Co., Dordrecht, Holland.

31 Harding S.A., Clapp C.E. and Larson W.E. 1984 Nitrogen availability and uptake from field soils five years after incorporation of sewage sludge. J. Environ. Qual. 14, 95-100.

32 Highiri R. 1974 Plant uptake of cadmium as influenced by cation exchange capacity, organic matter, zinc and soil temperature. J. Environ. Qual. 3, 180-183.

33 Hinesly T.D. and Jones R.L. 1974 Agricultural benefits and environmental changes resulting from long-term use of digested sewage sludge as fertilizer for field crops and as an amendment in reclamation of strip-mine spoil banks. Metropolitan Sanitary District of Greater Chicago, Chicago IL.

34 Hinesly T.D., Jones R.L. and Ziegler E.L. 1972 Effects on corn by application of heated anaerobically digested sludge. Compost Sci. 13, 26-30.

35 Hinesly T.D., Redborg K.E., Ziegler E.L. and Rose-Innes I.H. 1982 Effects of chemical and physical changes in strip-mined spoil amended with sewage sludge on the uptake of metals by plants. pp. 339-352. *In* Land Reclamation and Biomass Production with Municipal Wastewater and Sludge. Eds. W.F. Sopper, E.M. Leakes and R.K. Bastian. Pennsylvania State University Press, University Park PA.

36 Hohla G.N., Jones R.L. and Hinesly T.D. 1978 The effect of anaerobically digested sewage sludge on organic fractions of Blount silt loam. J. Environ. Qual. 7, 559-563.

37 Jacobowitz L.A. and Steenhuis T.S. 1984 Compost impact on soil moisture and temperature. BioCycle 25, 56-60.

38 Kagawa H. 1977 The significance of organic N compounds as the substance for iron-reduction metabolism in submerged paddy soils. Plant and Soil 47, 81-87.

39 Kaplan D.L., Hartenstein R., Neuhauser E.F. and Malecki M.R. 1980 Physicochemical requirements in the environment of the earthworm *Eisenia foetida*. Soil Biol. Biochem. 12, 347-352.

40 Keeney D.R., Lee K.W. and Walsh L.M. 1975 Guidelines for the application of wastewater sludge to agricultural land in Wisconsin. Bull. 88. Wisconsin DNR, Madison WI.

41 Kelling K.A., Peterson A.E., Walsh L.M., Ryan J.A. and Keeney D.R. 1977 A field study of the agricultural use of sewage sludge: I. Effect on crop yield and uptake of N and P. J. Environ. Qual 6, 339-345.

42 Kelling K.A., Walsh L.M., Keeney D.R., Ryan J.A. and Peterson A.E. 1977 A field study of the agricultural use of sewage sludge: II. Effect on soil N and P. J. Environ. Qual. 6, 345-351.

43 Khaleel R., Reddy K.R. and Overcash M.R. 1981 Changes in soil physical properties due to organic waste applications: a review. J. Environ. Qual. 10, 133-141.

44 King L.D. 1973 Mineralization and gaseous losses of nitrogen in soil-applied liquid sewage sludge. J. Environ. Qual. 2, 356-358.

45 King L.D. and Morris H.D. 1972 Land disposal of liquid sewage sludge: II. The effect on soil pH, manganese, zinc and growth and chemical composition of rye (*Secale cereale* L.) J. Environ. Qual. 1, 425-429.

46 King L.D. and Morris H.D. 1973 Land disposal of liquid sewage sludge: IV. Effect of soil phosphorus, potassium, calcium, magnesium and sodium. J. Environ. Qual. 2, 411-414.

47 King L.D., Rudgers L.A. and Webber L.R. 1974 Application of municipal refuse and liquid sewage sludge to agricultural land: I. Field study. J. Environ. Qual 3, 361-365.

48 Kirkham M.B. 1974 Disposal of sludge on land: Effect on soils, plant, and ground water. Compost Sci. 15, 6-10.

49 Kirkham M.B. 1980 Characteristics of wheat grown with sewage sludge placed at different soil depths. J. Environ. Qual. 9, 13-18.

50 Kladivko E.J. and Nelson D.W. 1979 Changes in soil properties from application of anaerobic sludge. J. Water Poll. Control Fed. 51, 325-332.

51 Korcak R.F. 1980 Effect of applied sewage sludge compost and fluidized bed material on apple seedling growth. Comm. Soil Sci. Plant Anal. 11, 571-585.

52 Leeper G.W. 1972 Reactions of heavy metals with soils with special regards to their application in sewage wastes. Dept. of the Army, Corps of Engineers. Contract No. DACW 73-73-C-0026, Washington DC.

53 Linden D.R., Clapp C.E. and Dowdy R.H. 1983 Hydrologic and nutrient management aspects of municipal wastewater and sludge utilization on land. pp. 79-101. *In* Workshop Proceedings: Utilization of Municipal Wastewater and Sludge on Land. Eds. A.L. Page *et al.* Univ. of California, Riverside CA.

54 Lunt H.A. 1959 Digested sewage sludge for soil improvement. Connecticut Agric. Exp. Sta. Bull. 622.

55 Magdoff F.R. and Amadon J.F. 1980 Nitrogen availability from sewage sludge. J. Environ. Qual. 9, 451-455.

56 Magdoff F.R. and Chromec F.W. 1977 Nitrogen mineralization from sewage sludge. J. Environ. Sci. Health A12, 191-201.

57 Magdoff F.R., Keeney D.R., Bouma J. and Ziebell W.A. 1974 Columns representing moundtype disposal systems for septic tank effluent: II. Nutrient transformations and bacterial populations. J. Environ. Qual. 3, 228-234.

58 Mays D.A., Terman G.L. and Duggan J.C. 1973 Municipal compost: Effect on crop yield and soil properties. J. Environ. Qual. 2, 89-91.

59 McCoy J.H. 1971 Sewage pollution of natural waters. pp. 46-64. *In* Microbial Aspects of Pollution. Eds. G. Sykes and F.A. Skinner. Academic Press, New York NY.

60 Miller R.H. 1973 Soil microbiological aspects of recycling sewage sludges and waste effluents on land. pp. 79-89. *In* Recycling Municipal Sludges and Effluents on Land. Nat. Assoc. St. Univ. and Land-Grant Coll., Washington DC.

61 Miller R.H. 1974 Microbiology of sewage sludge disposal in soil. Report No. EPA-670/2-74-074, NTIS, Springfield VA.

62 Miller R.H. 1974 Factors affecting the decomposition of an anaerobically digested sewage sludge in soil. J. Environ. Qual. 3, 376-380.

63 Mitchell M.J., Hartenstein R., Swift B.L., Neuhauser E.F., Abrams B.I., Mulligan R.M., Brown B.A., Craig D. and Kaplan D. 1978 Effects of different sewage sludges on some chemical and biological characteristics of soil. J. Environ. Qual. 7, 551-559.

64 Mitchell M.J., Parkinson C.M., Hamilton W.E. and Dindal D.L. 1982 Role of the earthworm, *Eisenia foetida*, in affecting organic matter decomposition in microcosms of sludge-amended soil. J. Appl. Ecol. 19, 805-812.

65 Morel J.L. and Guckert A. 1983 Influence of limed sludge on soil organic matter and soil physical properties. pp. 25-39. *In* The Influence of Sewage Sludge Application on Physical and Biological Properties of Soils. Eds. G. Catroux, P. L'Hermite and E. Suess. D. Reidel Publ. Co., Dordrecht, Holland.

66 Naghrajah S., Posner A.M. and Quirk J.P. 1970 Competitive adsorption of phosphates with polygalacturonate and other organic anions on kaolinite and oxide surfaces. Nature 228, 83-84.

67 Pagliai M. and Guidi G. 1981 Porosity and pore size distribution in a field test following sludge and compost application. pp. 545-552. *In* Characterization, Treatment, and Use of Sewage-Sludge. Eds. P. L'Hermite and H. Ott. D. Reidel Publ. Co., Dordrecht, Holland.

68 Pagliai M., Guidi G., LaMarca M., Giachetti M and Lucamante G. 1981 Effects of sewage sludges and composts on soil porosity and aggregation. J. Environ. Qual. 10, 556-561.

69 Pagliai M., LaMarca M. and Lucamante G. 1983 Micromorphological investigation of the effect of sewage sludges applied to soil. pp. 219-225 *In* Soil Micromorphology 1: Techniques and Applications. Eds. P. Bullock and C.P. Murphy. A.B. Academic Publ., Hertfordshire, England.

70 Palazzo A.J. 1983 Long-term plant persistance and restoration of acidic dredge soils with sewage sludge and lime. CRREL Report 83-28. US Army Corps of Engineers, Hanover NH.

71 Parsa A.A. and Lindsay W.L. 1972 Plant value in organic wastes. Iranian J. Agr. Res. 1, 60-71.

72 Patrick W.H. Jr. and Khalid R.A. 1974 Phosphate release and sorption by soils and sediments: Effect of aerobic and anaerobic conditions. Science 186, 53-56.

73 Pera A., Giovannetti M., Vallini G. and De Bertoldi M. 1983 Land application of sludge: effects on soil microflora. pp. 208-228. *In* The Influence of Sewage Sludge Application on Physical and Biological Properties of Soils. Eds. G. Catroux, P. L'Hermite and E. Suess. D. Reidel Publ. Co., Dordrecht, Holland.

74 Pomares-Garcia F. and Pratt P.F. 1978 Recovery of ^{15}N–labeled fertilizer from manured and sludge-amended soil. Soil Sci. Soc. Am. J. 42, 717-720.

75 Pratt P.F., Broadbent F.E. and Martin J.P. 1973 Using organic wastes as nitrogen fertilizer. Calif. Agric. 27, 10-13.

76 Premi P.R. and Cornfield A.H. 1969 Incubation study of nitrification of digested sewage sludge added to soil. Soil Biol. Biochem. 1, 1-4.

77 Robertson W.K., Lutrick M.C. and Yuan T.L. 1982 Heavy application of liquid-digested sludge on three ultisols: I. Effects on soil chemistry. J. Environ. Qual. 11, 278-282.

78 Rothwell D.F. and Hortenstine C.C. 1969 Composted municipal refuse: its effect on carbon dioxide, nitrate, fungi, and bacteria in Aredonda fine sand. Agron. J. 61, 837-840.

79 Rebhun M. and Manka J. 1971 Classification of organics in secondary effluents. Environ. Sci. Technol. 5, 606-609.

80 Ryan J.A., Keeney D.R. and Walsh L.M. 1973 N transformation and availabilty of anaerobically digested sewage sludge in soil. J. Environ. Qual. 2, 489-492.

81 Sabey B.R. 1980 The use of sewage sludge as a fertilizer. pp. 73-103. *In* Handbook of Organic Waste Conversion. Ed. M.W.M. Bewick. Van Nostrand-Reinhold Company, New York NY.

82 Sabey B.R. and Hart W.E. 1975 Land application of sewage sludge: I. Effect on growth and chemical composition of plants. J. Environ. Qual. 4, 252-256.

83 Schaumberg G.D., Levegue-Madore C.S., Sposito G. and Lund L.D. 1980 Infrared spectroscopic study of the water-soluble fraction of sewage sludge-soil mixtures during incubation. J. Environ. Qual. 9, 297-303.

84 Sheaffer C.C., Decker A.M., Chaney R.L. and Douglass L.W. 1979 Soil temperature and sewage sludge effects on corn yield and macro-nutrient content. J. Environ. Qual. 8, 450-454.

85 Sobhan-Ardakani M. 1971 Stability constants of metal polyelectric complexes occurring naturally in soil and sewage sludge. Ph. D. Thesis. Dept. of Agron. University of Illinois, Urbana IL.

86 Sommers L.E. 1977 Chemical compostion of sewage sludge and analysis of their potential use as fertilizer. J. Environ. Qual. 6, 225-232.

87 Stadelmann X. and Furrer O.J. 1983 Influence of sewage sludge application on organic matter content, microorganisms and microbial activities of a sandy loam soil. pp. 141-164. *In* The Influence of Sewage Sludge Application on Physical and Biological Properties of Soils. Eds. G. Catroux, P. L'Hermite and E. Suess. D. Reidel Publ. Co., Dordrecht, Holland.

88 Stark S.A. and Clapp C.E. 1980 Residual nitrogen available from soils treated with sewage sludge in a field experiment. J. Environ. Qual. 9, 505-512.

89 Terry R.E., Nelson D.W. and Sommers L.E. 1981 Nitrogen transformation on sewage sludgeamended soil as affected by soil environmental factors. Soil Sci. Soc. Am. J. 45, 506-513.

90 Tomati U., Grappelli A. and Galli E. 1983 Sludge effect on soil and rhizosphere biological activities. pp. 229-241. *In* The Influence of Sewage Sludge Application on Physical and Biological Properties of Soils. Eds. G. Catroux, P. L'Hermite and E. Suess. D. Reidel Publ. Co., Dordrecht, Holland.

91 Varanka M.W., Zdzislaw M.Z. and Hinesly T.D. 1976 The effect of digested sludge on soil biological activity. J. Water Poll. Control Fed. 48, 1728-1740.

92 Vigerust E. 1983 Physical properties in sewage sludge and sludge treated soils. p. 107-119. *In* The Influence of Sewage Sludge Application on Physical and Biological Properties of Soils. Eds. G. Catroux, P. L'Hermite and E. Suess. D. Reidel Publ. Co., Dordrecht, Holland.

93 Wade S.E., Bache C.A. and Lisk D.J. 1982 Cadmium accumulation by earthworms inhabiting municipal sludge-amended soil. Bull. Environ. Contam. Toxicol. 28, 557-560.

94 Webber L.R. 1978 Incorporation of nonsegregated, noncomposted solid waste and soil physical properties. J. Environ. Qual. 7, 397-400.

95 Whisler F.D., Lance J.C. and Linebarger P.S. 1974 Redox potential in soil columns intermittently flooded with sewage sludge. J. Environ. Qual. 3, 68-74.

96 Wollan E., Davis R.D. and Jenner S. 1978 Effects of sewage sludge on seed germination. Environ. Pollut. 17, 195-205.

97 Wong M.H. 1979 Sewage sludge as conditioner for improving soils affected by sulfur dioxide. Bull. Environ. Contam. Toxic. 23, 717-724.

98 Wong M.H., Cheung Y.H. and Cheung C.L. 1983 The effects of ammonia and ethylene oxide in animal manure and sewage sludge on the seed germination and root elongation of *Brassica parachinensis.* Environ. Poll. (Series A) 30, 109-123.

99 Zettel J. and Klingler J. 1983 Influence of sewage sludge application on microarthropods (collembola and mites) and nematodes in a sandy loam soil. pp. 167-169. *In* The Influence of Sewage Sludge Application on Physical and Biological Properties of Soils. Eds. G. Catroux P. L'Hermite and E. Suess. D. Reidel Publ. Co., Dordrecht, Holland.

100 Zhou Q. and Chen H.K. 1983 The activity of nitrifying and denitrifying bacteria in paddy soil. Soil Sci. 135, 31-38.

Section V

The use of peat and composts as container media

11. Peat and peat substitutes as growth media for container-grown plants

M. RAVIV, Y. CHEN and Y. INBAR

11.1. Historical review

People have probably used organic materials as an aid for plant culture since the eve of human history, but documentation is scarce. At the beginning of the 18th century we find a detailed description of soil improvement by organic materials (148). The author has distinguished between peat moss as an amendment for clay soils and animal dung as an ammendment for sandy soils. An elaborate procedure for the use of partially fermented cow manure in a frame covered with glass to form hotbeds has been described in the 18th century (86). This procedure enabled fast germination and growth of many species as a result of soil warming. During periods of cold weather the frames were kept closed to retain the heat produced by organic matter decomposition. Loudon, in his well-known "Encyclopedia of Gardening" listed in 1839 many organic materials used as plant growth media in greenhouses (77). Loudon also gave detailed recipes for their composting and stated that in many cases mixing different ingredients is preferable from the standpoint of plant growht. Loudon mentioned that peat *per se* did not possess any nutritional value but nonetheless, a grower could use only sphagnum peat, loam and sand to satisfy the needs of all plants.

The first motivation for growing plants in media other than soils *in situ* was probably the need for plants to be transferred from one place to another. Tree saplings are a good example. The early Egyptians are credited with being the first to transplant trees with a ball of earth. The Greek philosopher, Theophrastus, has discussed in his book *On The History of Plants* a few transplanting procedures (about 302 B.C.). It took, however, hundreds of years for growers to realize that growing plants in containers offers them many advantages.

11.2. General introduction

The concept of growing plants in containers is markedly different than growing plants in open soil. When pots are used the volume of medium from which the plant can absorb water and minerals is limited and is usually smaller than that available for plants growing in open soil. Moreover, under intensive culture with controlled temperatures and high nutritional levels, the stomata commonly stay open for longer periods of time enhancing water uptake and loss. In order to take full advantage of these conditions an adequate amount of easily available water must exist within the root zone. Whenever such a water regime is being maintained the problem of aeration arises. Since artificial growth media are usually porous and the root zone is homogenous, strict control of water and air contents is attainable. In contrast, many soils have relatively poor air porosity and their profile is heterogenous, therefore a proper simultaneous control over both water and air regimes is difficult to achieve (61).

The production of healthy, uniform plants is a basic requirement of modern agriculture. It is necessary to obtain a reporducible, pathogen-free medium, and delay its recontamination. Container-grown plants are easy to be graded frequently thus enabling the supply of uniform plants.

In order to meet the need for a sterilizable medium with proper air and water capacities growers use many types of organic and inorganic materials. The distinction between organic and non-organic materials is somewhat misleading and requires some clarification. In this review, we shall refer to organic materials as those materials which serve as substrate for biological processes and hence are subject to biological decomposition. Shredded polyethylene, polystyrene and polyurethane foams which are organic chemical products, are not subject to biological decomposition and shall be discussed as non-organic.

Although it is beyond the scope of this review, it is worth mentioning that nonorganic materials can be very useful consituents for plant growing.

A question frequently raised is whether one "ideal" growth media can be identified? The obvious answer is "no". The best medium for a specific case will vary according to numerous factors such as the type of plant material (seedlings, cuttings, etc.), species, climatic conditions, watering systems and regimes, container size, just to mention a few of the pertinent factors. Economical considerations must, of course be taken into account.

Plants can be supported and grown in many types of materials. In fact, plants can be grown and survive on any medium if the roots can penetrate the substrate. Obviously, survival is not the primary goal and continuous research is conducted in order to find the optimal substrate and growing conditions for various plants. For good germination, rooting and growth results, the following characteristics of the medium are required (51, 52, 64, 103, 147):

11.2.1. Physical properties

1. High retention of easily available water.

2. Adequate air supply.
3. Particle size distribution that will maintain the conditions mentioned above.
4. Low bulk density to provide light weighted substrate, but heavy enough to anchor the plant.
5. High porosity
6. Stable structure that will prevent shrinking (or swelling) of the medium.

11.2.2. Chemical properties

1. High cation exchange capacity.
2. Adequate available nutrient level.
3. Low salinity.
4. High buffering capacity and the ability to maintain constant pH levels.
5. Minimal decomposition rate of the substrate.

11.2.3. Other properties

1. Free from weed seeds, nematodes and various pathogens.
2. Reproductibility and availability.
3. Low cost.
4. Easy to mix.
5. Resistance to extreme physical, chemical and environmental changes.

In conclusion, it can be said that if the general background is understood and the specific conditions are known, an "ideal" solution can be worked out for each individual situation. This review is primarily concerned with the physical, chemical and biological characteristics of organic materials relevant to their function as growth media. The prominent organic materials which serve as constituents of growth media are also discussed.

11.3. Physical characteristics

The physical characteristics of a growth medium is of utmost importance. Once the medium is in the pot and a plant is growing in it, there is no practical way to improve its basic physical features. This is in contrast to the chemical status which can be affected by the growing techniques practiced by the grower (51).

For example, the daily water consumption of a mature container-grown Chrysanthemum plant in summer has been reported to be as high as 300 ml water/liter medium/day (16). This means that practically all the readily available water of the medium must be replenished frequently. If the soluble salts content of the water is high, as commonly found in arid regions, a thorough leaching of the substrate is also necessary to avoid high salinity levels. Under these conditions the air content and oxygen diffusion rate in the substrate are very important. Aeration is even more important when the decomposition process continues in the medium leading to biological competition for oxygen. The main physical parameters that determine and hence enable us to predict the air and water-holding capacity of a medium are:

— Particle and pore size distribution
— Bulk density
— Total porosity
— Moisture retention curve

These parameters are related to each other. The exact nature of these relationships vary and each parameter will be discussed separately.

Once those physical parameters are known there is a possibility to adopt proper irrigation and fertilization regimes to meet the requirements of various plants (24, 31).

11.3.1. Bulk and particle density

Particle density of organic matter ranges between 1.4 to 2.0 g/cm^3. The particle density of various organic substrates depends mostly on the ash content, Increases in ash content usually result in an increase in the particle density up to 2.0 g/cm^3. An average value of 1.5 g/cm^3 is commonly used on ash free basis (97).

Particle and bulk densities of some organic substrates are listed in Table 11.1.

Bulk density plays an important role as the containers are usually transported for marketing and weight has to be considered. In addition, roots anchorage should be considered as an important parameter: the taller the plant, the heavier should the substrate be or a larger container be used. Where wind is not a factor, bulk density may be as low as 0.15 g/cm^3. Crops which are grown outdoors must be grown in heavier mixes with bulk density ranging between 0.50 - 0.75 g/cm^3 (92, 142).

Table 11.1. Particle and bulk densities of some organic substrates (147).

Material	Particle density g/cm^3	Bulk density g/cm^3
Peat	1.55	0.05 - 0.20
Pine leaf mould	1.90	0.10 - 0.25
Bark	2.00	0.10 - 0.30

Bulk and particle density measurements are two parameters required for porosity calculations. Values given in the literature for the same substrate vary widely, probably due to the use of different measurement procedures. One may find that the bulk density of organic substrates ranges from 0.05 to 0.30 g/cm^3, while for mineral soils the values are in the range of 1.1 - 1.7 g/cm^3.

11.3.1.1. Particle and pore size distribution

Particle and pore size distribution correlate to each other although quantitative correlation is very difficult to draw. Particle size distribution is measured by sieving or by a sedimentation procedure whereas pore size distribution is calculated from water retention curves (24). Particle size affects plant growth mostly through pore size. Pore and particle size distribution determine the balance between air and water content of the substrate at any moisture content. As a general rule, the larger the particles, air content is higher and water content is lower for any given suction. Experiments with sphagnum peat have shown that the minimum air content should be about 15% by volume. At water tension of 10 cm, this value corresponds to particle size of about 1 mm. This size is considered a border line between fine and coarse particles (96, 99). Particles smaller than 0.5 - 0.8 mm in diameter, hold large amounts of water at a relatively high potential and aeration may be poor. The availability of the water at this suction is low. Small organic particles are more susceptible to decomposition than large ones. However, small particles contirbute to a high cation exchange capacity and increased surface area. In order to maintain aerated stable substrates it is recommended that the particles size should be above 0.5 - 1.0 mm (92, 98).

According to the particle size distribution, peat and other organic substrates are divided to fine, medium and coarse. Textural grades vary from one country to another. A comparison between ASTM and Scandinavian Standard exhibits immense differences and is presented in Table 11.2.

Table 11.2. ASTM and Scandinavian textural grades.

Grade	ASTM (D 2977-71) Particle size mm	Scandinavian standards(NS 28090 E) All particles under	Particles ⟨ 1 mm %
Coarse	⟩ 2.38	40 mm	⟨ 30
Medium	0.84 - 2.38	15 mm	⟨ 40
Fine	⟨ 0.84	6 mm	⟨ 70

In general terms, fine material is particularly suited for the production of seedlings, whereas medium and coarse materials should be selected for pot culture and rooting.

The pore size distribution can be calculated from the matric potential by the following equation (83)

$$r = {}^{2\gamma}/_{\rho gh}$$

where: γ = water surface tension, ρ = water density, g = gravitational constant, h - water tension (suction) and r = equivalent tubular pore radius. A pore diameter of 30 microns equivalent to 100 cm suction, is considered to separate the "non capillary" or macro-pores from the "capillary" pores. Macropores are filled with air during a significant portion of the growth cycle and thereby serve as pathway for oxygen and carbon dioxide diffusion (97).

A classification of pore size grades and their water economy are described in Table 11.3.

It can be concluded that in coarse materials where large pores are present low amount of water is held while the substrate is well aerated. On the other hand, fine material holds large amounts of hardly available water along with small amounts of air. Puustjarvi (99) defines the best substrate as a coarse to medium textured material with pore size distribution between 30 - 300 microns (equivalent to particle size ranging from 0.25 to 2.5 mm) so that there will be enough easily available water along with adequate air content.

As mentioned earlier, other values, derived from comparisons among materials exhibiting different particle size distribution were also present in the literature (92, 98). The apparent lack of accordance results, probably, from variations in the measurement methods but also may correspond to other inherent properties of the tested materials which may affect plant's performance.

Table 11.3. Pore size grades with regard to the water economy in peat (99).

Texture grade	Pore size microns	Matric potential mbar	Equivalent particle size mm	Duration of Water availability	Drainage
Very coarse	⟩ 300	⟨ 10	⟩ 2.6	High for very short time	Very fast
Coarse	300 - 100	10 - 30	2.6 - 0.9	High for short time	Fast
Medium	100 - 30	30 - 100	0.9 - 0.25	Easily available	Medium
Fine	30 - 3	100 - 1000	0.25 - 0.025	Low availability	Slow
Very fine	⟨ 3	⟩ 1000	⟨ 0.025	Almost insignificant	Very slow

11.3.2. Porosity and aeration

Total porosity in most organic substances is generally above 80% by volume. The pore space is divided between air filled and water filled pores. The air content of a substrate is defined as the proportion of the volume that contains air after it has been saturated with water and allowed to drain, usually at a water tension of 10 mbars (cm of water) (16, 34, 97).

Plant growth is inhibited and sometimes root rot occurs as a result of poor aeration caused by compaction or excessive watering, especially when substrates with inadequate pore space are used. Container substrates are usually shallow and plants may experience drainage problems to which close contact between the container wall and the substrate may also contribute. The uptake of water and nutrients is strongly inhibited in the absence of adequate levels of oxygen (92, 141). The oxygen is transferred into the roots by diffusion through the water layers surrounding them. Oxygen diffusion rate in water is 10^4 times lower than in air. Therefore, thickness of the water layer around the root is of great importance. If the texture and structure of the substrate are such that most of the pores remain filled with water after irrigation, oxygen supply will be severely restricted, CO_2 will accumulate, ethylene might be produced and as a result growth will be inhibited and sometimes the plant may wilt.

As mentioned earlier, the pore size distribution is the key to the water and air status in the substrate. The air content is measured at a constant matric suction, usually at 10 cm tension (34). Opinions in regard to the air content are different among various investigators: Bunt (15) and Joiner and Conover (64) require air content of above 10% by volume; Poole *et al.* (92) 5 - 30%; Puustjarvi (96) - 45%. Verdonck *et al.* (131) state that optimal growth was obtained at an air content of 20%. Penningsfeld (85) found that container-grown plants require minimum air content of 15%. In general the optimal air content is between 10 - 45% by depending upon the plant type, the container size and the substrate. Organic substrates that tend to decompose and require oxygen for the decomposition should be used in mixes with relatively high air content.

The container depth has an important effect, which is sometimes overlooked on the air content. The deeper the container, the higher is the air content. This phenomenon results from the water accumulation at the bottom of the container. When a small or shallow container is used, coarse substrates are preferable to maintain adequate aeration (115, 116).

11.3.3. Water holding capacity (water retention curve)

Water retention curves of soils are usually measured over a wide range of suctions (0 - 15 bars). Points frequently mentioned on these curves are "field capacity" (1/3 bar) and "wilting point" (15 bar). In container media narrower range of suction is usually refered to (0 - 100 mbar or cm). Points commonly refered to are at 10, 50 and 100 cm.

Plants grown in containers cannot be subjected to high water tensions without severe losses is growth rate, because of the limited volume of the medium. The drier the substrate the greater the energy required for water uptake.

In practice this means that suction should not exceed 100 cm. The minimal water tension shortly after saturation should be about 10 cm. If the particle size distribution is appropriate, this will result in adequate air content.

Moisture content at 10, 50 and 100 cm suction is thought to be usful in characterization of substrates for the following reasons: the substrate layer on which container plants are grown, generally has a thickness of 10 - 15 cm. When the plants are adequately watered the moisture content at the bottom layers is close to saturation, and gradually decreases from bottom to top of the container. Therefore, a minimum tension of 10 cm (equivalent to 10 cm height of the container) is required to obtain a minimum air content value. The next point of importance is the driest moisture conditions which will not inhibit plant growth. It was found in many experiments that water tension of 100 cm might be considered as a maximum (15, 34, 35, 98, 115).

The water retention curve is the best graphical presentation showing the amount of water retained by a substrate at various tensions. From this curve one can draw conclusions on the main physical properties of the substrate and hence to apply the proper water regime.

De Boodt and Verdonck (34) define several sections on a water retention curve of a substrate (on a percent of total volume vs. suction plot) as follows:

(a) Moisture content at 0 cm tension is equal to the total pore space (T.P.S.).
(b) Volume of solid material is presented on the ordinate at zero suction. Bulk density can be calculated based on this value and the dry weight.
(c) Volume of air in the substrate at any given suction.
(d) Volume of water at any given suction.

They introduced three new terms which are now widely adopted (all the values are measured on a volumetric basis). This concept and related definitions are also presented in Fig. 11.1.

(a) Volume percent of air after irrigation (AIR) - the difference in volume between total pore space and volumetric moisture content at 10 cm suction.
(b) Easily available water (E.A.W.) - the quantity of water released from the material when the suction is increased from 10 to 50 cm (usually 75 - 90% of the total available water).
(c) Water buffering capacity (W.B.C.) - the quantity of water released when the suction is increased from 50 cm to 100 cm (the water reserve in the substrate).

These investigators present values for an "ideal" substrate as follows: the total pore space (T.P.S.) should be 85%; AIR 10-30%; EAW 20-30%; and WBC 4-10% (percent of total volume).

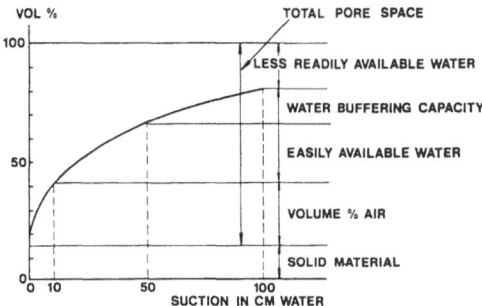

Fig. 11.1. Volume percent of solid, air and water in growth media *vs* suction (ref. (34)).

Puustjarvi (97) proposed a different procedure to calculate the optimum moisture regime in peat under the following conditions:

a. The air space of peat under intensive culture should be at least 50% by volume.

b. The maximum water tension at which no growth inhibition is likely to occur is 50 cm.

This leads to the following equation (using the water retention curve):

Optimal moisture regime $= WS_1 - WS_2$

where: Optimal moisture regime is the amount of easily available water.

WS_1 is the volumetric water content in peat when its air space is 50% by volume.

WS_2 is the volumetric water content of the peat when its water retaining force equals a suction of 50 cm.

In this equation both the minimum air content and the upper limit of the easily available water were taken into consideration. When considering different crops various minimum air space may be put in this equation according to the plant's need.

Table 11.4. Physical properties of some important organic substrates (34, 129, 135).

Substrate	B.D. g/cm^3	T.P.S.	AIR	E.A.W.	W.B.C.
			% of volume		
Pine leaf mould (fresh)	0.09-0.16	89-95	38-55	15-18	3-4
Pine leaf mould (old)	0.18-0.19	87-90	37-55	13-16	4-0
White peat	0.08-0.18	89-95	15-27	30-33	7-9
Black peat	0.11-0.20	86-92	14-17	16-25	4-6
Bark	0.12-0.27	81-95	38-61	3-6	0.5-2

Water retention and other physical properties of various organic substrates which are used in horticulture are described in a number of publications (24, 34, 35, 36, 63, 97, 129, 135). Some examples are presented in Table 11.4.

In practice, the choice of an optimal irrigation regime for a certain crop will depend on its physiological needs and the physical properties of the growth medium. Various growth media with different physical properties can be used for a certain crop as long as a proper irrigation regime is applied.

11.4. Chemical characteristics

The most important chemical properties of materials serving as components for growth media are: 1. cation exchange capacity. 2. buffer capacity. 3. pH. 4. quantity and availability of nutrients. 5. dgree and rate of decomposition. In general, these properties are linked together. The organic compounds which are used in growth media are the main contributors to the chemistry of the substrates, mainly due to the formation and presence of humic substances which consist the major end product of decomposition of organic materials. Organic materials which are used as substrates should maintain stable characteristics. This can be achieved either by composting or by the use of sphagnum peat moss which has been stabilized during its formation process. Stability in both cases is not absolute and any organic matter is subjected to further decomposition. Decomposition during the growing period results in the release of available nutrients such as NH_4^+, NO_3^-, PO_4^{-3}, SO_4^{-2} and many others, and the release of CO_2 to the atmosphere. Humic substances form complexes and chelates with micronutrients thereby improving their availability to plants. Another important characteristic of this group of compounds is high cation exchange capacity. As a result, nutrients which are applied as fertilizers may remain available for longer periods of time as compared to substrates without organic components.

11.4.1. Carbon/Nitrogen (C/N) ratio

The C/N ratio is traditionally used as an indicator for organic matter origin, maturity and stability. The main causes of damage to plants grown on immature organic matter is believed to be nitrogen starvation and lack of oxygen in the rhizosphere. This is caused by microorganisms which decompose raw organic materials and utilize nitrogen for the synthesis of cell proteins. Oxygen is consumed by the microbial activity. Nitrogen content in microorganisms' tissues varies from one systematical group to another. In general, the C/N ratio ranges from 5 to 30. A C/N ratio of less than 20 is considered desirable and serves as an indicator of mature and stable organic substrate (4, 47, 121). In order to prepare substrates from

organic materials such as bark, straw, sawdust etc. with high initial C/N ratio
(⟩ 100), nitrogen is provided during decomposition in the form of conventional
fertilizers, such as urea.

Puustjarvi and Robertson (98) recommend, for certain systems, to add nitrogen
to peat in order to prevent nitrogen starvation. Some problems may arise from
these additions: after a while decomposition rate slows down, the microorganisms
die and release of fixed nitrogen to the substrate solution takes place, thereby increasing the medium salinity. In order to avoid this process, leaching of the composted organic material should be performed before the growing season when nec-
cessary.

Accurate calculation of optimum nitrogen requirements and C/N ratios during
the composting of organic matter is difficult due to three factors (49): (a) some of
the carbon in the substrate is in the form of lignin and other compounds which
are resistant to microbial decomposition. This group of compounds decomposes
only slightly over a long period of time. (b) some of the nitrogen is in unavailable
forms e.g. as keratin-type proteins, and is not available during most of the composting process. (c) some nitrogen fixation takes place through the agency of bacteria of the *Azotobacter* spp. especially in the presence of phosphatic material.

The composting process results in softening of initial differences in C/N ratios
among various raw materials. Bark has a C/N ratio of ⟩ 100 which drops, after
composting to 30 - 40 (129). While in sewage sludge this ratio rises from 8.7 to
11.2 (21).

A new approach to the evaluation of the degree of maturity of compost has recently been developed in Japan. Carbon/organic Nitrogen ratio in the water extract has been proposed as a maturity index by Chanyasak and Kubota (21). They
state that since biochemical decomposition of organic matter is performed by microorganisms in the liquid phase of the material, the changes can be monitored
and detected in the water extract. These investigators found that for a well-mat-
ured compost, the ratio of C/N organic in the water extract has almost constant
values of 5 - 6 irrespective of the type of the raw material.

11.4.2. Cation exchange capacity

Cation exchange capacity (CEC) is defined as the sum of exchangeable cations
that a substrate can adsorb per unit weight. These cations are retained against the
leaching effect of water and usually are available to the plant. Organic matter, especially humic substances, contain activated carboxilic and phenolic groups which
are capable of releasing and adsorbing hydrogen ions, following the general equations:

$$RCOOH \rightarrow RCOO^- + H^+$$

Humic and fulvic acids behave like weakly acidic polyelectolytes and are amenable to examination by techniques based on the ionization of acidic functional groups. In addition to carboxylic groups, negative charges may arise from the presence of phenolic OH, enolic OH, imide (=NH) and some other groups. The occurrence of charged sites (*e.g.* COO⁻) accounts for the ability of the organic materials to retain cations in nonleachable forms (121). The dissociation of these weak organic acids is pH dependent. The higher the pH value, the higher is the C.E.C. During an exchange process the negatively charged organic ions are able to adsorb cations such as NH_4^+, K^+, Ca^{2+}, Mg^{2+}, Na^+ and others at relative ratios which correspond to their affinity with the adsorbing sites and concentration in the solution.

The capacity of organic substrates to adsorb metalic cations depends on the removal of hydrogen ions. This is achieved at an increasing extent as the pH rises. Puustjarvi (97) found in sphagnum peat at pH - 3.5 CEC of 50 meq/100g while at pH - 5.5 the CEC was 100 meq/100g. Helling *et al.* (54) found a linear corellation between the CEC of humic substances and the pH. They give the following equation:

CEC meq/100g organic C = – 59 + 51 pH

According to this equation, for an increase of one pH unit an increase in CEC of 51 meq/100g organic carbon is expected, or about 30 meq/100g organic matter. In clays Helling *et al.* (54) found an increase in CEC of only 4.4 meq/100g caly per one pH unit.

It may be concluded that CEC is mainly dominated by the presence of humic substances, which in turn depends on the degree of decomposition of the organic substrate.

Some organic substrates, such as peat, exhibit low initial pH level and low CEC. In order to increase the pH, lime is often added to the substrates and thus calcium and magnesium replace hydrogen ions. As a result, higher CEC is achieved and more cations may adsorb on the adsorbing sites.

Organic materials usually exhibit high CEC and buffering action against rapid cahnges in nutrients availability and pH. High CEC provides nutrient reservoirs, whereas materials with low CEC, such as most of the mineral substrates, retain small amounts of nutrients and require frequent application of fertilizers. Rapid changes in the substrate acidity or alkalninity can be prevented by using organic materials in substrate mixes (16, 85, 97, 131).

When measuring CEC of organic materials the pH at which the measurement has been performed should be stated (pH-7 is preferable). Results are usually presented in meq/100g (preferable on ash free basis). Other unit commonly used for container substrates is meq/volume unit.

11.4.3. Substrate pH

Most plants can survive at a wide range of substrate's pH (4 - 8) without suffering significant physiological disorders as long as all the nutrients are provided in available forms (7). However, growth rate and plant performance may be slowed down at extreme acid or alkaline conditions. The main effects of pH are on nutrient availability, CEC and biological activity. Under intensive cultivation it is recommended to maintain the substrate pH within a narrow range.

A prominent feature of humic substances is the buffering capacity over a wide pH range. This capacity is of considerable practical significance when a stable pH value is required. Verdonck (130), Puustjarvi (97) and Stevenson (121) studied the buffering capacity of organic substances by potentiometric titrations. From these titrations the investigators provided evidence that peat, leaf mould, humic acid and other organic materials, maintain higher buffering capacity than minerals such as clays and perlite.

The avialibility of many nutrients is affected by substrate pH. Lucas and Davis (78) examined the availability of 12 different nutrients at various pH levles in organic soils (⟩ 50% O.M.). They found that at pH range of 5.0 - 5.8 most of the nutrients maintained their maximum availability level. Above pH 6 difficulties in availability of Fe, P, Mn, B and Zn may arise, while below pH 5 deficiencies of many nutrients such as P, K, Ca, N, B, Cu, Mo may occur. If the pH of the substrate is beyond the recommended range, pH adjustment has to be applied. In the case of acidic substrates such as peat, addition of gypsum, lime or dolomite is required to achieve a pH increase. While on basic substrates a pH decrease may be achieved by addition of sulfur. The amount of lime or surfur added to the substrate depends on its original pH and CEC (79, 92).

11.4.4. Nutrients availability in organic substrates

In organic substrates such as peat, bark and leaf mould the amount of naturally occuring available nutrients is small. When using organic substrates originating from animals excreta or city refuse some nutrients may exhibit high levels depending on the origin of the compost and on the composting process.

In any case, for optimal plant growth additional nutrients should be added as base fertilizers and/or as a top dressing whenever organic substrates are used. The amount and frequency of fertilization depends on the substrate's characteristics such as CEC and on the irrigation regime. High CEC as in peat, increases the efficiency of addition of base fertilizers during the manufacturing process. When substrates of low CEC are used fertilizers are usually applied frequently through the irrigation system (fertigation). Fertigation seems to be the most promising approach for any substrate, but can be applied in lower concentration and frequency when substrates of high CEC are used. This technique is applied in a constantly increasing number of greenhouses all over the world.

Biochemicals which are produced by organic matter decomposition (*e.g.* aliphatic acids, sugar acids, amino acids and phenols) and humic substances have the ability to transform solid phase forms of micronutrients cations such as Fe^{3+}, Fe^{2+}, Mn^{2+} and Zn^{2+} into soluble metal complexes which are available to plants. In contrast, high-molecular weight and insoluble compounds such as lignin and humin function as a sink for polyvalent cations such as cadmium and lead, thereby reducing the toxicity danger of these heavy metals (121, 137).

Some peat substrates originally retain large amounts of DTPA extractable micronutrients which should be taken into consideration during the growing season (25).

When using organic substrates which retain high amount of cations and at the same time decompose and release nutrients into the solution, salinity problems may arise. These substrates should be thoroughly leached prior and sometimes during the growing period. The substrate solution concentration should be kept at levels which will not exceed 3.0 mS/cm. This limit should be lowered when sensitive plants are grown.

11.5. Biological characteristics

Other factors being constant, the physical and chemical characteristics are usually regarded as the sole parameters of importance with respect to the plant's performance in growth media (9, 85, 133). Nevertheless, in many instances non- or poorly-identified biological factors can greatly obscure the picture. This may have led Acock and Overcash (2) to state that "No method could be offered other than an empirical one for testing compatibility between growing medium and cultivar". Various cultivars may, indeed, reveal different requirements in relation to their growth medium. While these learnable needs are beyond the scope of this discussion the frustration manifested in the above-cited sentence can, in part, be relieved by improving our understanding of the biological effects exerted by the medium itself.

According to this strict definition, the forgoing discussion will not deal with the role of microorganisms in the medium and their interactions with plant's life.

11.5.1. Decomposition rate

All organic substrates, including the reltively stable ones are susceptible to continuous biological degradation. This process is enhanced by the favorable conditions prevailing in greenhouses. Various microbial and fungal populations are responsible for this process and their combined activity can lead to oxygen and nitrogen deficiencies, release of phytotoxic substances and medium shrinkage. Therefore, the net effect of organic matter decomposition in growth media is negative from the grower's standpoint and adequate precautions should be undertaken in order to minimize its damage to plants. Availability of biodegradable compounds such as carbohydrates, fatty acids and proteins is the prime factor that determines the decomposition rate. Contents of these compounds can be greatly reduced by proper composting procedures including pile rotation and by maintaining adequate water and available nitrogen contents. Growing conditions should also be considered. In general terms, the longer the growing period and the more intensive the growing practices, the greater the care that should be taken. Medium temperature and level of nitrogen nutrition are the most important factors during the growing period in relation to decomposition rate. If intensive cultivation for prolonged periods is to be expected then the use of more stable materials such as sedge peat and coarsely ground bark is recommended, while fast-growing plants thrive in less decay-resistant materials such as sphagnum peat and separated manure.

In any case, decomposition must be taken into consideration for assessing nutritional needs and in order to prevent soluble salts buildup. In extreme cases, medium top-dressing may be required. For example, roses grown in peat beds require an annual addition of about 10% by volume of fresh sphagnum peat under the conditions prevailing in Finland (Puustjarvi - personal communication).

11.5.2. Effects of decomposition products

Many of the biological effects of organic matter in soils and in growth media are attributed to humic and fulvic acids (HA, FA) which are the end-products of biological degradation of lignin and hemicellulose. A wide variety of plant functions both at the organ and at the cellular levels have been shown to be affected by HA and FA. Humic substances have also been reported as carriers of micronutrients to plants (27).

Toxicity, allelopathy and suppressivity as a general phenomena are the subject of another chapter in this volume (59) and, therefore, this paragraph will deal only with some specific examples of biological effects exerted by the medium which demonstrates both the complexity of the system and some beneficial possibilities which reside in it.

11.5.2.1. Enzymatic activity

A plethora of chemicals is liberated to the soil after the death and breakdown of plants and other organisms. Perhaps the most striking phenomenon resulting from this liberation is the existance of active enzymes in soils. Some other enzymes are secreted by living organisms. Proteinases and cellulases, for example, hydrolyze large macromolecules, turning them into available nutrients for these organisms.

Relatively little is known about the mode of action of enzymes in this cell-free environment, but generally, a pre-requisite to their activity is the adsorption of the enzyme molecules on clays or humus (17, 110).

Enzymatic activity has also been found in soilless organic materials such as peat (14). The distinction between enzymatic activity of living cells and that of cell-free enzymes can be done by applying ionizing radiation (Gamma, electron, x-rays) to a level which kills any cell but leaves the enzymes intact (18).

A wide variety of enzymes were detected in soils (110), in peat (14) and in manures (110). Among these, urease is one of the most abundant and agronomically important. Breakdown products of diverse organic vestiges can inhibit urease activity thereby reducing nitrogen loss from the root zone or from the containers (87, 109). Nitrogen fate in organic-based media should, therefore, be carefully assessed in order to optimize its utilization by selection of appropriate types and amounts of applied fertilizers.

11.5.2.2. Growth regulating activity

Many herbicides act, at lower concentrations, as growth-promoting substances. In a rather similar way the second facet of phytotoxicity phenomenon which is discussed elsewhere in this volume (59) can be described as growth regulation. There is little doubt that known plant hormones exist in soils. Levels of auxin which control cell growth and root initiation have been found to coincide with soil fertility (122). Addition of organic manures increased the extractable auxin concentrations (119). It is well-known, however, that various phenolics can interfere with auxin activity. Generally o-di hydroxy phenols enhance auxin activity while monophenols inhibit it (127). The existance of phenolic compounds in breakdown products of organic materials is well-decumented and attributed mainly to lignin decomposition (62, 143). Restricted attempts, however, have been carried out to elucidate the direct effects of realisitic concentrations of these substances on plant growth in soilless media and their interactions with auxin under these conditions. These effects have yet to be studied. Water extracts of sphagnum peat, sieved slurry produced by methanogenic fermentation of cow manure and of seaweed compost, separated by paper chromatography, revealed rooting promoting activity which cannot be associated with HA and FA and should probably be attributed to synergism between auxin (which was either produced by the plants or supplied exogenously) and phenolic compounds which were present in these extracts (101, Raviv and Medina - unpublished data).

Ethylene is another plant hormone which may be produced in organic media by microorganisms. It is almost inevitably produced under waterlogged conditions, leading to leaf epinasty and eventually to plant death (12).

Another indirect effect of organic materials may be the inactivation of applied growth regulators, probably via irreversible adsorption to humus particles (8, 65, 120).

11.6. Organic materials used as growth media

11.6.1. Peat

The most important and widely-used organic component of growth media is peat. Its value for gardening and plant production was recognized as early as the 18th century (86, 148). The origin and natural history of peat were thoroughly described in the begining of the 19th century (102, 117). Since that time peat was the subject of research conducted by botanists, pedologists, chemists, microbiologists and other scientists.

The flora of peat bogs and its bearing upon the physical and chemical properties of peat was discussed by Godwin and Clifford (45) and later by other authors (32, 50, 66, 97, 140). The horticultural potentials of peat was described in detail by many authors (e.g. 3, 31, 33, 41, 72, 97, 104).

In this review, peat which is not suitable for horticultural purposes such as peat formed under tropical conditions, will be ignored.

Peat is formed as a result of partial decomposition of plants under anaerobic or semianaerobic conditions. It appears that no organism is capable of decomposing lignin or sphagnol (the ligninlike substance of mosses) anaerobically, thus the main structure of peat-forming plants remains unaltered (44). Other typical conditions prevailing in peat bogs are low pH values and low levels of nutritional elements.

Peat bogs are formed under climatic conditions of high precipitation, low evaporation, low annual solar radiation and low summer temperatures. Peat is merely a general name for many types of decomposition products of plants. The actual nature of the peat is determined by such factors as botanical origin (97) and the climatic conditions prevailing during its formation (66, 71). It is assumed that the main changes leading to peat formation occur aerobically at the surface layer and once this layer is covered by fresh peat or young vegetation, little further change takes place. However, during extended periods of warmer weather which result in the lowering of the water table, more intense aerobic decomposition can take place.

The main peat types currently used in horticulture can be classified according to their botanical origin and degree of decomposition (97). The most important peat-forming plants are *Sphagnum* spp., other Bryales, sedges and many species of woody plants.

Sphagnum spp. is a primitive plant and continues growth by apical buds while the lower plant's parts die. Plants belonging to the genus *Sphagnum* are characterized by a unique transport system. The lignified cells which consitute this system remain mechnically stable after the plant's death. The cell walls of the transport system are porous, thus enabling vertical and horizontal capillary water and solutes movement whether the plant is alive or dead.

The cell wall-forming material is also characterized by a high surface activity which explains the fact that sphagnum peat has high cation exchange capacity (CEC) even before the breakdown process takes place.

A side effect, related to the above-mentioned property, is that cations movement in the "soil solution" of peat is extremely slow (110), suggesting the need for adequate fertilizing practices. Sphagnum tissues are decay-resistant (44). This resistance is believed to result, in part, from the complexity of the cell wall composition. In addition to sphagnol, the cell wall contains cellulose, high in xylan units and a high level of pectin. Other peat-forming plants are less important for horticulture. Sedge peat has a lower water-holding capacity and CEC than sphagnum peat and serves mainly when good aeration is more important than high water-holding capacity, for example, mist propagation. Another important property of sedge is its relative stability under greenhouse (aerobic) conditions even as compared to sphagnum peat. It has been found that the initial decomposition rate of sphagnum peat is twice as high as that of sedge peat (during the first month). When the decompostion rate reaches a constant level, sphagnum peat decomposes three times faster than sedge peat (74).

Peat which formed as a result of *Bryales* (Hypnum peat moss) or of woody plant degradation must reach a stable form and in this stage it contains a high conventration of humic acids (HA) in a collodial form. Drying out in this case is hardly reversible due to the hydrophobic nature of the resulting clods. This situation can be prevented by a process of freezing and thawing which breaks down the clods and creates a granular structure. Liming of the peat is also effective in forming a granular structure due to the precipitation of the non-soluble humate salts.

Sphagnum and sedge peats are characterized by relatively low content of HA. Typical values of HA in peat *in situ* are 5-25% (82) and for Dutch fine horticultural sphagnum peat - 5% (Raviv and Medina, unpublished data).

In general terms the importance of peat as a component in growth media stems from its inherent characteristics of which the most important ones are high water-holding capacity, high air capacity, low bulk density (which reduces shipment cost) and high CEC.

Despite all the above-mentioned advantages of peat, an extensive search for peat substitutes is undertaken in many parts of the world.

The main reason is the high cost of peat, particularly in countries without local peat resources. In Israel, for example, the cost of peat can comprise 10-15% of the total cost for pot plant production. Since high quality horticultural peat resources are relatively limited, the high price of peat is considered an economical burden in other countries as well (30). Other factors leading to a search for peat substitutes are listed herewith: (i) in many instances peat-based media proved to be conducive for phytopathogenic fungi (59). To overcome this disadvantage, costly procedures of media sterilization and/or of fungicide drenches are needed. (ii) peat formation is a unique phenomenon in nature since it represents a break in the carbon cycle. This intense carbon accumulation (leading finally to coal formation) is completely arrested by peat removal which subsequently leads to the exposure of deeper layers of peat to fast aerobical degradation, otherwise impossible. In the short term, peat resources can be regarded, therefore, as non-renewable. Furthermore, even on a year-by-year basis peat bogs serve as an important sink for atmospheric carbon dioxide (13). In an era of an increasing CO_2 emission to the atmosphere, due to fossil fuel burning as well as the gradual decrease in sinks efficacy due to oceans pollution and forest clearings, every important CO_2 sink should be preserved.

11.6.2. Peat substitutes

Peat substitutes can be defined from a practical standpoint as materials which can be added to a growth medium and constitute a significant fraction of its volume, thereby avoiding or reducing the need for peat in the medium. However, some of the organic materials are very different from peat in their physical and chemical nature, consequently, the strict control over moisture, air and nutrition level, attainable with peat-based media, is no longer possible. The following description will be confined to materials which, in principal, can be defined as peat substitutes, judging from their physical and chemical characteristics. The following parameters will be used as guidelines for determining the suitability of various materials as a replacement for peat: (i) Bulk density $\langle 0.4$ g/cm^3; (ii) Pore volume $\rangle 70\%$ V/V; (iii) Organic matter $\rangle 80\%$ W/W; (iv) CEC $\rangle 20$ meq/100g.

These limits were set as a result of accumulated experience by the authors and according to data compiled from the literature. Corresponding figures for most peat types are well within these limits (16, 97).

11.6.2.1. Bark

Perhaps the most important contemporary organic substitute for peat is bark. The role of bark in nature is to protect the inner part of the tree against desiccation and against phytopathogenic organisms. Since bark is the outer layer of the stem it is also exposed to other weathering agents. During evolution, it became an extremely decay-resistant tissue. By taking advantage of this feature, bark, which formerly presented an expensive disposal problem to wood, timber and paper industries, became a valuable horticultural product.

Bark is stripped off the logs in the sawmill by debarking machines and is then ground, usually by means of hammer mills, to pass an appropriate screen.

The botanical description of bark used for horticultural purposes is generally poor due to the use of mixed barks and to partial or even erroneous definitions. An important source of misunderstanding is the frequent use of common names instead of botanical names and the use of the genus name without mentioning the species. Examples for these ambiguities are "Red fir" for either *Abies magnifica* or *A. nobilis* and "White fir" for *Abies amabilis* or *A. concolor*. Many contradictory results in the literature may be explained by these botanical differences and it is therefore suggested that a proper use of botanical names should be the rule.

The species used for horticultural bark are commonly defined as softwood (coniferous) or hardwood (dicotyledoneous).

Important softwood trees for bark production are *Pinus* spp. such as *P. ponderosa, P. monticola, P. pinea, P. maritima, P. sylvestris* and *P. taeda, Pseudotsuga menziesii* (Douglass fir), *Picea* spp. (spruce), *Cypressus macrocarpa* (Monterey cypress), *Sequoia sempervirens* (redwood), *Larix* spp. (larch, *Cedrus* spp.*Tsuga* spp. (hemlock) and *Abies* spp. (fir).

Important hardwood trees for bark production are *Quercus* spp. (oak), *Fagus* spp. (Beech) and *Betula* spp. (Birch). However, their higher C/N ratio necessitates a longer and better controlled composting period.

As with peat, bark is merely a general name for numerous types of products. The horticultural quality of the end-product depends primarily upon its botanical origin, particle size distribution and type and duration of the composting process. Particle size distribution is affected by grinding, composting and sieving procedures.

Although some researchers reported successful use of non-composted softwood bark (73, 94), in most cases, problems such as continued breakdown and hence microbial competition for nitrogen and oxygen (16), phytotoxicity (1, 43), allelopathy (70, 124) and too low water-holding capacity dictates the use of composted bark. Various bark composting methods have been recently reviewed (16, 53, 59, 112) and are beyond the scope of this article.

The possible horticultural outlets for bark were thoroughly explored and reviewed (1, 5) during the last two decades and a comprehensive bibliography is also available (89).

Some prominent characteristics of bark determine its possible uses. Bark has low water-holding capacity, high air capacity (9, 93) and in some cases the capacity to release water-soluble fungistatic agents (60, 114). Degradation rate of bark is slow. Low water-holding capacity can be corrected by mixing bark with other materials such as peat (9, 29, 73), compost (132), non-organic materials (94, 146) or by maintaining an adequate water regime (146). Mixing with fine components will also solve problems of a limited capillary rise which may interfere with bottom irrigation. Bark is excellent for mulching purposes in outdoor gardens, for propagation media both under high humidity conditions and under intermittent mist (76, 88, 90) and as a growth medium for many species of epyphitic plants whose natural habitat is simulated by bark (5) . Bark can also serve successfully as a growth medium for various vegetables such as greenhouse tomatoes (30, 138), flower crops such as chrysanthemum (123, 146), pot plants (91, 132), nursery stock (42, 67, 105) and woody ornamentals (28). Large quantities of bark serve as a bulking agent during composting heavier organic wastes such as city refuse and sewage sludge (144) although its increasing cost promoted lately the search for cheaper bulking agents (56). Bark can also be used to improve the air status of materials like black peat (136). Some inherent difficulties are, however, associated with the use of bark in growth media. Most of these problems can be successfully treated if the grower is aware of their possible occurrence. As mentioned, the water-holding capacity of bark is poor. However, due to its good air porosity, frequent watering is not likely to cause oxygen deficiency. By the use of automatic irrigation controllers, an adequate moisture regime can be easily maintained (146). Whenever a very close physical contact between the medium and the plant material is essential, as in the case of small seed germination, difficulties can be expected (68) probably due to lack of water at the immediate vicinity of the seed.

Barks are also poor in nutrient and their low CEC's do not enable nutrient accumulation in the medium. Constant liquid or slow release fertilization is the straightforward solution to this problem.

In some cases, high quantities of manganese released from bark had caused severe toxicity symptoms in plants (112). Manganese toxicity can be overcome by raising the pH of the medium to near neutral (78) and/or by supplying higher quantities of available Fe. It has been shown that high Fe/Mn ratios reduce Mn uptake (69, 113). Growth regulators applied as drenches to pot plants have been shown to be less effective in bark-containing media (8, 65). This phenomenon resembles the well-known effect of organic matter in soil on herbicide efficacy (120). There is no clear-cut solution for this problem, other than increasing application dosage.

11.6.2.2. Sawdust and woodchips

Unlike bark, sawdust and woodchips originate from the inner parts of the tree and are less decay resistant. The lignin content is lower and C/N ratio is higher as compared to bark (46). As a result, both the nitrogen and oxygen demands during decomposition are higher than those of bark and longer composting period is usually required for stabilization (81). Rate and extent of wood waste decomposition may be higher (6) and the process usually results in high (70-80°C) temperatures.

The use of wood waste in growth media without previous composting is not advisable not only due to fast breakdown in the medium, but also because of severe phytotoxicity problems (80, 150).

In general, wood wastes are obtained from the same species as bark. In some countries (especially Australia), *Eucalyptus* spp. serve as the main source of sawdust. Composted wood wastes are characterized by low air porosity (as compared to sphagnum peat or bark), low content of easily available water (as compared to peat) (9, 93), low CEC (46), and very high C/N ratio (46). The range of bulk densities reported among sawdust of various origins is 0.14 (46) - 0.23 (93) g/cm^3. Contradictory results concerning the utilization of wood wastes as peat subsitutes can be found in the literature. This suggests that further evaluation is necessary to determine clear and safe recommendation.

Sources of variability as reported by different researchers may be caused by a number of factors: the existence of residual phytotoxicity (80, 150), the use of different species (43), other components of the media affecting the air and water status and the duration of the composting period which may affect nitrogen and oxygen availability to roots (55, 150).

Good results with sawdust as a peat substitute were obtained with foliage plants, seedlings and flower plants (126, 151). Negative results, however, were reported in other cases (94, 150).

Forced aeration or the use of other means of supplying oxygen to the medium can prevent the appearance of oxygen-dificiency symptoms (11) but for commercial operations this procedure is probably not economically feasible.

As a result of experience and many years of research work, Poole *et al*. recommended not to exceed the proportion of 20% sawdust in the medium (92).

An unequivocal outlet for wood waste is its use as bulking agents in composting of sewage sludge and other municipal wastes (144). In this system, wood wastes are used to control the initial required C/N ratio and can help in reducing excessive water content of the raw materials. The end product is richer in organic material, more friable and has lower bulk density. Another efficient outlet may be the addition of wood wastes to manures prior to the composting process. This addition can be made in an earlier stage, just by increasing the amounts of shavings and/or sawdust which serve as animal litter. This reduces undesired anaerobic fermentation and the cost of the additional operation of adding and mixing the wood waste before composting can be avoided.

11.6.2.3. Sewage sludge and municipal composts

Although sewage sludge and municipal composts are not always within the definitions given earlier for peat substitute, these materials will be discussed because they are sometimes utilized. Their physical and chemical properties differ greatly from those of peat and consequently, the grower's ability to control water, air and nutrition in the media is limited considerably by introducing these materials to the growing mixtures. Another serious drawback stems from the high content of heavy metals in many of these products. This is not to say there is no room for these products in agriculture and horticulture. There are many examples of careful and successful utilizations of these materials (37, 48, 106, 149), but one should be aware of their inherent limitations. Some of the most important are: considerable heterogenetity of products (22) and even between different batches of the same product (56), existence of undesired materials such as glass and plastics (22), phytotoxicity problems (resulting from fatty acids (23, 39) and unidentified factors (152)), low CEC (106), low content of organic matter (30), high contents of toxic elements to both plants and animals (20), high salinity levels (20), high bulk density and low air capacity (149).

As a result of the above the use of these materials should be considered for only non-edible products, which require low levels of control. Good examples may be forest seedlings and flowering bedding plants.

11.6.2.4. Treated animal excreta

At present, animal excreta poses an enormous environmental problem throughout the world. In India alone an estimated 1335 million tons of wet manure were produced annually in the mid-seventies (19). The traditional way of treating this resource is to spread it on soil prior to soil cultivation. This system is far from economical and leads to considerable waste of energy and nutrients, without eliminating some of the undersirable problems such as leaching of nitrates to ground waters, rivers and lakes. Some recently suggested improvements may form a basis for a new approach towards manure utilization, and one of the potential products of these processes is likely to be a peat substitute.

It appeares that fractionation of the manure may play a key role in making its utilization more efficient. The operation *per se* is technologically feasible (84). It has been shown that most fermentable materials of manure remain in the liquid fraction (139) and the fibrous fraction reduces the efficiency of biogas convertors (75). These findings are corroborated by the known effect of the C/N ratio on the efficiency of anaerobic digestion (58). Since most of the nitrogen, which appears in the fiber fraction, is a part of the non-digestible ligneous compounds, its removal is likely to increase this ratio and hence improve the fermentation efficiency.

The fibrous fraction will compost spontaneously if kept wet and can serve as a peat substitute (30, 38). In developing countries, this fraction can also serve, after drying, as an energy source by direct burning with a heat-release value comparable to wood (19).

After fermentation, the resulting slurry of the liquid phase can be applied to fields as liquid fertilizer, free of pathogenic organisms and odor problems. In Israel a two-stage stabilization process was tested by conducting the fractionation after the anaerobic digestion (25, 26). The fibrous fraction that resulted after aerobic composting, was comparable to sphagnum peat both physically, chemically (25, 95) and in its horticultural performance (95, 101). Since the anaerobic reactions are, under realistic conditions, not less energy-efficient than the aerobic reactions (111), a prompt and continuous processing of fresh manure separated to liquid and fibrous fractions either prior to or after anaerobic fermentation seems for the time being to present an optimal solution for both maintaining good sanitary status for the animals and for the profitable utilization of their excreta.

11.6.2.5. Other organic materials

The utilization of other organic materials is common in various countries according to local availability and demand. Many different sources can be utilized as horticultural substrates either with or without composting. The transportation costs are high for most of these materials which limits their use to the production area hence does not attract widespread interest. The most promising and commonly used materials are straw balls (125, 145), leaf mould (107, 108), rice and peanut hulls (10, 40) and seaweed composts (100, 118) and grape/marc. Examples for exotic organic materials of local interest and limited availability are coconut fibre dust (134) and macadamia husks (128).

Acknowledgement: This literature review was partially supported by a grant from the National Council for Research and Development, Israel and the European Community.

11.7. References

1 Aaron J.R. 1976 Conifer bark: its properties and uses. Forestry Commission Record No. 110., U.K.

2 Acock M.C. and Overcash J.P. 1983 Effect of growing medium and cultivar on the container culture of pecan seedlings. HortScience 18, 330-331.

3 Adamson A. 1975 Crop production on peat-vegetables. pp. 97-110. *In* Peat in Hoticulture. Eds. D.W. Robinson and J.G.D. Lamb. Academic Press, London.

4 Alexander M. 1977 Introduction to Soil Microbiology. John Wiley & Sons Inc. (2nd ed.).

5 Allen M.L. 1973 Agricultural and horticultural uses for bark. *In* Bark utilization Symposium Proc. Ed. E.L. Ellis. pp. 88-95. School of Forestry, Univ. of Canterbury, Christchurch, New Zealand.

6 Allison F.E. and Clover R.G. 1960 Rates of decomposition of short-leaf pine sawdust in

soil at various levels of nitrogen and lime. Soil Sci. 89, 194-201.

7 Arnon D.I. and Johnson C.M. 1942 Influence of hydrogen ion concentration on the growth of higher plants under controlled conditions. Plant Physiol. 17, 525-539.

8 Barnett J.E. 1982 Chrysanthemum height control by Ancymidol, PP 333 and EL 500 dependent on medium composition. HortScience 17, 896-897.

9 Beardsell D.V., Nichols D.G. and Jones D.L. 1979 Physical properties of nursery potting-mixtures. Sci. Hort. 11, 1-8.

10 Bilderback T.E., Fontano W.C. and Johnson D.R. 1982 Physical properties of media composed of peanut hulls, pine bark and peat moss and their effect on azalea growth. J. Am. Soc. Hort. Sci. 107, 522-525.

11 Bowen P.A. 1983 The effect of oxygen fumigation of sawdust medium on the yield and yield-components of greenhouse cucumbers. Sci. Hort. 20, 131-136.

12 Bradford K.J. and Yang S.F. 1981 Physiological responses of plants to waterlogging. HortScience 16, 25-29.

13 Bramryd T. 1980 The role of peatlands for the global carbon dioxide balance. I.P.S. 6th Int. Peat Congress, Minnesota, 1980, pp. 9-11.

14 Brown K.A. 1981 Biochemical activities in peat sterilized by gamma irradiation. Soil Biol. Biochem. 13, 469-474.

15 Bunt A.C. 1974 Physical and chemical characteristics of loamless pot plant substrates and their relation to plant growth. Proc. Symp. Artificial Media in Hort., 1973, Ghent, pp. 1954-1964.

16 Bunt A.C. 1976 Modern Potting Composts. George & Unwin Ltd., London.

17 Burns R.G. 1982 Enzymatic activity in soil: location and a possible role in microbial ecology. Soil Biol. Biochem. 14, 423-427.

18 Cawse P.A. 1975 Microbiology and biochemistry of irradiated soils. pp. 213-267 In Soil Biochemistry, Vol. 3. Eds. E.A. Paul and A.D. McLaren. Marcel Dekker, Inc., New York.

19 Chandra S. and Seckler D.W. 1980 Note on fractionation of animal dung. Ind. J. Agric. Sci. 50, 93-95.

20 Chaney R.L., Munns J.B. and Cathey H.M. 1980 Effectiveness of digested sewage sludge compost in supplying nutrients for soilless potting media. J. Am. Soc. Hort. Sci. 105, 485-492.

21 Chanyasak V. and Kubota H. 1981 Carbon/Organic Nitrogen ratio in water extract as measure of composting degradation. J. Ferment. Technol., 59(3), 215-219.

22 Chanyasak V. and Kubota H. 1983 Source separation of garbage for composting. Bio Cycle 24(2), 56-58.

23 Chanyasak V., Katayama A., Hirai F.M., Mori S. and Kubota H. 1983 Effect of compost maturity on growth of Komatsuma (*Brassica rapa* var. *Pervidis*) in Neubauer's pot. I. Comparison of growth in compost treatments as controls. Soil Sci. Plant Nutr. 29, 239-250.

24 Chen Y., Banin A. and Ataman Y. 1980 Characterization of particles and pores, hydraulic properties and water-air ratios of artificial growth media and soil. 5th Int. Cong. on Soilless Culture, Wageningen, pp. 63-82.

25 Chen y., Inbar Y., Raviv M. and Dovrat A. 1984 The use of slurry produced by methanogenic fermentation of cow manure as a peat substitute in horticulture - physical and chemical characteristics. Acta Horticulturae 150, 553-561.

26 Chen Y., Inbar Y. and Raviv M. 1984 Slurry produced by methanogenic fermentation of cow manure as a peat substitute in horticulture. Proc. 2nd Intl. Symp. Peat in Agr. and Hort. pp. 297-317.

27 Chen Y. and Stevenson F.J. 1986 Soil organic matter interaction with trace elements. *In* The role of Organic Matter in Modern Agriculture. Eds. Y. Chen and Y. Avnimelech Martinus Nijhoff Publ., The Hague. pp 73-116

28 Chrustic G.A. and Wright R.D. 1983 Influence of liming rate on holly, azalea and juniper growth in pine bark. J. Am. Soc. Hort. Sci. 108, 791-795.

29 Conover C.A. and Poole R.A. 1981 Effect of soil compaction on physical properties of potting media and growth of *Picea pubescens* Liebm. "Silver Tree" J. Am. Soc. Hort. Sci. 106, 604-607.

30 Cull D.C. 1981 Alternatives to peat as container media. Organic resources in the U.K. Acta Horticulturae 126, 69-81.

31 Dasberg S. and Feigin A. 1978 The effect of irrigation, fertilization and organic matter on roses grown in four soils in greenhouses. Sci. Hort. 9, 181-188.

32 Davis J.H. 1946 The peat deposits of Florida: their occurrence, development and, uses. Fla. Geol. Bull. No. 30.

33 Davis J.F. and Lucas R.E. 1959 Organic soils: their formation, distribution and management. Bull. 425, Dept. Soil Sci., Michigan State Univ.

34 De Boodt M. and Verdonck O. 1972 The physical properties of the substrates used in horticulture. Acta Horticulturae 26, 37-44.

35 De Boodt M., Verdonck O. and Cappaert I. 1974 Determination and study of the water availability of substrates for ornamental plant growing. Acta Horticulturae 35, 105-111.

36 De Boodt M., Verdonck O. and Cappaert I. 1974 Method for measuring the water release curve of organic substrates. Acta Horticulturae 37, 2054-2060.

37 De Vleeschauwer D., Verdonck O. and De Boodt M. 1980 The use of town refuse compost in horticultural substrates. Acta Hort. 99, 149-155.

38 De Vleeschauwer D., Verdonck O. and De Boodt M. 1982 The use of chicken and piggery manure in composts. Acta Horticulturae 126, 105-111.

39 De Vleeschauwer D., Verdonck O. and Van Assche P. 1981 Phytotoxicity of refuse compost. Bio Cycle 22(1), 44-46.

40 Einert A.E. and Baker C.E. 1973 Rice hulls as a growing medium component for cut tulips. J. Am. Soc. Hort. Sci. 98, 556-558.

41 Gallagher P.A. 1975 Peat in protected cropping. pp. 133-146. *In* Peat in Horticulture. Eds. D.W. Robinson and J.G.D. Lamb. Academic Press, London.

42 Gartner J.B., Meyer Jr. M.M. and Saupe D.C. 1971 Hardwood bark as a growing media for container-grown ornamentals. Forest Prod. J. 21(5) 25-29.

43 Gartner J.B. Still S.M. and Klett J.E. 1974 The use of bark waste as a substrate in horticulture. Acta Hort. 37, 2003-2013.

44 Given P.H. and Dickinson C.H. 1975 Biochemistry and microbiology of peats pp. 123-212. *In* Soil Biochemistry, vol. 3. Eds. E.A. Paul and A.D. McLaren. Marcel Dekker, Inc., NY.

45 Godwin H. and Clifford M.H. 1930 Studies of the post-glacial history of British vegetation. Phil. Trans. Royal Soc., London. Ser B 229, 323-406.

46 Goh K.M. and Haynes R.R. 1977 Evaluation of potting media for commercial nursery production of container-grown plants. 1. Physical and chemical characteristics of soil and soilless media and their constituents. New Zealand J. Agric. Res. 20, 363-370.

47 Golueke G.C. 1977 Biological reclamation of solid wastes. Rodale Press, Emmaus.

48 Gouin F.R. 1982 Using composted waste for growing horticultural crops. Bio Cycle 23 (1), 45-47.

49 Gray K.R. Sherman K. and Biddlestone A.J. 1971 Review of composting. Part 2 - The

Practical Process. Process Biochemistry (6) 10, 22-28.

50 Hammond R.F. 1975 The origin, formation and distribution of peatland resources. pp. 1-22. *In* Peat in Horticulture. Eds. D.W. Robinson and J.G.D. Lamb. Academic Press, London.

51 Hanan J.J., Holley W.D. and Goldsberry K.L. 1978 Soils and soil mixtures. Greenhouse Management, Springer-Verlag pp. 255-322.

52 Hartmann H.T. and Kester D.E. 1983 Propagation structure. Media fertilizers, soil mixtures and containers. *In* Plant propagation, Principles and Practices. 4th ed. Prentice-Hall Inc., Englewood Cliffs, N.J. pp. 18-59.

53 Haug R.T. 1980 Compost engineering principles and practice. Ann Arbor Science, Ann Arbor, Michigan. 655 pp.

54 Helling C.S., Chester G. and Corey R.B. 1964 Contribution of organic matter and clay to soil cation-exchange capacity as affected by the pH of the saturating solution. Soil Sci. Soc. Am. Proc. 28, 517-520.

55 Hicklenton P.R. 1983 Flowering, vegetative growth and mineral nutrition of pot chrysanthemum in sawdust and peat-lite media. Sci. Hort. 21 189-197.

56 Higgins A.J. 1983 Reducing costs for bulking agents. Bio Cycle 24(5), 34-38.

57 Hillel D. 1982 Introduction to soil physics. Academic press, New York, N.Y.

58 Hills D.J. 1979 Effects of carbon: nitrogen ratio on anaerobic digestion of dairy manure. Agric. Wastes 1. 268-278.

59 Hoitink H.A.J. and Kuter G.A. 1986 Organic residues and composts as growth media for plants. *In* The role of Organic Matter in Modern Agriculture. Eds. Y. Chen and Y. Avnimelech. Martinus Nijhoff Publ., The Hague. pp 289-306.

60 Hoitink H.A.J. VanDoren Jr. D.M. and Schmittenner A.F. 1977 Suppression of *Phytophthora cinnamomi* in a composted hardwood bark potting medium. Phytopathol. 67, 561-565.

61 Holley W.D. 1967 Inert media to replace soil. Colo. Flo. Gro. Ass. Bull. 205, pp. 1-2.

62 Jalal M.A.F. and Read D.J. 1983 The organic composition of calluna heathland soil with special reference to phyto- and fungitoxicity. I. Isolation and identification of organic acids. Plant and Soil 70, 257-272.

63 Johnson P. 1968 Horticultural and agricultural uses of sawdust and soil amendments. Tech. Bull. Pub. by P. Johnson, 1904 Cleveland Ave. National city, Calif.

64 Joiner J.N. and Conover C.A. 1965 Characteristics affecting desirability of various media components for production of container-grown plants. Proc. Soil Crop Sci. Soc. of Fla. 25, 320-328.

65 Joiner J.N. *et al.* 1977 Effects of Ancymidol and N, P and K on growth appearance and foliar content of *Dieffenbachia picta* "Exotica". HortScience 12, 238.

66 Kingsley W.G. 1935 Biological and physiological conditions in peat bogs. M.Sc. thesis, Cornell Univ., Ithaca, N.Y.

67 Klett J.E. Gartner J.B. and Hughes T.D. 1972 Utilization of hardwood bark in media for growing woody ornamental plants in containers. J. Am. Soc. Hort. Sci. 97, 448-450.

68 Klingman G.L. and Jaster S.W. 1982 A comparison of seeding mixes for bedding plant production. HortScience 17, 735-736.

69 Kohno Y. and Foy C.D. 1983 Manganese toxicity in bush bean as affected by concentration of manganese and iron in the nutrient solution. J. Plant Nutrition 6, 363-386.

70 Krogstadt O. and Solbraa K. 1975 Effects of extracts of crude and composted bark from spruce on some selected biological systems. Acta Agric. Scand. 25, 306-312.

71 Kurbatov I.M. 1963 The question of the genesis of peat and its humic acids. Vol. 1: 133-137. Second Int'l. Peat Congress, Ed. R.A. Robertson. Leningrad. H.M.S.O. Edinburgh.

72 Lamb J.G.D. 1975 Crop production on peat-hardy nursery stocks. pp. 111-118. *In* Peat in Horticulture. Eds. D.W. Robertson and J.G.D. Lamb. Academic Press, London.

73 Lemaire F., Dartiques A. and Riviere L.M. 1980 Properties of substrates with ground pine bark. Acta Hort. 99, 67-80.

74 LeVesque M., Jacquin F. and Polo A. 1980 Comparative biodegradability of a sphagnum and sedge peat from France. pp. 584-590. *In* Proc. of the 6th Int'l Peat Congress, Duluth, Minnesota.

75 Lo K.V. *et al.* 1983 The effect of solids-separation pretreatment on biogas production from dairy manure. Agric. Wastes 8, 155-165.

76 Lorenzo P. *et al.* 1982 Effects of physical media properties on *Codiaeum variegatum* rooting response. Acta Hort. 126, 293-302.

77 Loudon J.C. 1839 An Encyclopedia of Gardening. Longman, London.

78 Lucas R.E. and Davis J.K. 1961 Relationship between pH values of organic soils and the availabilities of 12 plant nutrients. Soil Sci. 92, 177-182.

79 Lucas R.E., Rieke P.E., Shickluman F.C. and Cole A. 1975 Lime and fertilizer requirement for peat. *In* Peat in Horticulture. Eds. D.W. Robinson and J.G.D. Lamb. Academic Press Inc. (London) Ltd., pp. 119-132.

80 Maas E.F. 1981 How toxic is cedar sawdust to plants? Research Reviews. Res. Station, Agassiz, British Columbia, No. 10.

81 MacDonald W.A. and Dun S. 1953 Sawdust composts in soil improvement. II. Pot culture studies with compost mixtures of sawdust and manure steam treated composts and miscellaneous mixtures. Plant and Soil 4, 235-247.

82 Manskaya S.M. and Drozdova T.V. 1968 Geochemistry of organic substances (Transl. by I Shapiro and I. Berger). Pergamon, Oxford.

83 Marshal T.C. 1958 A relation between permeability and size distribution of pores. J. Soil Sci. 9, 1-8.

84 Pain B.F., Hepherd R.A. and Pittman R.J 1978 Factors affecting the performance of four slurry separating machines. J. Agric. Eng. Res. 23, 231-242.

85 Penningsfeld F. 1978 Substrates for protected cropping. Acta Horticulturae 82, 13-22.

86 Perfect T. 1759 The Practice of Gardening. Baldwin, London.

87 Perucci P., Giusquiani P.L. and Scarponi L. 1982 Nitrogen losses from added urea and urease activity of a clay-loam soil amended with crop residues. Plant and Soil 69, 457-469.

88 Pokorny F.A. and Perkins H.F. 1967 Utilization of milled pine bark for propagating woody ornamental plants. Forest Products J. 17(8), 43-48.

89 Pokorny F.A. 1982 Horticultural uses of bark: a bibliography. Research Rpt. College of Agric., Experiment Station, Georgia Univ., No. 402.

90 Pokorny F.A. and Austin M.E. 1982 Propagation of blueberry by softwood terminal cuttings in pine bark and peat media. HortScience 17, 640-642.

91 Poole R.T. and Waters W.E. 1972 Media for potted foliage plants. Fla. Foliage Grower 9 (7), 5-7.

92 Poole R.T., Conover C.A. and Joiner J.N. 1981 Soils and potting mixtures. *In* Foliage Plant Production. Ed. J.N. Joiner. Prentice-Hall Inc., Englewood Cliffs, N.J. pp. 179-202.

93 Prasad M. 1979 Physical properties of media for container-grown crops. I. New Zealand peats and wood wastes. Sci. Hort. 10, 317-323.

94 Prasad M. 1980 Evaluation of wood wastes as a substrate for ornamental crops watered by capillary and drip irrigation. Acta Hort. 99, 93-103.

95 Putievski E., Raviv M. and Chen Y. 1983 Development and regeneration ability of lemon

balm and marjoram on various growth media. Biol. Agric. Hortic. 327-333.

96 Pusstjarvi V. 1974 Physical properties of peat used in horticulture. Acta Horticulturae 37, 1222-1229.

97 Puustjarvi V. 1977 Peat and its uses in horticulture. Turveteollisuusliitto Ry. Pub. 3, Helsinki, 1977.

98 Puustjarvi V. and Robertson R.A. 1975 Physical and chemical properties. *In* Peat in Horticulture. Eds. D.W. Robinson and J.G.D. Lamb. Academic Press Inc. (London) Ltd., pp. 23-38.

99 Puustjarvi V. 1982 Peat and Plant Yearbook 1981-1982.

100 Ramanathan K.M. *et al.* 1964 A note on the composting of seaweeds. Madras Agric. J. 51, 451-457.

101 Raviv M., Chen Y., Geler Z., Medina S., Putievski E. and Inbar Y. 1984 Slurry produced by methanogenic fermentation of cow manure as a growth medium for some horticultural crops. Acta Horticulturae 150, 563-573.

102 Rennie R. 1807. Essays on the Natural History and Origin of Peat Moss. A. Constable & Co., Edinburgh.

103 Richards S.J., Warneke J.E., Marsh A.W. and Aljibury F.K. 1964 Physical properties of soil mixes used by nurseries. Calif. Agr. 18(5), 12-13.

104 Rockwell F.F., and Breitenbucher W.G. 1928 Gardening with Peat Moss. Atkins & Durbrow, Inc., New York.

105 Rouse R.E. 1982 Evaluation of 5 media and 5 fertilizer treatments for container citrus. Rio Grande Valley Hort. Soc. 35, 167-171.

106 Sanderson K.C. and Martin C.W. 1974 Performance of woody ornamentals in municipal compost medium under nine fertilizer regimes. HortScience 9, 242-243.

107 Sawhney B.L. 1976 Leaf compost for container-grown plants. HortScience 11, 34-35.

108 Shanks J.B. 1974 The use of several industrial and municipal by-products as components of plant growth media. Maryland Florist. No. 192.

109 Sivapalan K. *et al.* 1983 Hunified phenol-rich plant residues and soil urease activity. Plant and Soil 20, 143-146.

110 Skujins J.J. 1967 Enzymes in soil: pp. 371-414. *In* Soil Biochemistry. Eds. A.D. McLaren and G.H. Paterson. Marcel Dekker Inc., New York.

111 Sobel A.T. and Muck R.E. 1983 Energy in animal manures. Energy Agric. 2, 161-176.

112 Solbraa K., Sant M.D. and Selmer-Olson A.R. 1983 Composting soft and hardwood barks. Bio Cycle 24, 44-48, 59.

113 Somer I.I. and Shive J.W. 1942 The iron-manganese relation in plant metabolism. Plant Physiol. 17, 582-602.

114 Spencer S. and Benson D.M. 1982 Pine bark hardwood bark compost and peat amendment effects on development of *Phytophthora* spp. and lupine root rot. Phytopathol. 72, 346-351.

115 Spomer L.A., 1974 Two classroom exercises demonstrating the pattern of container soil water distribution. HortScience 9(2), 152-153.

116 Spomer L.A., 1975 Soil aeration and plant growth. Illinois State Florists Assoc. Bull. No. 363.

117 Steele A. 1826 The natural and agricultural history of peat moss. W & D Laing, Edinburgh.

118 Stephenson W.A. 1968 Seaweed in agriculture and horticulture. Faber & Faber, London.

119 Stevenson F.J. 1967 Organic acids in soil. pp. 147-169. *In* Soil Biochemistry. Eds. A.D. McLaren and G.H. Peterson. Marcel Dekker, Inc., New York.

120 Stevenson F.J. 1972 Role and function of humus in soil with emphasis on adsorption of

herbicides and chelation of micronutrients. BioScience 22, 643-650.

121 Stevenson F.J. 1982 Humus Chemistry. John Wiley & Sons Inc.

122 Stewart W.S. and Anderson M.S. 1942 Auxins in some American soils. Bot. Gat. 103, 570-575.

123 Still S.M. 1977 Comparison of chrysanthemum growth in pine bark or commercial soil-less mixes. HortScience 12, 573-574.

124 Still S.M., Dirr M.A. and Gartner J.B. 1976 Phytotoxic effects of several bark extracts on mung bean and cucumber growth. J. Am. Soc. Hort. Sci. 101, 34-37.

125 Stokes D.A. and Tinley G.H. 1981 Cucumber: thermal screen and growing media investigations. Acta Hort. 118, 135-145.

126 Thomas M.B. and Matheson S.D. 1981 The influence of slow release nitrogen, liquid fertilization and media on the growth of *Fatsia japonica*. Ann. J. Roy. New Zealand Inst. Hort. 9, 38-45.

127 Tomaszewski M. and Thimann K.V. 1966 Interactions of phenolic acids, metallic ions and chelating agents in auxin-induced growth. Plant Physiol. 41, 1433-1454.

128 Trochoulias T. 1983 Macadamia husks for use in nursery potting media: a promising alternative to peat. Aust. Hort. 81, 95-98.

129 Verdonck O. 1984 Reviewing and evaluation of new materials used as substrates. Acta Horticulturae 150, 467-473.

130 Verdonck O., Cappaert I. and De Boodt M. 1973 The rooting of Azalea in inert substrates. Int. Working Group on Soilless Culture, I.W.O.S.C. Proceedings, Sossari, 1973 pp. 143-148.

131 Verdonck O., Cappaer I. and De Boodt M. 1974 The properties of the normally used substrates in the region of Ghent. Acta Horticulturae 37, 1930-1944.

132 Verdonck O., De Vleeschauwer D. and De Boodt M., 1980 The use of bark-sludge composts as a substrate for ornamental plants. Acta Hort. 99, 119-129.

133 Verdonck O., De Vleeschauwer D. and De Boodt M. 1982 The influence of the substrate on plant growth. Acta Horticulturae 126, 451-458.

134 Verdonck O., De Vleeschauwer D. and Penninck R. 1983 Cocofibre dust, a new growing medium for plants in the tropics. Acta Hort. 133, 215-220.

135 Verdonck O., Penninck R. and De Boodt M., 1984 The physical properties of different horticultural substrates. Acta Horticulturae 150, 155-160.

136 Verdure M. 1982 Improvement of physical properties of black peat. Acta Hort. 126, 131-142.

137 Verloo M.G. 1980 Peat as a natural complexing agent for trace elements. Acta Horticulturae 99, 51-56.

138 Verschambre D. *et al.* 1982 Utilization de l'ecorce de pin maritime comme substrat en culture legumière, Acta Hort. 126, 37-43.

139 Ward G.H. and Seckler D.W. 1975 Recycling the protein of animal waste: protein to support animal protein to support animal protein production. World Review Animal Production 11, 21-23.

140 Walker D. and Walker P.M. 1961 Stratigraphic evidence of regeneration in some Irish bogs. J. Ecol. 49, 169-185.

141 Waters W.E. 1970 The chemical, physical and salinity characteristics of 27 soil media. Proc. Fla. State Hort. Soc. 83, 482-488.

142 White J.W. 1974 Criteria for selection of growing media for greenhouse crops. Florists'
     Rev. Vol. 155, Issue No. 4009, pp. 28-30, 73-75.
143 Whitehead D.C. Dibb H. and Hartley R.D. 1983 Bound phenolic compounds in water ex-
     tracts of soils, plant roots and leaf litter. Soil Biol. Biochem. 15, 133-136.
144 Wilson G.B. *et al.* 1980 Manual for composting sludge by the Beltsville, aerated-pile
     method. Municipal Environmental Research Lab., U.S. EPA, Cincinnati, Ohio.
145 Wilson G.C.S. 1978 Modified straw bale technique for cucumbers. Acta Hort. 82, 75-77.
146 Wilson G.C.S. 1982 Bark composts for pot chrysanthemums. Acta Hort. 126, 95-104.
147 Wilson G.C.S. 1984 The physico-chemical properties of horticultural substrates. Acta
     Horticulturae 150, 19-32.
148 Wooldridge J. 1719 Systema Horti-culturae or the Art of gardening. Freeman, London.
149 Wooton R.D., Gouin F.R. and Stark F.C. 1981 Composted, digested sludge as a medium
     for growing flowering annuals. J. Am. Soc. Hort. Sci. 106, 46-49.
150 Worrall R.J. 1978 The use of composted wood waste as a peat substitute. Acta Hort. 82,
     79-86.
151 Worrall R.J. 1981 Comparison of composted hardwood and peat-based media for the
     production of seedlings, foliage and flowering plants. Sci. Hort. 15, 311-319.
152 Zucconi F., Pera A., Forte M. and Bertoldi M. 1981 Evaluating toxicity of immature
     compost. Bio Cycle 22(2), 54-57.

12. Effects of composts in growth media on soilborne pathogens

H.A.J. HOITINK and G.A. KUTER

12.1. Introduction

Properties of organic matter that affect its utilization in container media can be conveniently divided into three major categories, *i.e.* chemical, physical and biological. Changes in any one of these frequently have significant effects on the overall characteristics of the medium. For example, changes in particle size due to decomposition of an organic component may change aeration and, therefore, affect growth of the plant, the microflora and the interaction of these biological components.

Sphagnum peat for many years was the principle organic component used in media. It resists decomposition and its physical and chemical properties are well recognized and do not change considerably throughout the production cycle of plants. Media of predictable and relatively stable properties can by formulated with sphagnum peats (2, 8, 11, 23, 75).

During the past two decades various composted organic wastes prepared from tree barks, solid waste, sewage sludges and crop residues have partially replaced peat in container media (37, 38, 49, 51, 57, 65, 77, 89, 93, 95). The recycling of these wastes has been adopted in many parts of the world. Several advantages are associated with the use of composts. For example, some of the compost-amended media suppress the major soilborne plant pathogens (21, 39, 46, 62, 79, 82, 83, 91). This has eliminated the need for sterilization of such media, as well as reduced fungicide use (26).

A distinct disadvantage of most composts is that they continue to decompose after potting. Inadequate techniques are available for measuring their stability. Therefore, the history of each type of compost needs to be known so that media of predictable qualities can be formulated.

Compost type, maturity level as well as the physical and chemical properties of the container media in which they are used, affect plant growth and soilborne plant pathogens and the diseases they cause. In this article the effects of composting process and effects of utilizing composts in container media on soilborne plant diseases are emphasized.

Y. Chen and Y. Avnimelech (eds.), The Role of Organic Matter in Modern Agriculture.
ISBN 90-247-3360-X.
© *1986, Martinus Nijhoff Publishers, Dordrecht.*

12.2. The composting process

Several recent articles and books review composting (36, 44, 73). Therefore, this discussion is limited to some general definitions and specific factors affecting preparation of composts for use in container media. Composting is a form of waste stabilization that requires special conditions, particularly of moisture and aeration, to yield thermophilic temperatures. Although there is no generally accepted technical definition, this article defines composting as "the biological decomposition of organic constituents in wastes under controlled conditions", as proposed by Golueke (36).

Control of environmental conditions distinguishes composting from natural rotting or putrefaction such as occurs in open dumps, manure heaps or field soil. The process, which is largely aerobic, involves both thermophilic and mesophilic microorganisms. Basically, the process can be divided into three phases:1) an initial phase of 1-2 days during which readily degradable compunds are decomposed; 2) a thermophilic phase possibly lasting months, during which largely cellulose is degraded; and 3) stabilization, a period when temperatures decline, decomposition rates decrease and mesophilic microorganisms recolonize the compost.

Although temperature has traditionally been used to monitor composting, the process is most readily controlled by regulating heat loss through manipulation of air flow (34). Microbial decay of organic matter results in considerable heat production. When heat is allowed to accumulate, high temperatures ($>$ 60° C) inhibitory to continued microbial activity are reached rapidly. Recent research on composting of hardwood bark and municipal sludge in a full scale (108 m^3) aerated reactor (Paygro System) has demonstrated the effects of using air flow to control the composting process (48). For experimental purposes the reactor was modified so that air flow through the compost could be regulated by temperature feedback. Precise measurements of air flow rates could be obtained. Rates of microbial activity (measured by monitoring carbon dioxide output) and drying were greater when the compost was maintained at low (38-60°C) rather than at high temperatures (60 - 70°). These results agree with those obtained for a variety of wastes composted in small-scale composters (12, 13, 52, 53, 54) and those obtained in an aerated pile study (34).

Several types of composting systems are available. This subject was reviewed recently (44). Windrows must be high in free air space to enhance natural airflow. Various systems with forced aeration have been developed. The most widely used composting system in the U.S. is the Beltsville aerated-pile system (94), in which air is exhausted by negative-pressure ventilation through tubes placed under the compost pile (33, 70). Heat accumulates in the pile, which is detrimental. An improved system was developed at Rutgers University in which air is forced through an identical ducting system but with positive-pressure ventilation (34). Heat, therefore, is removed more efficiently which allows for lower process temperatures and higher rates of decomposition.

Aerobic composting implies decomposition of organic substrates in the presence of oxygen. The main products of aerobic metabolism are carbon dioxide, water and heat. The amount of ammonia released through deamination varies with the process, the waste and bulking agent being composted. For example, large amounts of ammonia are released during composting of low C/N ratio wastes such as municipal sludge, particularly if inadequate or inappropriate bulking agents are used.

Even in the most highly aerated systems some anaerobic metabolism occurs in micro habitats after wastes are first mixed for composting. Metabolic-end products of anaerobic composting are methane, carbon dioxide and numerous intermediates, such as low molecular weight organic acids and various alcohols. This process releases significantly less energy per unit organic matter decomposed than during the aerobic process. It has a high odor potential and the toxic end product is referred to as "sour". Some anaerobically produced composts remain toxic for months or years (6). Composts produced for container media, therefore, must be produced under predominantly aerobic conditions.

The rate of aeration required to prevent anaerobic metabolism varies with the product being composted. For example, most pine barks are highly resistant to decomposition (1) and generally are not aerated by forced ventilation during composting in windrows. On the other hand, most hardwood barks (1, 12, 13, 81) and also some softwood barks such as spruce bark (80) are high in cellulose content and should be aerated throughout composting.

Two types of problems, one that is obvious and the other more subtle, are caused by "sour" composts. In some climatic regions mature composts can become anaerobic after a heavy snowfall or rain. This frequently occurs when inadequate energy is left in the system to evaporate excessive precipitation that soaks into the pile. Such compost in the base of a windrow develops a slightly sour odor that may not be detected when the windrow is turned and the compost is utilized. Although plants vary in sensitivity to this toxicity, root injury may occur in media amended with these composts resulting in poor performance of crops. This can be avoided by placing windrows under a roof during the later phase of composting or by placing compost in very large piles during the curing phase (6 m high). A disadvantage of high piles is the excessively high temperature (> 80°C) that develops resulting in pyrolysis and kill of beneficial microorganisms.

An acute toxicity problem may develop in the spring when the sugar content in tree bark is high. Highly toxic bark (pH 2.5) often is generated at sawmills at this time of year because operators store the bark in excessively large piles. Other sour composts have been generated by placing high temperature compost in sealed containers (covered trucks) during transport for 12-24 hr resulting in fermentative metabolism. All these problems can be avoided with adequate ventilation.

Wood residues, such as barks and sawdusts, are amended with nitrogen and, usually, phosphorus (12, 49, 51, 58, 80, 89) and water before composting. The amount of N added generally ranges from 1 to 1.5 kg/m^3. The pH of bark (4.5-5.2) is below the optimum range (6.5-8.5) for the composting process (12,13,54). Addition of ammonium nitrogen and/or poultry manure raises the pH and therefore largely results in higher rates of decomposition than that obtained when ammonium nitrate is used as the sole source of N (12, 13). Urea, poultry manure and other high nitrogen wastes (*i.e.* municpal sewage sludges), therefore, are preferred sources of N. However, the pH of compost should remain below 7.4 in aerated compost systems to curtail excessive ammonia loss. Mixtures of ammonium nitrate and urea nitrogen or urea and other organic N sources should be used to avoid this problem (46). Plant response in composts prepared with poultry manure frequently is better than that in composts prepared with synthetic N sources (14). Pig manures also have been used with success (28).

The optimum moisture content (on a wet weight basis) for composting varies with the product to be composted. The initial moisture content may range from 60-70%. To avoid low oxygen tensions, water should not drain out of or restrict airflow up through the base of the compost pile. Generally this means that moisture levels should be below 60%. Moisture levels below 50% can reduce rates of decomposition (12, 13, 53). Highly porous wood wastes which are difficult to wet can frequently dry out during composting and may need to be rewetted.

During composting of sewage sludges the lack of drying may present a problem, particularly in systems where wood chips are used. This is partially caused by the large amount of water liberated by the respiration process. Chips are wetted thoroughly shortly after they are mixed with sludge. Near the end of the curing process moisture moves outward from the center and is then evaporated from the surface of the chip. The rate of water movement limits drying in this system. In systems where small particle size bulking agents are used, *i.e.*, sawdusts, higher surface area to nonexposed volume ratios exist. The relatively larger surface area promotes increased microbial and enzyme activity and, therefore, increased heat output and drying potential.

12.3. Maturity of composts

Stabilization implies oxidation or conversion of organic materials to a more refractory (*i.e.* stable) form. The more stable compounds that remain after composting are still degradable, but at a much reduced rate compared to the raw material. The question is how much should composts be stabilized before their utilization in container media. This depends on the percentage volume of compost to be incorporated into a container medium and the rate of oxygen uptake by the compost. All compost-amended container media must provide adequate levels of O_2 in the root environment. Oxygen consumption rates, therefore, should not exceed the sum of oxygen needs for continued aerobic decomposition of the compost and oxygen uptake by the root system.

Several procedures for testing stability or maturity of composts have been proposed. Negative plant responses in composts could be due to high cellulose content resulting in N immobilization, allelopathic chemicals, toxic levels of metabolites of anaerobic metabolism, high salt content and, possibly, other factors. Some examples of proposed maturity level assays are: starch content, cellulose content, cation exchange capacity, C/N ratio, carbon/organic nitrogen ratio in water extracts of compost, the presence of indicator microorganisms and others (16,

17, 19, 20, 36, 41, 42, 43, 90). The most widely used methods are plant bioassays. Several of these have been proposed (17, 18, 80, 97, 98). However, a wide range of responses occur among plants to composts of various maturity levels. Generally, all plants are inhibited by immature composts. On the other hand, growth frequently improves upon contact with mature composts.

A disadvantage of most assays for maturity is the time required to complete an assay and its cost. A simple and inexpensive procedure has been developed that may eventually solve this problem if paired with other assays (90). In this procedure, a given amount of compost is incubated in a coulometer under standard conditions. The total amount of CO_2 produced provides a measure of compost stability. The procedure therefore is based on older procedures which measured the rate of respiration (36, 44).

Two approaches have been proposed recently for estimating compost maturity: the cation exchange capacity and the water extract organic carbon/organic nitrogen ratio. Both were developed in Japan. In municipal refuse compost the cation exchange capacity (CEC) increases with maturity. A highly significant negative correlation was established between the CEC and C/N ratio of city refuse compost. The CEC values for municipal refuse composts, which were considered to have been sufficiently matured for utilization were greater than 60 meq/100 g ash-free compost (42, 43).

The C/N ratio itself cannot be regarded as a reliable indicator of compost maturity since raw materials used as bulking agents vary widely in lignin/cellulose ratio and, therefore, biodegradability. The organic carbon/organic nitrogen ratio in water extracts of various types of composts is a function of compost maturity. Many components of compost dissolve in water and interact with plant roots and microorganisms. Enzymes degrade insoluble components which affect nutrition of the biomass including the plant. It is not surprising therefore, that the carbon/ organic N ratio of water extracts of various types of well matured composts, with almost constant values of 5-6, was indicative of maturity (16, 17, 18, 19). Obviously, this procedure is not effective for low C/N ratio wastes such as sewage sludges which after mixing of raw sludge and bulking agents already have low ratios in the extract.

12.4. Chemical properties

The lignin/cellulose ratio of wastes not only affects the composting process but also the utilization of their composts. Pine bark for example, with its high lignin content frequently is not composted before utilization in container media. Pine bark media must be amended with essential micro nutrients as well as a source of Ca and Mg for growth of most crops (10). Hardwood barks and sewage sludges on the other hand, which undergo significant levels of decomposition during composting, generally are utilized without addition of essential micro nutrients or sources of Ca and Mg. Elemental S is added to these composts to lower their pH (49, 58). Frequently, manganese is released at excessively high levels by many composted barks; in Norway, limits have been set for Mn levels (81). Concentrations of exchangeable Mn [with $Mg(NO_3)_2$] must be below 200 ppm.

Composts prepared from wood industry wastes frequently are not adequately stabilized before utilization and still immobilize N during utilization. Composts from sewage sludges, on the other hand, release significant levels of N during crop production thus requiring lower levels of N fertilization for some time after planting (15, 37, 38). Nitrogen fertility affects severity of several soilborne plant diseases, including those caused by *Phytophthora* and *Fusarium* spp (32, 78). Any investigation on effects of organic components on plant growth and severity or suppression of diseases caused by these pathogens must consider N nutrition.

Treatment processes may also affect utilization of organic wastes. For example some municipal sewage treatment plants utilize lime during the dewatering process and others use organic polymers. The lime-dewatered sludges present variable and high pH problems after composting not associated with polymer-dewatered sludges.

The acidity of the organic component may affect disease. Low pH of Sphagnum peats and pine bark theoretically could have beneficial side effects for some plants because *Phytophthora* root rot (*Phytophthora cinnamomi* Rands) for example is inhibited at a pH ⟨ 4.0 (5, 82, 83). This low pH reduces sporangium formation, zoospore release and motility (5). In nurseries however, the pH in container media rapidly rises to the range of severe disease development (pH ⟩ 5.7). Furthermore, growth of most plants is inhibited at a pH of 4.0.

Allelopathy has been described as chemical warfare between plants. Tree barks in particular may be sources of chemicals that inhibit plant growth. This inhibition cannot be explained by excessive levels of N-immobilization due to the high cellulose content of these barks. Allelopathic chemicals responsible for this effect in both softwood as well as hardwood bark, apparently are destroyed within a few weeks of composting (59, 86). Plant and microbial bioassays have been developed for testing bark composts for these inhibitors.

In addition to allelopathic chemicals, natural-product fungicides (inhibitors of fungal plant pathogens) occur in some composts (35, 50, 79, 82, 83, 84). Literature on this subject is summarized in Table 1. The age or maturity levels of bark and sawdust composts for example affects this activity. Composted hardwood tree bark (CHB) (mostly from oak) that is less than 1 yr old, releases components into the water phase of the container medium that inhibit production of sporangia and zoospores by *Phytophthora* spp. (50, 83, 84). Water extracts from 2-yr-old CHB are not inhibitory even though the CHB media from which the extracts are prepared still suppress *Phytophthora* root rot. This suggests a change in the mechanism of suppression with age of the compost which is discussed further in the section below on biological properties. The chemical identity of the inhibitors of *Phytophthora* spp. in CHB has been explored. The most toxic components in 6- month old CHB (prepared mostly from oak) were ethyl esters of hydroxy-oleic acids (4, 96), which are known constituents of oak bark. Extracts prepared from green (fresh) pine barks also have inhibitory effects on *Phytophthora cinnamoni*, whereas extracts from composted pine bark do not (35, 79). Inhibitors in barks do not

Table 12.1 Summary of literature on effects of water extracts prepared from various tree barks on sporangia production and zoospore release by *Phytophthora cinnamomi*

Bark species	Reference[1]	Bark age[2]	
		Fresh	Aged
Hardwoods (*Quercus* spp.)	50, 83, 84	+[3]	–
Pinus radiata	35	+	0
Pinus radiata	79	0	–
P. pinaster	79	+	–
Pinus spp.	83, 84	+	0

[1] See references for identity of citations.

[2] Bark age was classified as fresh and aged.
Hardwood bark extracts were not inhibitory after 1 yr of composting. Age or maturity levels of the pine bark composts were not available.

[3] += toxic;– = nontoxic; 0 = not tested.

affect plant pathogens equally. For example, growth of *Rhizoctonia solani* Kuhn as well as *Rhizoctonia* damping-off are not suppressed in fresh hardwood bark, even though it is toxic to *Phytophthora* spp. (67, 68, 69). Fresh pine also is largely conductive to *Rhizoctonia* damping-off (67).

Toxicity caused by organic acids in refuse compost has been examined in detail recently (18, 20, 29, 98, 99). Percentage emergence in composts of various maturity levels is negatively correlated with organic acid concentrations in compost extracts. Low molecular weight fatty acids (acetic, propionic and butyric acids) are most toxic. In one study after 4 months of curing of windrowed compost the acetic acid concentration was (300 ppm. This is below the level where it inhibits germination of cress (29). In another study on sewage sludge the carbon contributed by the low molecular weight fatty acid content was as high as 15%of the total organic carbon content of the raw material (20). In cured compost, when odors were not noticeable, fatty acids also were below detectible levels (20, 29). In practice, toxicity caused by organic acids as mentioned in the previous section is avoided by producing compost in properly aerated systems so that anaerobic pockets are avoided. Furthermore, in climatic regions with high levels of precipitation, stabilized composts should be stored under a roof to avoid excessive moisture levels and subsequent fermentation.

12.5. Physical properties

The suitability of organic components for utilization in container media is determined primarily by their effects on hydraulic conductivity, water retention and air capacity. Bulk density is also an important physical property of container media. The optimum bulk density differs for various production systems. For example, high bulk densities are required for support in the production of tall plants in containers to that they do not blow over during storms.

The air capacity of the medium has a direct effect on the oxygen levels in the root environment. Although oxygen availability is actually a function of diffusion rates, air capacity is the standard property measured. Plants vary in their responses to oxygen levels in container media. For chrysanthemum, optimal oxygen levels are obtained with an air capacity of 15% in a 10-cm tall column (71). Specific guidelines for a variety of crops are not available however.

Air capacity not only directly affects plant growth but also has an impact on root rot severity. For example, observations made in nurseries indicate that *Phytophthora* root rots are most prevailant in media with an air capacity $\langle$ 15% (15 cm tall column). Media that contain tree barks typically have air capacities $\rangle$ 25% (46, 50, 83) and percolation rates $\rangle$ 2.5 cm/min and thus suppress root rots. Addition of silica sand to a pine bark medium reduced air capacity to 15% and negated suppression of *Phytophthora cinnamoni* (83).

Investigations into the mechanism of suppression of *Phytophthora* root rots in bark-amended media therefore need to relate differences to physical properties.

Composts still undergoing high rates of decomposition promote low O_2 concentrations, particularly in the base of large containers, and reduce plant growth. Also, constant decomposition decreases particle size and air capacity which compounds the problem. Thus pine bark, which consists largely of lignin and is not readily biodegradable, was one of the first "recycled organic wastes" to be used widely in container media. It retains its original physical properties for several years in container media. Sphagnum peat is frequently added to imporve its water holding capacity. In most hardwood barks 25-45% of the total carbon is decomposed rapidly (1). Media amended with composted hardwood bark, therefore, have to be amended with expanded neutral aggregates to maintain air capacity. Sphagnum peat is added to these media to increase the percentage volume of stabile organic matter and reduce O_2 uptake in the root zone. Each type of compost needs to be tested in media over a 1-2 yr growing period to develop adequate guidelines for its successful utilization.

During the past decade, procedures have been proposed for determining physical properties of container media (22, 27, 40, 74). In practice, the choice of an irrigation regime for a given crop usually is based on its physiological needs and the hydraulic characteristics of the container medium. As noted above, the fate of water mold plant pathogens, such as *Phytophthora* spp. and *Pythium* spp., also needs to be considered. For example, release of zoospores from sporangia of several *Phytophthora* spp. is inhibited at negative tensions $\langle$ 5 mb (31). Therefore, irrigation systems could conceivably be designed to introduce water into container media in a way so as to expose roots for the shortest possible time with water held in the root environment at low tensions.

12.6. Biological properties

Published literature on the incidence of soilborne diseases in container media is reminiscent of earlier work on those diseases in field soils with organic amendments (9, 63). A recent text book reviews this subject (24). Various mechanisms of biological control were proposed. In field soils, interpretation of results often is difficult due to effects of previous cropping practices. In container media, chemical and physical properties can be standardized which is an advantage. In addition, the history of the introduced organic component frequently can be controlled as well, thus emphasizing differences between treatments caused by biotic factors.

During the past decade a significant body of research has been published by horticulturalists on utilization of organic components in container media (2, 8, 10, 11, 15, 22, 23, 40, 49, 58, 61, 71, 74, 80). Although knowledge on stability or maturity of composts and their microbiology still is inadequate, some information is now available on contributions of biological components to the observed disease suppressive effects. Since various types of peats also are used in container media, some of the pertinent literature on peat will be discussed as well.

Essentially all reports on media containing various types of peats show that they are conducive to disease development, particularly if steamed first. Examples of diseases that are not suppressed in Sphagnum peat media are those caused by *Phytophthora* spp. (50, 84), *Pythium* spp. (39, 76), *Rhizoctonia solani* (67, 85), *Fusarium* spp. (21), nematode diseases (64) and others. Recent Finnish reports indicate however that some sources of Sphagnum peat are suppressive (87, 88). This apparent contradiction may be related to differences in levels of decomposition of peat sources. The surface layer known as "light peat" is less decomposed, high in cellulose, low in lignin and may contain some living plant parts at harvest time (55, 75). The deeper layer is more decomposed, low in cellulose and is classified as "dark peat".

The low pH of Sphagnum peat severely limits its microbial population diversity (30, 55). Fungi present at the surface of the bog in "light peat" either are not present or in low numbers further down the profile in "dark peat". Facultative anaerobic bacteria occur at the surface in low numbers and increase down the profile. During the preparation of media the pH is usually raised from that of peat (3.1-4.0) to 5.0-6.0 which further complicates any interpretation of the data on microbiology of peats in the literature.

Some "light peat" sources harbor high population levels of various antagonists, particularly *Trichoderma viride* Pers. ex Gray and *Streptomyces* spp. (55) and are suppressive to various diseases (87) including *Pythium* and *Rhizoctonia* damping-off and tomato *Fusarium* wilt. The suppressive effect may last 3-6 weeks (88). Other "light peat" sources are conducive however and do not harbor these antagonists (87, 88). The depth from which Sphagnum peat is harvested, therefore, apparently affects the quality of peat as well as the fate of plant pathogens. In summary it appears that some of the "light peat" sources, all of which contain high levels of cellulose, may harbor antagonists and suppress some soilborne plant pathogens. The stabilized "dark Sphagnum" peats apparently do not have these properties.

It has been proposed to add specific pathogen-suppressive field soils to container media and render them suppressive (25, 72). Soil-amended container media have disadvantges however because a) suppressive soils frequently affect only one pathogen and may harbor others which are not suppressed (24), b) herbicide residues in soils may cause problems in production, c) the bulk density of soil-amended media frequently is too high, d) clay and silt particles may migrate through media to the base of the medium resulting in inadequate drainage and finally e) top soil is not a renewable resource. For these and other reasons the concept of utilizing suppressive soils in media is not utilized widely.

One of the striking differences observed in nurseries between media which contain only peat in the organic component, versus those which also are amended with composts, is the total absence or decreased losses caused by soilborne plant pathogens. In some compost-amended media, suppressiveness is high enough so that fungicide drenches are not used (26, 46). The observed disease control could be due to 1) kill of plant pathogens during composting of the media and 2) suppression of the pathogen and/or the disease after composting and during crop production. Each will be discussed separately.

During composting in large windrows, pathogen kill can be explained on the basis of temperature-time exposure (47, 92, 97). In one study, effects of heat treatment during the self-heating phase of composting and that of possible microbial antagonism during the maturing phase on survival of the pathogen, were examined separately. The pathogen of cabbage club root was killed during heating but survived maturing (92). Other studies show that pathogens are killed during the maturing process. For example *R. solani* propagules lose viability if added to mature compost but only if specific antagonists recolonize the compost after peak heating (68, 69). This mechanism of pathogen kill during composting apparently occurs in typical backyard compost heaps where temperatures attained are sublethal to the pathogen (97). Tobacco mosaic virus is not readily inactivated during composting, even if temperatures exceed 60°C (46). In summary, the sparse literature available suggests that the composting process, if performed properly, can be managed so as to yield a pathogen-free product. Some of the more heat-resistant pathogens may survive or not be inactivated.

Suppression of *Phytophthora* root rots in container media amended with tree barks, as discussed before, can be due to 1) physical factors associated with drainage (50, 83), 2) release of inhibitors of *Phytophthora* spp. into the water phase of the medium (50, 79, 84), 3) low pH (5), 4) low levels of N nutrition (78), and other factors. To evaluate the contribution of biotic factors involved in suppression of *Phytophthora* root rot, media amended with composted hardwood bark of two maturity levels were heated to 60°C for 5 days. Desorption curves of the media were nearly identical (50), the pH was 6.0 and high levels of slow release nutrients were applied. Heating destroyed the suppressive effect in the medium amended with 1-yr-old composted hardwood bark but not in that amended with 6-mo-old compost (Fig. 1.). This suggests that heat stable factors (inhibitors) were

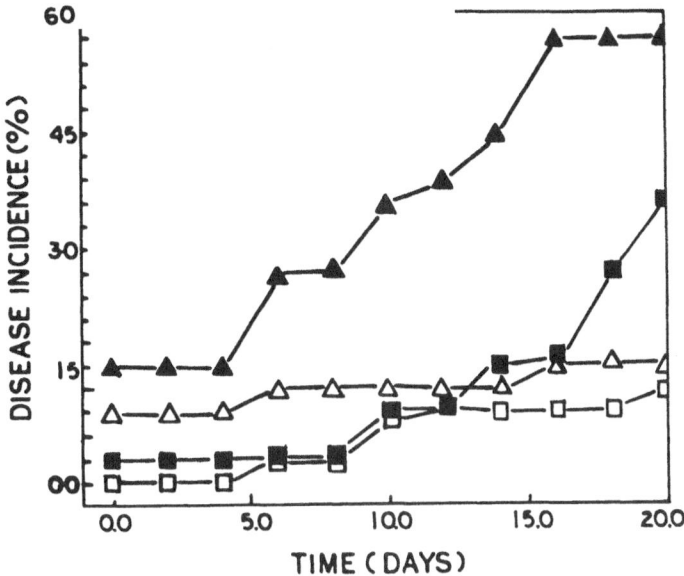

Fig. 12.1. Effects of heat treatment and maturity level of composted hardwood tree bark on severity of lupine *Phytophthora* root rot. Closed symbols represent heated samples (5 days, 60° C) □ = 6 mo-old and △ = 1 yr-old compost-amended media. Media were inoculated with one chlamydospore/10 CC. Areas under the disease progress curves for heated and non-heated 1-yr and 6-mo-old treatments are 7.1, 2.4, 2.0 and 1.4, respectively. LSD(*p* = 0.05) = 3.3.

involved in suppression in the 6-mo-old compost and heat labile factors in the 1-yr-old compost. Green composted hardwood bark releases inhibitors of *Phytophthora* spp. into the water phase of the container medium, whereas 1-yr-old and older composts do not. Therefore, it appears that a shift in mechanism of suppression from chemical to biotic factors occurs with increase in age of composts prepared from hardwood tree bark.

It is generally accepted that suppression is due to biotic factors if 1) it is negated by heating and 2) if it can be restored by adding small amounts of suppressive soil back to the heated conductive sample (24). Both small amounts (⟨ 10%, v/v) and a specific antagonist (*Trichoderma harzianum* Rifai of *P. cinnamomi*) restored the suppressive effect to the heated 1-yr-old composted hardwood bark-amended medium. This proved that suppression observed in media amended with 1- and 2-yr-old composted hardwood bark was due to biotic factors.

Not all composts prepared from hardwood tree barks are suppressive to *P. cinnamomi*. In the mid-70's, papermills adopted whole tree chippers in their harvesting program. The bark produced at mills from this process was approximately 60% wood. The bark proper contained 40-50% cellulose (46). This mixture, after composting for 3 months, still resulted in high levels of N-immobilization and was conducive to *P. cinnamomi*. It has been reported that Douglas fir sawdust increases *Phytophthora* root rot of strawberry, but Douglas fir bark provided control (reviewed in 46). The initial cellulose content of the parent material as well as the age of the compost, therefore, has a significant effect on *Phytophthora* root rots.

Pine barks used in the Southeastern U.S., Australia and Europe are low in readily available carbon and release inhibitors of *Phytophthora* spp. after potting (1, 79). Although concise guidelines are not available, the published literature suggests that media amended with older or aged pine bark are suppressive because of their physical properties related to drainage. Possibly, inadequate amounts of biodegradable carbon are available in these processed pine barks to support adequate levels of antagonists after composting. In Australia, fungicides are recommended for control of *P. cinnamomi* in pine bark media.

Some "batches of pine barks" available commercially in the U.S., harbor high *Trichoderma* population levels and these media are suppressive. However, the parent organic matter may have contained wastes other than bark. For example, sawdust was landfilled with bark at most sawmills until recently. Aged bark removed from these sites may have properties different from the processed milled pine barks available today.

Diseases other than *Phytophthora* root rots are affected by biotic factors in media amended with composts. Examples are *Fusarium* wilt of flax (21), *Rhizoctonia* damping-off (60, 68, 69), and *Pythium* damping-off (Y. Hadar, The Hebrew University, Rehovot, Israel, *personal communication*). As previously mentioned, *R. solani* is not affected by the inhibitors present in composted hardwood bark that affect *Phytophthora* spp. In fact *R. solani* colonizes fresh hardwood bark. The role of biotic factors in suppression of soilborne plant diseases in media amended with composted hardwood bark was therefore examined further with *R. solani*. A radish bioassay (45) specifically developed for assigning quantitative parameters to suppressiveness was used for this purpose. Media amended with green composted hardwood bark were conducive or mildly suppressive.

Suppressiveness increased with maturity of compost. Furthermore, only the outer low temperature layer of windrows was suppressive. Compost removed from the high temperature ($>$ 55°C) center of the windrow was conducive but could be made suppressive by the artifical introduction of specific antagonists (*Trichoderma* spp. and *Gliocladium virens* Miller, Giddent and Foster) of *R. solani* (60, 69). Similar knowledge has now also been developed for media amended with composted municipal sludge (Authors', *unpublished*).

Addition of a *T. harzianum* antagonist to green composted hardwood bark does not induce high levels of suppression, in spite of the development of high population levels of the antagonist in the compost (69). Induction of suppression of *Rhizoctonia* damping-off by this antagonist therefore depends on 1) chance of recolonization of the compost by the antagonist and 2) the maturity level of compost. It is of interest that propagules of *R. solani* survive in *T. harzianum*-colonized green compost, but not in similar but mature compost. The age of the compost, therefore, affects the activty of the antagonist.

A detailed examination of fungal populations isolated from suppressive and conducive media has revealed that they differ quantitatively (60). Principal components analysis demonstrated relationships among fungal taxa based on their relative densities in suppressive and conducive media. Factor loadings produced principal components in which predominant fungal taxa associated with suppressive compost were well separated from the taxa in conducive compost. Although no single taxon predominated in all suppressive media, *Trichoderma* spp. were among the most abundant taxa associated with suppressiveness. However, other antagonists played a role as well. The ultimate aim of this field of research is to develop procedures for production of composts that suppress a wide range of plant pathogens. Some of the technology appears available to accomplish this but a significant body of knowledge on the ecology of antagonists of pathogens remains to be uncovered.

The aim of plant pathologists in various parts of the world is to develop practical alternatives to some of the more hazardous and costly chemical disease control procedures. Most of this research is limited to the fate of single antagonists in soil, on seed or on plant parts. Very few systems have actually been applied successfully in agriculture. A strain of *Agrobacterium radiobacter* is used widely for control of crown gall (56, 57, 66). By artifically increasing the ratio of antagonist to pathogen, the percentage infected plants has been reduced effectively. The population level of the antagonist unfortunately decreases after application. Procedures which maintain high levels of this antagonist have yet to be developed. For example, the role of organic matter in maintaining a desirable balance of these microorganisms is unknown.

A recent example of successful use of antagonists in greenhouses is control of several carnation diseases with a benomyl resistant mutant of *T. harzianum* (3) which is effective against *Rhizoctonia* stem rot and *Pythium* root rot of carnation. Growers in Colorado produce carnations in ground beds which are frequently amended with composted cow manure before steaming. The antagonist, *T. harzianum,* is introduced after steam treatment which may affect its efficacy. This system, therefore, in principle may not differ from amendment of container media with suppressive antagonist-recolonized compost.

Aerated steaming has been proposed for treatment of peat and soil media. If applied properly, at least in theory, it eradicates plant pathogens possibly without killing some of the beneficial microorganisms (2, 7). No attempt is made to review this subject in detail here. However an important feature is overlooked frequently, *i.e.,* the "residual effect" (the length of time that the substrate remains suppressive). Even though peat and soil-amended container media can be suppressive (87, 88), the effect apparently lasts only a few weeks (88). This is supported by performance data in greenhouses and nurseries which has resulted in recommendation for application of fungicide drenches.

Incorporation of suppressive composts in media with appropriate chemical and physical properties has successfully solved soilborne disease problems in nursery and greenhouse crops in several parts of the world. Some of the container media not only suppress the diseases, but eradicate pathogens as well. Production in compost-amended media therefore has advantages because fungicide treated systems, generally do not eradicate pathogens. Even though some production probems have been solved, many questions remain unanswered. For example, incorporation of excessively high levels of compost into container media increases fungus gnat problems. Incorporation of small amounts of inadequately matured composted municipal sludge also may have this effect. Techniques for monitoring maturity of composts, including production of pathogen suppressive composts, and the fate of insect pests need to be evaluated further. Such research also may yield principles which eventually can be applied successfully to field agriculture.

12.7. References

1 Allison F.E. and Murphy R.M. 1962 Comparitive rates of decomposition in soil of wood and bark particles of several hardwood species. Soil Sci. Soc. Am. Proc. 26, 463-466.

2 Baker K.F., Ed. 1957 The U.C. System for producing healthy containergrown plants. Calif. Agr. Expt. Sta. Manual 23.

3 Baker R. 1983 State of the art: plant diseases. Natl. Interdiciplinary Biological Control Conf. Ed. S.L. Battenrield pp. 14-22 *In* Proc. 15-17 Feb. 1983. Las Vegas, NV. CSRS/USDA. Washington DC.

4 Basham L.M. 1980 Extraction and partial purification of non-volatile fungal inhibitors from composted hardwood bark. M.S. Thesis, The Ohio State University. 42 p.

5 Blaker N.S. and MacDonald J.D. 1983 Influence of container medium pH on sporangium formation, zoospore release, and infection of rhododendron by *Phytophthora cinnamomi*. Pl. Dis. 67, 259-263.

6 Bollen W.B. and Lu K.C. 1970 Sour sawdust and bark - its origin, properties, and effect on plants. U.S. Dept. Agr. PNW 108.

7 Bollen G.J. and Van der Pol-Luiten B. 1975 Mesophilic heat-resistant soil fungi. Acta Bot. Neerl. 24, 254-255.

8 Boodley J.W. and Sheldrake R.S. 1972 Cornell peat-like mixes for commercial plant growing. Cornell Univ. Plant Sci. Info. Bull. 43.

9 Broadbent P., Baker K.F. and Waterworth Y. 1971 Bacteria and actinomycetes antagonistic to fungal root pathogens in Australian soils. Aust. J. Biol. Sci. 24, 925-944.

10 Brown E.F. and Pokorny F.A. 1975 Physical and chemical properties of media composed of milled pine bark and sand. J. Am. Soc. Hortic. Sci. 100, 119-121.

11 Bunt A.C. 1976 Modern potting composts. A manual on the preparation and use of growing media for pot plants. The Pennsylvania State Univ. Press. Univ. Park and London, 277 p.

12 Cappaert I., Verdonck O. and DeBoodt M. 1976 Composting of bark from pulp mills and the use of bark compost as a substrate for plant breeding. Part I. The effect of physical paramemeters on the rate of composting of bark. Compost Sci. 17, 6-9.

13 Cappaert I., Verdonck O. and DeBoodt M. 1976 Composting of bark from pulp mills and the use of bark compost as a substrate for plant breeding. Part II. The effect of physical parameters on the composting rate of bark. Growth experiments with bark compost. Compost Sci. 17, 18-20.

14 Castro Gomez R.J.H. and Park Y.K. 1983 Conversion of cane bagasse to compost and its chemical characteristics. J. Ferment Technol. 61, 329-332.

15 Chaney R.L., Munns J.B. and Cathey H.M. 1980 Effectiveness of digested sewage sludge compost in supplying nutrients for soilless potting media. J. Am. Soc. Hortic. Sci. 105, 485-492.

16 Chanyasak V., Hirai M. and Kubota H. 1982 Changes of chemical components and nitrogen transformation in water extracts during composting of garbage. J. Ferment. Technol. 60, 439-446.

17 Chanyasak V., Katayama A., Hirai M.F., Mori S. and Kubota H. 1983 Effects of compost maturity on growth of komatsuna (*Brassica rapa* var. *pervidis*) in Neubauer's pot. I. Comparison of growth in compost treatments with that in inorganic nutrient treatments as controls.

Soil Sci. Plant Nutr. 29, 239-250.

18 Chanyasak V., Katayama A., Hirai M.F., Mori S., and Kubota H. 1983 Effects of compost maturity on growth of komatsuna (*Brassica rapa* var. *peridis*) in Neubauer's pot. II. Growth inhibitory factors and assessment of degree of maturity by org.-C/org.-N ratio of water extract. Soil Sci. Plant Nutr. 29, 251-259.

19 Chanyasak V. and Kubota H. 1981 Carbon/organic nitrogen ratio in water extract as measure of composting degradation. J. Ferment. Technol. 59, 215-219.

20 Chanyasak V., Yoshida T. and Kubota H. 1980 Chemical components in gel chromatographic fractionation of water extract from sewage sludge compost. J. Ferment. Technol. 58, 533-539.

21 Chef D.G., Hoitink H.A.J., and Madden L.V. 1983 Effects of organic components in container media on suppression of *Fusarium* wilt of chrysanthemum and flax. Phytopathology 73, 279-281.

22 Chen Y. Banin A., and Ataman Y. 1980 Characterization of particles and pores, hydraulic properties and water-air ratios of artifical growth media and soil. Proc. Fifth Internat. Cong. on Soilless Culture, Wageningen, pp. 63-82.

23 Conover C.A. 1967 Soil mixes for ornamental plants. Florida Flower Grower 4, 1-4.

24 Cook R.J. and Baker K.F. 1983 The nature and practice of biological control of plant pathogens. The American Phytopathological Society, St. Paul, Minnesota, 539 p.

25 Couteaudier Y. and Alabouvette C. 1981 *Fusarium* wilt diseases in soilless cultures. Acta Hortic. 126, 153-158.

26 Daft G.C., Poole H.A. and Hoitink H.A.J. 1979 Composted hardwood bark: A substitute for steam sterilization and fungicide drenches for control of poinsettia crown and root rot. HortScience 14, 185-187.

27 DeBoodt M., Verdonck O. and Cappaert I 1974 Method for measuring the water release curve of organic substrates. Acta Hortic. 37, 2054-2062.

28 DeVleeschauwer D., Verdonck O. and DeBoodt M. 1981 The use of chicken and piggery manure in composts. Acta Hortic. 126, 105-119.

29 DeVleeschauwer D., Verdonck O. and Van Assche P. 1981 Phytotoxicity of refuse compost. BioCycle 22, 44-46.

30 Dickinson C.H. and Maggs G.H. 1974 Aspects of the decomposition of Sphagnum leaves in an ombrophilous mire. New Phytol. 73, 1249-1257.

31 Duniway J.M. 1979 Water relations of water molds. Annu. Rev. Phytopathol. 17, 431-460.

32 Englehard A.W. and Woltz S.S. 1973 *Fusarium* wilt of chrysanthemums: Complete control of symptoms with an integrated fungicide-limenitrate regime. Phytopathology 63, 1256-1259.

33 Epstein E., Willson G.B., Burge W.D., Mullen D.C. and Enkiri M.K. 1967 A forced aeration system for composting waste water sludge. J. Water Pollut. Control Fed. 48, 688-694.

34 Finstein M.S., Miller F.C., Strom P.F., MacGregor S.T. and Psarianos K.M. 1983 Composting ecosystem management for waste treatment. Bio Technology 1, 347-353.

35 Gerritson-Cornell L. and Humphreys F.R. 1977 Results of an experiment on the effects of Pinus radiata bark on the formation of sporangia in *Phytophthora cinnamomi* Rands. Phyton. 36, 15-17.

36 Golueke C.G. 1972 Composting. A study of the process and its principles. Rodale Press, Emmaus, Penna., 110 p.

37 Gouin F.R. 1982 Using composted waste for growing horticultural crops. BioCycle 23, 45-47.

38 Gouin F.R. and Walker J.M. 1977 Deciduous tree seedling response to nursery soil amended with composted sewage sludge. HortScience 12, 45-47.

39 Gugino J.L., Pokorny F.A. and Hendrix Jr. F.F. 1973 Population dynamics of *Pythium irregulare* Buis. in container-plant production as influenced by physical structure of media. Plant and Soil 39, 591-602.

40 Hanan J.J., Holley W.D., and Goldsberry K.L. Eds. 1978 Soils and soil mixes. *In* Greenhouse Management. Springer-Verslag, Berlin.

41 Harada Y. and Inoko A. 1980 The measurement of the cation-exchange capacity of composts for the estimation of the degree of maturity. Soil Sci. Plant Nutr. 26, 127-134.

42 Harada Y. and Inoko A. 1980 Relationship between cation-exchange capacity and degree of maturity of city refuse composts. Soil Sci. Plant Nutr. 26, 353-362.

43 Harada Y., Inoko A., Tadaki M., and Izawa T. 1981 Maturing process of city refuse compost during piling. Soil Sci. Plant Nutr. 27, 357-364.

44 Haug R.T. 1980 Compost engineering principles and practice. Ann Arbor Science, Ann Arbor, Mich., 655 p.

45 Henis Y., Ghaffar A., and Baker R. 1978 Integrated control of *Rhizoctonia solani* damping-off of radish: Effect of successive plantings, PCNB, and *Trichoderma harzianum* on pathogen and disease. Phytopathology 68, 900-907.

46 Hoitink H.A.J. 1980 Composted bark, a lightweight growth medium with fungicidal properties. Pl. Dis. 64, 142-147.

47 Hoitink H.A.J. Herr L.J., and Schmitthenner A.F. 1976 Survival of some plant pathogens during composting of hardwood tree bark. Phytopathology 66, 1369-1372.

48 Hoitink H.A.J. and Kuter G.A. 1983 Factors affecting composting of municipal sludge in a bioreactor. Final report to USEPA Ultimate Waste Disposal Section Cincinnati OH., 100 p.

49 Hoitink H.A.J. and Poole H.A. 1980 Factors affecting quality of composts for utilization in container media. HortScience 15, 171-173.

50 Hoitink H.A.J., VanDoren, Jr., D.M., and Schmitthenner A.F. 1977 Suppression of *Phytophthora cinnamomi* in a composted hardwood bark potting medium. Phytopathology 67, 561-565.

51 Isomaki O. 1974 Using possibilities of barking waste in sawmill industry. Specially using a soil improver and substrate for plants. Acta Forest. Fenn. 140, 1-79.

52 Jeris J.S. and Regan R.W. 1973 Controlling environmental parameters for optimum composting. I Experimental procedures and temperature. Comp. Sci. 14, 10-15.

53 Jeris J.S. and Regan R.W. 1973 Controlling environmental parameters for optimum composting. II Moisture, free air space and recycle. Comp. Sci. 14, 8-15.

54 Jeris J.S. and Regan R.W. 1973 Controlling environmental parameters for optimum composting. III Effect of pH, nutrient storage and paper content. Comp. Sci. 14, 16-22.

55 Kavanagh T. and Herlihy M. 1975 Microbiological aspects. *In* Peat in Horticulture. Eds. W.D. Robinson and J.G.D. Lamb pp. 39-49. Acad. Press London, 170 p.

56 Kerr A. 1974 Soil microbiological studies on *Agrobacterium tumefaciens* and biological control of crowngall. Soil Sci. 118, 168-172.

57 Kerr A. 1980 Biological control of crowngall through production of agrocin 84. Plant Dis. 65, 24-25, 28-30.

58 Klett J.E., Gartner J.B. and Hughes T.D. 1972 Utilization of hardwood bark in media for growing woody ornamental plants in containers. J. Amer. Soc. Hortic. Sci. 97, 448-450.

59 Krogstadt O. and Solbraa K. 1975 Effects of extracts of crude and composed bark from spruce on some selected biological systems. Acta Agric. Scand. 25, 306-312.

60 Kuter G.A., Nelson E.B., Hoitink H.A.J. and Madden L.V. 1983 Fungal populations in container media amended with composted hardwood bark suppressive and conducive to *Rhizoctonia* damping-off. Phytopathol. 73, 1450-1456.

61 Louvet J. 1982 The relationship between substrates and plant diseases. Acta Hortic. 126, 147-152.

62 Lumsden R.D., Lewis J.A. and Millner P.D. 1983 Effect of composted sewage sludge on several soilborne pathogens and diseases. Phytopathol. 73, 1543-1548.

63 Lumsden R.D., Lewis J.A., and Papavizas G.C. 1983 Effect of organic amendments on soilborne plant diseases and pathogen antagonists. Pages 51-70. *In* Environmentally Sound Agriculture, Ed. W. Lockeretz. Praeger Press, New York, 320 p.

64 Malek R.B. and Gartner J.B. 1975 Hardwood bark as a soil amendment for suppression of plant parasitic nematodes on container grown plants. HortScience 10, 33-35.

65 Millner P.D., Lumsden R.D. and Lewis J.A. 1982 Controlling plant disease with sludge compost. BioCycle 23, 50-52.

66 Moore L.W. and Cooksey D.C. 1981 Biology of *Agrobacterium tumefaciens:* Plant interaction. pp. 15-46. *In* The Biology of Rhizobiaceae. Supplement to International Review of

Cytology. Ed. K. Tiles. Suppl. 13, Academic Press, New York.

67 Nelson E.B. and Hoitink H.A.J. 1982 Factors affecting suppression of *Rhizoctonia solani* in container media. Phytopathology 72, 275-279.

68 Nelson E.B. and Hoitink H.A.J. 1983 The role of microorganisms in the suppression of *Rhizoctonia solani* in container media amended with composted hardwood bark. Phytopathology 73, 274-278.

69 Nelson E.B., Kuter G.A. and Hoitink H.A.J. 1983 Effects of fungal antagonists and compost age on suppression of *Rhizoctonia* damping-off in container media amended with composted hardwood bark. Phytopathology 73, 1457-1462.

70 Parr J.F., Epstein E., and Wilson S.B. 1978 Composting sewage sludge for land application. Agriculture and Environment 4, 123-137.

71 Paul J.L. and Lee C.I. 1976 Relation between growth of chrysanthemums and aeration of various container media. J. Am. Soc. Hortic. Sci. 101, 500-503.

72 Perrin R. and Camporata p. 1982 Etude de la receptivite de substrats organiques aux maladies de fontes de semis provoques par *Pythium* spp. et *Rhizoctonia solani* Kuhn. Acta Hortic. 126, 159-168.

73 Poincelot R.P. 1974 A scientific examination of the principles and practice of composting. Comp. Sci. 15, 24-31.

74 Prasad M. 1979 Physical properties of media for container-grown crops. I New Zeland peats and wood wastes. Scientia Hortic. 10, 317-323.

75 Puustjarvi V. 1977 Peat and its use in horticulture. Turveteollisuusliitto, Helsinki, 161 p.

76 Robertson G. I. 1973 Occurrence of *Pythium* spp. in Zealand soils, sands, pumices, and peat, and on roots of container-grown plants. N. Z.J. Agric. Res. 16, 357-365.

77 Sanderson K.C. 1980 Use of sewage-refuse compost in the production of ornamental plants. Hortic. Sci. 15, 173-178.

78 Schmitthenner A.F. and Canaday C.H. 1983 Role of chemical factors in development of Phytophthora diseases. pp. 186-196. *In* Phytophthora Its Biology, Taxonomy, Ecology, and Plant Pathology. Eds. D.C. Erwin, S. Bartnicki-Garcia and P.H. Tsao. The American Phytopathological Society, St. Paul, Minnesota, 392 p.

79 Sivasithamparam K. 1981 Some effects of extracts from tree barks and sawdust on *Phytophthora cinnamomi* Rands. Aust. Plant Pathol. 10, 18-20.

80 Solbraa K. 1979 Composting of bark III Experiments on a semi-practical scale. Medd. Norsk inst. Skogforsk. 34, 387-439.

81 Solbraa K., Sant M.D., Selmer-Olson A.R. and Gislerod H.R. 1983 Composting soft and hardwood barks. Biocycle 24, 44-48 & 59.

82 Spencer S. and Benson D.M. 1981 Root rot of *Aucuba japonica* caused by *Phytophthora cinnamomi* and *P. citricola* and suppressed with bark media. Pl. Dis. 65, 918-921.

83 Spencer S. and Benson D.M. 1982 Pine bark, hardwood bark compost, and peat amendment effects on development of *Phytophthora* spp. and lupine root rot. Phytopathology 72, 346-351.

84 Spring D.E., Ellis M.A., Spotts R.A., Hoitink H.A.J. and Schmitthenner A.F. 1980 Suppression of the apple collar rot pathogen in composted hardwood bark. Phytopathol. 70, 1209-1212.

85 Stephens C.T., Herr L.J., Hoitink H.A.J. and Schmitthenner A.F. 1981 Suppression of *Rhizoctonia* damping-off by composted hardwood bark medium. Pl. Dis. 65, 796-797.

86 Still S.M., Dirr M.A. and Gartner J.B. 1976 Phytotoxic effects of several bark extracts on mung bean and cucumber growth. J. Am. Soc. Hortic. Sci. 101, 34-37.

87 Tahvonen R. 1982 The suppressiveness of Finnish light coloured Sphagnum peat. J. Scient. Agric. Soc. Finl. 54, 345-356.

88 Tahvonen R. 1982 Preliminary experiments into the use of *Streptomyces* spp. Isolated from peat in the biological control of soil and seed-borne diseases in peat culture. J. Scient. Agric. Soc. Finl. 54, 357-369.

89 Uemura S. 1973 Production and application of composts from wood residues. Wood Industry 28, 237-248.

90 Usui T., Shoji A. and Yusa M. 1983 Ripeness index of wastewater sludge compost. BioCycle 24, 25-27.

91 Van Assche C. and Uytterbroeck P. 1981 The influence of domestic waste compost on plant diseases. Acta. Hortic. 126, 169-178.

92 Wijnen A.P., Volker D. and Bollen G.J. 1983 De lotgevallen van pathogene schimmels in een composthoop (The fate of pathogenic fungi in a compost heap). Gewasbescherming 14, 5.

93 Will H. 1973 Kann mullkompost bei der pflanzenanzucht verwendet werden. Der Ergebsgartner 27, 1629-1630.

94 Willson G.B. and Dalmat D. 1983 Sewage sludge composting in the U.S.A. BioCycle 24, 20-23.

95 Winkler H. 1972 Erste ergebnisse sur verwendung von stadtkompost bei sierpflanzen. Arch. Gartenbau 20, 591-599.

96 Wilton J.H. 1982 Studies of natural products from *Liriodendron tulipifera, Simmondsia californica* and hardwood bark compost. Ph. D. Diss. The Ohio State University, 367 p.

97 Yuen G.Y. and Raabe R.D. 1979 Eradication of fungal plant pathogens by aerobic composting. Phytopathology 69, 922.

98 Zucconi F., Forte M., Monaco A. and de Bertoldi M. 1981 Biological evaluation of compost maturity. BioCycle 22, 27-29.

99 Zucconi F., Pera A., Forte M. and de Bertoldi M. 1981 Evaluating toxicity of immature compost. BioCycle 22, 54-57.